▼▼▼▼▼▼▼▼▼▼▼▼▼▼▼▼▼▼▼▼▼▼▼▼▼

Fundamentals of Geological and Environmental Remote Sensing

Robert K. Vincent
Bowling Green State University

D0068827

Prentice Hall, Upper Saddle River, New Jersey 07458

Library of Congress Cataloging-in-Publication Data

Vincent, Robert K.
    Fundamentals of geological and environmental remote sensing /
  Robert K. Vincent.
      p.   cm.
    Includes bibliographical references and index.
    ISBN 0-13-348780-6
    1. Earth sciences—Remote sensing.   2. Environmental monitoring—
  Remote sensing.   I. Title.
  QE33.2.R4V56   1997
  550'.28—dc21   96-49767
                                                          CIP

Executive Editor: *Robert McConnin*
Production Supervision and Interior Design: *Joanne E. Jimenez*
Assistant Managing Editor: *Shari Toron*
Manufacturing Buyer: *Benjamin Smith*
Cover Designer: *Bruce Kenselaar*
Photo Editor: *Carolyn Gauntt*

On the cover: Wind River Basin and Range, Wyoming, as seen in a color ratio image of the LANDSAT MSS sensor (see Color Plate 4 and Chapter 5, p. 116). Soils rich in ferric oxide appear orange and vegetation appears blue-green (aqua). Water and shadows are black. (Courtesy of GeoSpectra Corp.)

 © 1997 by Prentice-Hall, Inc.
Simon & Schuster/A Viacom Company
Upper Saddle River, NJ 07458

Printed in the United States of America

10 9 8 7 6 5 4 3 2 1

ISBN 0-13-348780-6

PRENTICE-HALL INTERNATIONAL (UK) LIMITED, LONDON
PRENTICE-HALL OF AUSTRALIA PTY. LIMITED, SYDNEY
PRENTICE-HALL CANADA INC. TORONTO
PRENTICE-HALL HISPANOAMERICANA, S.A., MEXICO
PRENTICE-HALL OF INDIA PRIVATE LIMITED, NEW DELHI
PRENTICE-HALL OF JAPAN, INC., TOKYO
SIMON & SCHUSTER ASIA PTE. LTD., SINGAPORE
EDITORA PRENTICE-HALL DO BRASIL, LTDA., RIO DE JANEIRO

This book is dedicated to my loving wife, Dinah K. Vincent, who helped with the unsung chores of publishing, to the former employees of GeoSpectra Corporation, who performed so well, and to the late Dr. Graham R. Hunt, my favorite Australian, who gave me a spectroscopist's view of the world.

Prentice Hall Series
in Geographic
Information Science

KEITH C. CLARKE,
Series Editor

Avery/Berlin, *Fundamentals of Remote Sensing and Air Photo Interpretation, 5th Edition*

Clarke, *Analytical and Computer Cartography*

Clarke, *Getting Started with Geographic Information Systems*

Jensen, *Introductory Digital Image Processing: A Remote Sensing Perspective, 2nd Edition*

Peterson, *Interactive and Animated Cartography*

Star/Estes, *Geographic Information Systems*

Tomlin, *Geographic Information Systems and Cartographic Modeling*

Vincent, *Fundamentals of Geological and Environmental Remote Sensing*

Contents

▼ ▼ ▼ ▼ ▼ ▼ ▼ ▼ ▼ ▼ ▼ ▼ ▼ ▼

Preface

▼ ▼ ▼ ▼ ▼ ▼ ▼ ▼ ▼ ▼ ▼ ▼ ▼ ▼ ▼

Understanding of any book can be enhanced by at least some knowledge of the author's background. I have had a nontypical background for a geology professor, starting with my university degrees, which are in math (B.A. from Louisiana Tech University), physics (M.S. from the University of Maryland and B.S. from Louisiana Tech University), and geology (Ph.D. from the University of Michigan). I was a principal investigator on both LANDSAT I and Skylab Earth resources experiments in the early 1970s as a research geophysicist at Willow Run Laboratories of the University of Michigan, which became the Environmental Research Institute of Michigan (ERIM) in 1973. From 1974–1993, I was an industrial remote sensing scientist and entrepreneur, having founded GeoSpectra Corporation (geological remote sensing) and BioImage Corporation (medical image processing) in Ann Arbor, Michigan. In August, 1993, I began my teaching career as an associate professor in the Department of Geology at Bowling Green State University, Bowling Green, Ohio.

Although remote sensing has great utility in many scientific disciplines, its importance in geological and environmental applications is unsurpassed. Inexpensive computer hardware and software now are commercially available that can be used to fully exploit the multispectral aspects of satellite and aircraft remote sensing data. The nonmultispectral aspects of remote sensing, which are like the rind of an orange, had to carry the justification for existence of remote sensing during the early years of the technology. It is now time to taste the meat of the orange, the multispectral applications. The knowledge base that this book provides should be particularly helpful in appreciating the various roles that multispectral remote sensing can play.

This book was written primarily for senior-level undergraduate and graduate students studying geology and environmental science, but also for geologists and environmental scientists in industry, who should find Appendix C particularly useful. Since 1993, I have been using the first eight chapters of this book as a text for a course in Geological Remote Sensing (Geology 540/440 at Bowling Green State University) and the last two chapters, along with Cambridge University Press' *Atlas of Satellite Observations Related to Global Change* (edited by R. J. Gurney, J. L. Foster, and C. L. Parkinson 1993), as texts for a course in Environmental Remote Sensing (Geology 680). The first six and last two chapters could be used as an environmental remote sensing text alone in geography, earth science, and other environmentally oriented departments. Although one year of calculus, petrology, and mineralogy are recommended as prerequisites for the geological remote sensing course, the equations have been simplified such that all integrals are in Appendix A, and much of the petrological and mineralogical information needed for the course is included in this book. For environmental remote sensing, only calculus is a recommended prerequisite, and only then if the student is interested in understanding the more explicit equations in Appendix A.

The first six chapters deal with physical, chemical, and image processing backgrounds for remote sensing, with an emphasis on geological considerations. However, these chapters should prove useful for remote sensing specialists from many scientific disciplines, including military remote sensing specialists, who must understand the natural terrain in order to increase contrast between manmade objects and the background around them. Chapters 5 and 6 are about multispectral and spatial image processing methods, respectively. The digital photogrammetry section of Chapter 6 is one of the first ever in this new field, and should be helpful to photogrammetrists and Geographic Information Systems (GIS) database users alike.

The remaining chapters deal with applications of remote sensing to various subjects. Chapter 7 is devoted to the use of geological remote sensing in exploration for precious metals (including disseminated gold deposits), base metals, rare earth minerals, and industrial minerals. Chapter 8 is about petroleum and ground water exploration. Both of these chapters include several case studies, some from my own twenty years of industrial experience, which will particularly benefit geology students and extractive industry consultants. Agricultural and forestry remote sensing specialists may find the geobotanical remote sensing parts of chapters 7 and 8 interesting.

Chapter 9 deals with engineering geology and environmental studies, but will be useful to a variety of remote sensing practitioners, from solid-waste landfill operators to transportation and pipeline route planners. GIS is discussed in several of these applications, as it comes up in case studies. Chapter 10 deals with global monitoring, and is the most speculative chapter because the subject is new and rapidly changing. Plate tectonic geophysicists and oceanographers should also find Chapter 10 interesting. We live in exciting times for both of those fields of study, as well as for remote sensing, which I believe is a critical, enabling technology for environmental monitoring.

ACKNOWLEDGMENTS

I would like to thank my students at BGSU and my family, who nurtured me during the writing of the book. I also wish to thank my reviewers for their excellent suggestions. They include: Gwynn Suits, University of Michigan, Ronald W. Maars, University of Wyoming, Alesander F. H. Goetz, University of Colorado, Mark Jakubauskas, University of Oklahoma, James W. Merchant, University of Nebraska-Lincoln, Richard Beck, Miami University, H. R. Hopkins, Consulting Geologist, Bellaire, Texas, W. Terry Lehman, EOSAT, Lanham, Maryland, Michael Abrams, Jet Propulsion Laboratories, and John R. Jensen, University of South Carolina. In particular, my thanks to Dr. John Salisbury of Johns Hopkins University, also a reviewer, who was my first manager as a professional scientist (at Air Force Cambridge Research Laboratories in the late 1960s) and has been my longest-term friend and colleague in remote sensing. His work with reflectance spectra of rocks and minerals, many of which are included in this book, has been a great contribution to this field.

1

Introduction
and Overview

▼ ▼ ▼ ▼ ▼ ▼ ▼ ▼ ▼ ▼ ▼ ▼ ▼ ▼ ▼ ▼

FROM THE HUMAN EYE
TO THE MULTISPECTRAL SCANNER

Geology and environmental studies are largely observational sciences. Historically, the eyes of a field scientist have been the most important sensors available for characterizing objects on Earth's surface. When a scientist visually observes his or her surroundings, both visible color and spatial information (such as size, texture, three-dimensional shape, and contextual location) are perceived all at once by the eye–brain combination. The information is presented to the logical parts of the brain in co-registered form, such that different types of color and spatial information simultaneously overlay one another for any portion of the scene.

As good as the eye–brain observational system of a field scientist may be, however, there is much more information available about Earth's surface than the un-aided human can observe, process, or accurately retrieve. First, any ground-based sensor is limited to small areas of observation. Second, human depth perception is relative, rather than absolute, leading to misperceptions of how objects are spatially configured on Earth's surface. Third, there are many other forms of electromagnetic waves (also called electromagnetic radiation) besides visible sunlight that contain useful information about surface objects. Electromagnetic waves are virtually massless packets of energy called photons. Some of these photons are visible light, which means that they have wavelengths in the 0.40–0.67 μm range (one μm is one-millionth of a meter) that can be seen by the human eye; however, most are invisible. Fourth, human memory is sufficiently faulty so that when the field scientist draws surface

1

features in the form of maps, these artistic representations interpolate results of surface observations unevenly across the mapped area in a manner that is often more beautiful than accurate.

Remote sensing, which is the technology of determining characteristics of a distant object from electromagnetic waves emanating from and reflecting off the object, augments the sensory perception and archival capabilities of the field observer in at least one, and sometimes all, of those four ways. The first remote sensing device for recording images from electromagnetic waves was the black-and-white photograph, which recorded images from a spectral band of wavelengths that covered the ultraviolet and blue portion of the visible spectrum (Reeves et al. 1975). Louis Daguerre (Slama et al. 1980) produced the first tangible photographic results as positive images on a metal plate in 1837. In 1858, a French photographer/balloonist named Tournachon (Slama et al. 1980) first achieved aerial photography from a hot air balloon. This feat was followed in the early part of this century by aerial photography from airplanes, first by Wilbur Wright on April 24, 1909, over Centocelli, Italy (Reeves et al. 1975). The aerial photo was able to replicate much of what the field scientist's eyes could see, though at smaller scales and larger instantaneous fields of view than the surface-bound observer's eyes could behold during a field trip.

By 1904, Henry Fourcade, a British forester in South Africa, had produced a topographic map using stereophotographs (Slama et al. 1980), which quantified elevation measurements in a way that depth perception of the unaided human eye could not. Orthophotographs were developed in the 1950s (Reeves et al. 1975). Unlike normal aerial photos, they could be overlaid directly onto maps because distortions caused by elevation differences of Earth's undulating surface in the photographed scene had been removed. In the last decade, digital photogrammetry was developed (Vincent et al. 1987; Ackerman 1994; Jensen 1995); this advancement automated the extraction of elevation data from digitized stereo images by computers and created the first digital orthophotographs. Digital orthophotographs integrate the technologies of geographic information systems (GIS) and remote sensing through the accurate digital overlaying of digitized maps with distortion-free remote sensing images.

The *multispectral* (meaning more than one spectral band or wavelength region) aspect of remote sensing began with the development of color photographic film. Kodak introduced it to the commercial world in 1935 as Kodachrome (Reeves et al. 1975), as a film that covered three visible spectral bands (blue, green, and red). But color film presented problems when applied to aerial photography, one of the most important being the reduction of detail by atmospheric haze, which produces much more scattering of light in the visible blue wavelength region (approximately 0.43–0.49 μm) than at longer wavelengths. In the late 1930s, the U.S. Army Air Corps was active in the military adaptation of infrared film, sensitive to wavelengths between the lower limit of the visible red (0.60 μm) and 0.90 μm, for haze penetration, a practice that had become well established by 1941 (Reeves et al. 1975). During World War II, Kodak invented color infrared film (CIR) for the purpose of detecting the difference between natural vegetation (shown as red in a CIR photo) and camouflage that looked like vegetation in the visible wavelength region (Reeves et al. 1975). Multispectral cameras, with photographic film filtered to various narrow spectral bands in the 0.4–1.0 μm region, were developed after the war and reached an apex in the 1960s, with the first satellite multispectral photograph of Earth, taken by Apollo 9 (Reeves et al. 1975). Even though it permitted photographers to record images at wavelengths that were invisible to the human eye, infrared film was and still is limited to the small fraction of the infrared spectrum between 0.67 μm and 1.0 μm.

Collection of image data at wavelengths longer than 1.0 μm requires remote sensing technology that departs from photographic film. In the late 1950s and 1960s, a multispectral scanning device for recording images of Earth's surface in many different wavelength regions, even some longer than 1.0 μm, was developed under U.S. Army sponsorship at Willow Run Laboratories of the University of Michigan in Ann Arbor, Michigan by Dr. Gwynn Suits, Marvin (Mike) Holter, and a supporting team of engineers and scientists. More than a dozen different spectral bands were simultaneously recorded on analog tapes by an airborne multispectral scanner, and custom-built analog multispectral data processors were used in the laboratory to play back images of the data in various combinations of spectral bands. The multispectral scanner literally looked at the world in a new light that included both visible and invisible (ultraviolet and infrared wavelengths) electromagnetic radiation.

In 1972, the first Earth Resources Technology Satellite (ERTS I) was launched by NASA with a digital multispectral scanner (MSS) aboard. ERTS I, later renamed LANDSAT I, enriched the field of geology tremendously because it gave the geologist a synoptic view (with data collected from vast areas within a very short period of time). Earth's surface was seen from space in four different wavelength regions of light, two in the visible and two in the reflective infrared region. The image for each spectral band was broken up into picture elements (called pixels) that were approximately squares of 80 meters each side on the ground. With the advent of LANDSAT I, the applications phase of the remote sensing era began in earnest. Whereas multispectral scanner data prior to 1972 had been in the hands of only a few researchers at a handful of institutions and had been collected from aircraft for only tiny portions of the globe, the Open Skies policy of NASA towards LANDSAT data opened up the availability of Earth resources satellite data for almost any place on Earth (excluding some of the polar regions that were physically unavailable) to any researcher in the world whose government permitted it.

In the entire history of science, there has never been an event equal to the advent of LANDSAT I for the peaceful sharing of scientific data. For data from a satellite for any place on the globe to be available to anyone was a revolutionary idea for the Cold War era. Typical of revolutionary ideas, however, the Open Skies policy of LANDSAT caused a few uneasy times for the early participants in remote sensing commercialization, as the following recollection of a 1970s experience will attest.

I was trying, through a local interpreter in a developing country, to convince the military officer at the desk in front of me to purchase a LANDSAT petroleum and ground water exploration study, when he unexpectedly stated that he was ready to buy any satellite data of his country that was already in my possession. I explained that the data would be ordered after the study began, and that the satellite would collect the data on one of its passes every 18 days over his country. He suddenly leapt to his feet and shouted something in a bellicose manner. My interpreter looked at me, thought better of translating the angry words, then turned and answered something back without consulting me. There was a pregnant pause, whereupon both the military officer and my interpreter burst forth in laughter, as I tried to hide both my relief and confusion with a grin. Later, as we drove away from the office complex, I asked my interpreter what had transpired at that tense moment, and he started laughing so hard that it became difficult for him to drive. He then said, "When I interpreted what you said about the satellite passing over this country every 18 days, he shouted, 'Who gives the satellite that permission?' I thought for a second and said 'Nobody. The satellite doesn't need your permission. If you don't like it, shoot it down.' Then he laughed." No wonder experienced, remote sensing practitioners are somewhat scarce.

Later satellites of the LANDSAT series continued to include an MSS multispectral scanner with the same four spectral bands as LANDSAT I. Thus, data from those bands have been continuously collected since 1972, constituting an archived data set that is extremely important for all scientific disciplines, but especially for environmental monitoring. In the 1980s, the Thematic Mapper (or TM) multispectral scanner with seven spectral bands was added to the LANDSAT series. All TM bands have a 30-meter pixel size except for the thermal infrared band, which has a 120-meter pixel.

Colorplates CP1, CP2, and CP3 provide an example of how an observer's perspective changes from vantage points on the ground (looking East), on a high-flying aircraft and on an orbiting LANDSAT satellite, for an area in the southwest corner of the Wind River Basin, Wyoming, where a stream intersects the uplifted Chugwater Formation of Triassic age. The ground photo covers roughly 2 km $\times$ 2 km in area, the original aerial photo covers approximately 9.23 km $\times$ 9.23 km, and the original LANDSAT image covers approximately 185 km $\times$ 185 km, or about 400 times as much area as the aerial photo and about 8500 times as much as the ground photo. This scale implies that it would take no less than 400 aerial photos of 1:40,000 scale mosaicked together to cover the area in one LANDSAT frame. The aerial photos would be collected at different times of day and possibly on different days, so that differences in illumination complicated the mosaicking process, increasing its expense. In contrast, an entire LANDSAT frame is collected within 20 seconds, with no artifact boundaries within the image to confuse the interpreter. This is just one comparison that demonstrates the important contribution to geological and environmental remote sensing made by the synoptic coverage of LANDSAT.

The wavelength limits of LANDSAT MSS and TM spectral bands are given in Appendix B. Also in Appendix B are the spectral bands and pixel sizes of several other Earth resources satellites that have or will be orbited in the 1990s. This explosive increase in the availability of multispectral satellite data in the last decade of this millennium almost ensures that remote sensing will be one of the most important technologies of the next century.

When an observer modifies the term *remote sensing* with the adjective *multispectral,* it is equivalent to asking the question, What is the composition of the observed object? The change in reflectance or emittance of an object with a change in wavelength is controlled in the ultraviolet and visible wavelengths by the object's trace element content, and in the reflective infrared and thermal infrared wavelength regions it is controlled by the object's bulk composition. When applied to geology, chemical and/or mineral composition of the observed object is an even more important attribute than size or shape, though such spatial attributes are also useful.

Rocks are composed of minerals, and no two minerals have both the same chemical composition and crystalline structure, a fact that implies that no two minerals will have the same spectral reflectance at all wavelengths. Thus, geological remote sensing truly is multispectral. For geology, remote sensing offers greater compositional discrimination of exposed rocks and soils, easier accessibility to the study area, and better synoptic view than would be possible from a geologist in the field. However, the field geologist is still indispensable for interpretation and verification of remote sensing results, particularly for inferring what is at depth after remote sensing has observed the surface.

Geology is not the only discipline that benefits from multispectral remote sensing. Botanical remote sensing for total biomass and vegetative stress also benefits and requires fewer spectral bands than does geological remote sensing; recent studies (Martin and Aber 1993; Curran and Kupiec 1995) have shown that remote sensing for canopy chemistry (discussed in Chapter 3) can be successful if data from a plethora of narrow spectral bands (hyperspectral data) are available. Despite the fact that both fresh and salt water are opaque to most of the infrared portion of the electromagnetic spectrum, there are many multispectral oceanographic applications, including bathymetry and biological productivity. Broad-band thermal infrared measurements are very useful for sea surface temperature (SST) mapping. Atmospheric remote sensing requires perhaps the most demanding multispectral remote sensing methods, since there are many spectrally narrow, gaseous absorption bands located throughout the ultraviolet, visible, and infrared wavelength regions.

The previous omission in this chapter of wavelength regions longer than thermal infrared, called the microwave and radio wave regions, is due to the fact that such long wavelength radiation is not as sensitive to the chemical composition as it is to the physical shape and size of an object. It does not imply that those longer wavelengths are not important for geological and environmental remote sensing, because they certainly are. Although microwaves, along with infrared waves, can be observed passively as heat given off by objects, they can also be artificially beamed down to the object and the backward-reflected part of the beam can be detected to yield images of the object. This kind of active microwave sensing is called radar. It was developed during World War II primarily to detect flying aircraft, day or night and in all weather, since radar beams can be sent through either clouds or darkness. Side-looking radar (SLAR) was created to image terrain under any weather or time of day conditions (Reeves et al. 1975); in the 1960s, synthetic aperture radar (SAR) was developed at Willow Run Laboratories of the University of Michigan as a device that could remove the range-dependence of radar images and make those images more geometrically similar to aerial photos. Over land, radar is important for such disparate uses as mapping flood extent while clouds are still overhead; for geological structure mapping and digital elevation model extraction in perennially cloudy regions where higher-resolution visible stereo images are unavailable; and for determining slight elevation changes caused by earthquakes (Peltzer and Rosen 1995). Over the ocean, radar is becoming increasingly important for ocean circulation and current mapping, sometimes being used as a simple altimeter instead of an imager, as well as for mapping sea ice and oil slicks floating on the ocean surface. Listed in Appendix B are two recent SAR satellites, ERS-1 and RADARSAT.

ORGANIZATION OF PRESENTED MATERIAL

The material in this book is organized with the basics first and the applications last, as discussed in the Preface. The interaction of electromagnetic radiation with natural materials on Earth's surface is the first topic, followed by the spectral properties of important Earth materials and how these properties are related to the chemical composition of those materials. Spectral radiance available from Earth's surface and its measurement by different types of commercially available electro-optical, multi-spectral sensors is the next topic.

The next two topics deal with how remote sensing data are processed to yield useful information. Several multispectral image processing methods that can provide geologically important compositional information are discussed. Appendix C contains LANDSAT Thematic Mapper (TM) and MSS ratio codes (explained in the appendix) that were calculated for more than a hundred mineral laboratory reflectance spectra from 0.4–2.5 μm wavelengths that are contained in the Jet Propulsion Laboratory (JPL) Pilot Land Data System (Grove et al. 1992). A few vegetative spectra were added for good measure. These TM ratio codes can be used to determine which spectral ratio image combinations are best to employ for the enhancement of any particular mineral, and what color the mineral of interest is expected to be in the resulting color-ratio image. Three important spatial processing methods are then described: digital photogrammetry, which yields elevation data and digital orthophotos that are important for the production of geographic information system (GIS) databases; geophysical imaging, which yields information about underground geologic structures that have subtle inhomogeneities in topography, magnetic susceptibility, and density; and automatic linear feature extraction, which can be used for the objective mapping of faults in Earth's crust.

The remaining topics are all devoted to applications. Remote sensing provides databases for GIS manipulation, which combine into a powerful management tool for a wide spectrum of uses. However, nowhere is that tool more cost-efficient than in resource exploration and extraction, especially now that transportation costs to market have become such a large component of commodity prices. Remote sensing costs to society are, in my estimation, more than justified by the cost savings in exploration of minerals, petroleum, and ground water alone. Multispectral data reveal so much new mineral information otherwise invisible to the eye that remote sensing may actually moderate commodity prices of some minerals for which the technology is most potent, such as gold, by providing greater supply. The great benefit of remote sensing petroleum exploration is as a means to reduce the areal extent and cost of collecting and processing seismic data, which is over 100 times more expensive than remote sensing on a per unit area cost basis.

The use of remote sensing for geological engineering and environmental applications is discussed in the last part of this book with GIS applications included in some, but not all case studies. An example of a new type of GIS/remote sensing application is highlighted in the section on pipeline route selection, whereby a surface water run-off algorithm is used by Feldman et al. (1994) to determine the lowest-cost path of the pipeline. The use of multispectral remote sensing for the imaging of methane and other gases is presented within the context of solid-waste landfill monitoring.

The uses of multispectral remote sensing for plate tectonic research and global environmental monitoring are intertwined in this book because they share a global scope and may even have some cause and effect relationships. Environmental monitoring of the planet is broadened to include the subjects of geological hazards and anthropogenic contributions of gases to the atmosphere.

Remote sensing is a mature technology in terms of data collection and image processing. However, it is far less mature in environmental and engineering geology applications than in previously better-funded applications, such as defense mapping, mineral exploration, and petroleum exploration. More research is needed in the environmental and engineering areas before the full potential of remote sensing for these important applications can be realized.

Cited References

ACKERMAN, F. 1994. Digital Elevation Models-Techniques and Application, Quality Standards, Development. In *Proceedings of the Symposium on Mapping and Geographic Information Systems, Athens, Georgia,* vol. 30, no. 4, 421–432. International Society for Photogrammetry and Remote Sensing.

CURRAN, P. J., and J. A. KUPIEC. 1995. Imaging Spectrometry: A New Tool for Ecology. In *Advances in Environmental Remote Sensing,* ed. F. M. Danson and S. E. Plummer, 71–88. New York: John Wiley & Sons.

FELDMAN, S. C., R. E. PELLETIER, W. E. WALSER, J. C. SMOOT, and D. AHL. 1994. Integration of Remotely Sensed Data and Geographic Information System Analysis for Routing of the Caspian Pipeline. In *Proceedings of the Tenth Thematic Conference on Geologic Remote Sensing,* vol. 2, 206–213. Ann Arbor: Environmental Research Institute of Michigan.

GROVE, C. I., S. J. HOOK, and E. D. PAYLOR II. 1992. *Laboratory Reflectance Spectra of 160 Minerals, 0.4 to 2.5 Micrometers.* JPL Publication 92-2. Pasadena, Calif.: Jet Propulsion Laboratory.

JENSEN, J. R. 1995. Issues Involving the Creation of Digital Elevation Models and Terrain Corrected Orthoimagery Using Soft-Copy Photogrammetry. *Geocarto International* 10, no. 1: 5–21.

MARTIN, M. E., and J. D. ABER. 1993. Measurements of Canopy Chemistry with 1992 AVIRIS Data at Blackhawk Island and Harvard Forest. In *Proceedings of the Fourth Annual JPL Airborne Geoscience Workshop,* vol. I, AVIRIS Workshop, ed. R. O. Green, 113–116. Pasadena, Calif.: Jet Propulsion Laboratory.

PELTZER, G., and P. ROSEN. 1995. Surface Displacement of the 17 May 1993 Eureka Valley, California, Earthquake Observed by SAR Interferometry. *Science* 268, no. 5215: 1333–1336.

REEVES, R. G., A. ANSON, and D. LANDEN, eds. 1975. *Manual of Remote Sensing.* Falls Church, Va.: American Society of Photogrammetry.

SLAMA, C. C., C. THERUER, and S. W. HENRIKSEN, eds. 1980. *Manual of Photogrammetry.* 4th ed. Falls Church, Va.: American Society of Photogrammetry.

VINCENT, R. K., M. A. TRUE, and D. V. ROBERTS. 1987. Automatic Extraction of High-Resolution Elevation Data Sets from Digitized Aerial Photos and Their Importance for Energy Mapping. In *National Computer Graphics Association's Mapping and Geographic Information Systems 1987 Proceedings,* 203–210. San Diego, Calif.: NCGA.

2

Interaction of Electromagnetic Radiation with Natural Materials on Earth's Surface

▼ ▼ ▼ ▼ ▼ ▼ ▼ ▼ ▼ ▼ ▼ ▼ ▼ ▼ ▼

ELECTROMAGNETIC RADIATION AND THE SEARCH FOR COMPOSITIONAL INFORMATION

As electromagnetic waves interact with materials on the earth's surface, they can yield important information about the chemical (and in the case of rocks and soils, the mineralogical) composition of those materials. The discussion of the interaction of electromagnetic radiation with vegetation and water will be deferred to the following chapter, where spectra of natural materials are presented. Excepting the immediately following discussion about the electromagnetic spectrum, this chapter will focus primarily on the interaction of electromagnetic radiation with exposed soils and rocks.

Figure 2.1 (after the Hughes Santa Barbara Research Center Wall Chart of 1991) shows the entire spectrum of electromagnetic waves. The shortest wavelength photons (less than 1 Å, or 0.0001 μm) are gamma rays, which are emitted or absorbed by changes in the energy state of neutrons or protons in the nucleus of an atom. Thus, gamma rays yield information only about the nucleus. Alpha, beta, and neutron rays also offer information about the nucleus, but they are particles in motion, not photons of negligible mass.

X-rays, the second shortest wavelength category of electromagnetic radiation (from 0.0001–0.01 μm), consist of photons emitted or absorbed by the inner-shell electrons of atoms. Ultraviolet (0.01–0.4 μm), visible (0.4–0.67 μm), and reflective infrared (0.67–1.0 μm as near infrared, plus 1.0–3.0 μm as short wavelength infrared, or SWIR) photons are emitted or absorbed primarily by changes in the energy states

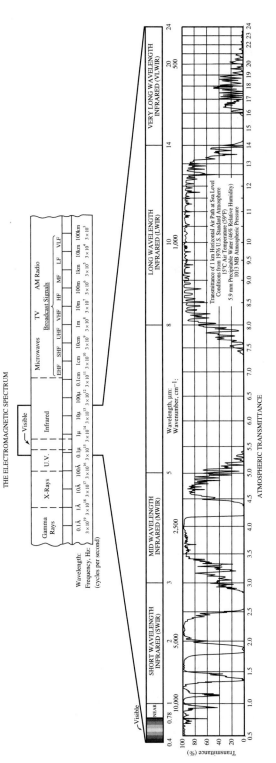

Figure 2.1 Spectrum of electromagnetic waves. (After the Hughes Santa Barbara Research Center Wall Chart 1991). (*Image courtesy of Santa Barbara Research Center.*)

of electrons in the outer shells of transition metal ions that are located at sparsely distributed sites in the crystalline structure of a solid. Therefore, the visible and infrared wavelength regions dominated by reflected sunlight (0.4–3.0 μm) contain information about the presence of even trace quantities of transition metals. Transition metals, which include iron, nickel, zinc, copper, manganese, chromium, titanium, vanadium, cobalt, and scandium, are the pigments of nature because of their partially occupied 3d or 4s electronic energy levels. (The exception is zinc, which has both of these energy levels fully occupied.) Ferric iron ions (Fe^{3+}), which color rocks and soils various shades of red, yellow, orange, purple, and rusty brown, are the most common pigments. Ferrous iron ions (Fe^{2+}) and cuprous ions (Cu^{2+}) create colors of green and blue in rocks and soils.

Of course, there are other colors created that cannot be seen by the naked eye, as will be discussed in more detail in the next chapter. Some outer-shell electronic transition absorption bands occur in the reflective infrared region (0.67–3.0 μm). The 1.4–4.0 μm wavelength region also contains overtone and combinational bands of some primary vibrational bands that occur in the 2.5–14.0 μm wavelength region. Solar radiance peaks in the visible light region and declines rapidly toward 3 μm. Earth's surface at room temperature emits maximum thermal infrared radiation near 10 μm, and the magnitude of the emission also declines toward 3 μm. Thus, at room temperature (300 K), 3 μm is taken to be the approximate cross-over point between the reflective infrared and thermal infrared wavelength regions. We will discover in Chapter 4, that 4 μm is a more accurate cross-over point.

Figure 2.1 shows the 3.0–5.0 μm wavelength region as the midwavelength infrared (MWIR) and the 8.0–14.0 μm wavelength region as the long wavelength infrared (LWIR). The 5.0–8.0 μm wavelength region is not named because Earth's atmosphere absorbs almost all electromagnetic radiation in this region. The very long wavelength thermal infrared region (VLWIR) actually extends from 14 μm to 1000 μm (0.1 cm).

In the thermal infrared wavelength region, it is primarily the twisting, rotational, and vibrational motions among ions of a compound that absorb or emit photons. Therefore, thermal infrared (MWIR and LWIR) photons yield information about molecules (inter-atomic bond strengths), whereas the previously discussed shorter wavelength regions yield information about elements (intra-atomic electronic transitions). The thermal infrared region is especially useful for geological remote sensing because multiple, medium-width thermal infrared spectral bands provide information about the bulk mineralogy of rocks and soils, while a single, broad-band thermal infrared spectral band can tell us whether they are hot or cold. Figure 2.2 summarizes relationships between spectral region and physical origin of compositional information that is available through multispectral remote sensing.

The microwave region begins at a wavelength of 1000 μm (0.1 cm) and extends upward to about 10 meters (through the TV wavelengths of Figure 2.1); radio waves are above 10 meters in wavelength. X-band and L-band radar waves are near 3.0 cm and 27 cm, respectively, in wavelength. All of the wavelength regions longer than the thermal infrared region yield very little information about composition, but can yield much about temperature and terrain roughness or particle size (such as whether the particles are boulder, gravel, sand, or clay). Passive or active (radar) mi-

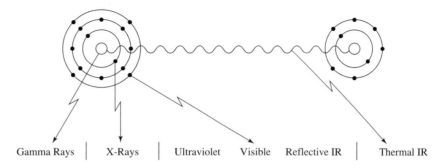

Figure 2.2 Relationship between spectral region and types of compositional information available from multispectral remote sensing.

crowaves can "see" through thick clouds that remote sensors from other wavelength regions cannot penetrate.

The principal point of the foregoing discussion is that the upper ultraviolet through the thermal infrared wavelength regions contain most of the information that can be obtained by remote sensing methods about the chemical and mineral composition of materials at Earth's surface. However, because of atmospheric absorption, not all of that wavelength region is accessible to us. Absorption of photons by gaseous constituents of the atmosphere dictates that remote sensing of Earth's surface be performed within the confines of what are called atmospheric windows—spectral regions of relatively low atmospheric absorption. Those atmospheric windows (Figure 2.1) permit remote sensing in the ultraviolet, visible, and reflective infrared region from about 0.3–2.50 μm and also in the thermal infrared regions from 3.0–5.0 μm, 8.0–14.0 μm, and 16.0–21.0 μm. The exceptions include a few water-absorption bands in the reflective infrared, a narrow carbon dioxide band near 4.2 μm, an ozone band near 9.4 μm, and numerous water bands in the 16.0–21.0 μm wavelength region.

THE EFFECT OF SINGLE-GRAIN INDICES OF REFRACTION AND ABSORPTION ON THE REFLECTANCE PROPERTIES OF A BEACH

Imagine a beach on a beautiful tropical island. What controls the amounts and colors of light, broadly defined as electromagnetic radiation of any wavelength, that reflects off the beach surface? To answer this question thoroughly, we must consider the multispectral nature of light and how it interacts with a single grain of beach sand.

Figure 2.3 shows interaction of electromagnetic radiation with the top layer of sand grains on the beach. When an incident ray of electromagnetic radiation strikes an air/grain interface, part of the ray is reflected and part of it is transmitted into the sand grain. The solid lines in Figure 2.3 represent the incident rays, and dashed lines 1, 2, and 3 represent rays reflected from the surface but that have never penetrated a sand grain. The latter are called specular rays by Vincent and Hunt (1968), or surface-scattered rays by Salisbury and Wald (1992); these rays result from first-surface reflection from all grains encountered. For a given reflecting

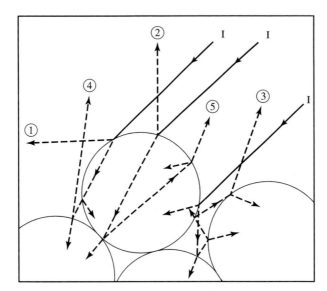

Figure 2.3 Interaction between electromagnetic radiation and the top layer of particles comprising a mat surface. Solid lines (I) represent incident rays, lines 1–3 represent specular rays, and lines 4 and 5 are volume rays. (After Vincent and Hunt 1968.) *(Courtesy of "Applied Optics," Optical Society of America.)*

surface, specular rays are all reflected in the same direction, such that the angle of reflection (the angle between the reflected rays and the normal, or perpendicular to the reflecting surface) equals the angle of incidence (the angle between the incident rays and the surface normal).

The measure of how much electromagnetic radiation is reflected off a surface is called its reflectance, which is a number between 0 and 1.0. A measure of 1.0 means that 100% of the incident radiation is reflected off the surface, and a measure of 0 means that 0% is reflected. In the case of first-surface reflection, this measure is called the specular reflectance, which will be designated here as $r^s(\lambda)$. The λ in parentheses indicates that specular reflectance is a function of wavelength. The reason that $r^s(\lambda)$ is a function of wavelength is that the complex index of refraction of the reflecting surface material is dependent on wavelength. The term *complex* means that there is a real and imaginary part to the index of refraction. Every material has a complex index of refraction, though for some materials at some wavelengths, only the real part of the complex index of refraction may be nonzero.

For a sand grain with complex index of refraction $N(\lambda) = n(\lambda)[1 - ik(\lambda)]$, the specular reflectance is expressed by Fresnel's equation (Jenkins and White 1957), as follows:

$$r^s(\lambda) = \frac{[n(\lambda) - 1]^2 + n^2(\lambda)k^2(\lambda)}{[n(\lambda) + 1]^2 + n^2(\lambda)k^2(\lambda)} g(\theta, \phi) \qquad \text{(Eqn. 2.1)}$$

where

$r^s(\lambda)$ = specular reflectance of one reflecting grain ($0 \le r^s(\lambda) \le 1$)

$n(\lambda)$ = intrinsic spectral index of refraction of the grain

$k(\lambda)$ = intrinsic spectral index of absorption of the grain

$g(\theta, \phi)$ = a generally nonzero function of the angle of incidence (ϕ) and angle of observation (θ) with respect to the macroscopic surface

The final reflectance of a specular ray bouncing off multiple grains of sand is simply the multiplicative product of specular reflectances from all of the encountered air/grain interfaces. For instance, if the specular reflectance of three grains for a particular wavelength of electromagnetic radiation were 0.9, 0.8, and 0.7, respectively, the final reflectance of a specular ray bouncing off all three grains would be $(0.9)(0.8)(0.7) = 0.504$. The specular reflectance of the beach surface, $R^s(\lambda)$, is the average of all the individual specular ray reflectances.

Rays of electromagnetic radiation that have been transmitted through some portion of one or more grains are called volume rays. These are shown as dashed lines 4 and 5 in Figure 2.3. The equation for the volume reflectance, $r^v(\lambda)$, of a sand grain is complicated because it depends on both the transmittance of the grain and the interface reflectance off the top of that grain and the underlying grain(s). More information on this subject can be found elsewhere (Vincent and Hunt 1968; Hapke 1981, 1984, 1986, and 1993; Salisbury and Wald 1992). For purposes of this text, it is sufficient to say that the spectral transmittance of a single grain is the greatest controlling factor on the size of $r^v(\lambda)$, as long as the grain size is appreciably larger than the wavelength of incident electromagnetic radiation so that interference effects can be ignored. This observation is certainly true of sand-sized particles with respect to visible and infrared light. The spectral transmittance of a sand grain is controlled by the index of absorption, $k(\lambda)$, as described by the Beer-Lambert law:

$$T(\lambda) = e^{\frac{-4\pi k(\lambda)t}{\lambda}}$$ (Eqn. 2.2)

where

$T(\lambda)$ = spectral transmittance of the grain
t = thickness of the grain

The average of $r^v(\lambda)$ for all the grains in the beach from which electromagnetic radiation is reflected is defined as the volume reflectance of the beach, $R^v(\lambda)$. The total reflectance of the beach, $R^T(\lambda)$, is the averaged sum of the specular and volume reflectances, as follows:

$$R^T(\lambda) = \frac{[R^s(\lambda) + R^v(\lambda)]}{2}$$ (Eqn. 2.3)

The dependencies of $R^s\lambda$ and $R^v(\lambda)$ on $k(\lambda)$ are markedly different, as is demonstrated in Figure 2.4 for the case of a uniform grain size and varying wavelength. As $k(\lambda)$ approaches infinity, Equation 2.1 shows that $r^s(\lambda)$, and, therefore, $R^s(\lambda)$, approaches a maximum of 1.0, while Equation 2.2 shows that $r^v(\lambda)$, and, therefore, $R^v(\lambda)$, approaches 0, because no light is transmitted through any grain. In lesser extremes, this implies that $R^s(\lambda)$ increases and $R^v(\lambda)$ decreases with increasing $k(\lambda)$, as Figure 2.4 shows. $R^T(\lambda)$ is an averaged blend of the specular and volume reflectances.

When mineral indices of refraction or absorption are listed in handbooks, they are usually for the sodium line emission wavelength (0.5893 μm, or 5893 Å). In reality, however, the refractive and absorptive indices of a grain vary with the wavelength of light impinging upon that grain. The λ subscripts in the above equations are

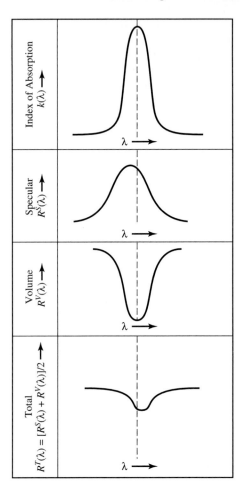

Figure 2.4 Absorption coefficient, specular reflectance, volume reflectance, and total reflectance vs. wavelength for a spectral feature. (After Vincent and Hunt 1968.) *(Courtesy of "Applied Optics," Optical Society of America.)*

reminders of that effect. In fact, all of the information about chemical or mineral composition discussed in the previous section is embedded in those λ subscripts, and we must "think multispectrally" in order to gain compositional information about any object under our scrutiny.

Let us return to the example of our beautiful tropical island beach. The principal constituent of sand on most beaches is quartz. Its great resistance to chemical weathering makes quartz the most common silicate mineral brought to the sea by rivers and streams. In tropical and subtropical seas, however, the most common constituent of beach sand can be calcite (calcium carbonate), which is a product of carbonate reef erosion. Therefore, our beach could be chiefly composed of either or both minerals. For either quartz or calcite grains of sand that have undergone significant weathering, each granular face is likely to be smooth.

The spectral index of refraction at most wavelengths for most materials is much larger than the spectral index of absorption. For instance, throughout the visible spectrum (0.40–0.67 μm), $n(\lambda)$ for both quartz and calcite is typically around 1.5, whereas $k(\lambda)$ is typically around 10^{-8} in magnitude. For air in the visible wavelength region, $k(\lambda)$ is even smaller than 10^{-8}, and $n(\lambda) = 1.0$ (by definition). These relationships explain why both the numerator and denominator of $r^s(\lambda)$ in Equation 2.1 have a 1 representing the index of refraction of the medium surrounding the grain.

Because the intrinsic spectral index of absorption of the grain $k(\lambda)$ is so low in the visible wavelength region for quartz and calcite, neither material absorbs very much for sand-sized and smaller grains ($t \leq 250$ μm), which makes them transparent in that wavelength region. Most of the light that enters a grain from one side can pass all the way through with little loss, exiting from the other side and reflecting off another grain of sand. This sequence implies that most of the reflected rays coming off the beach surface in the visible wavelength region are volume rays that have passed through one or more sand grains. Therefore, the volume reflectance component dominates the total reflectance in Equation 2.3 for the visible wavelength region, regardless of whether the sand grains are quartz or calcite.

However, quartz and calcite are not transparent at all wavelengths. In the thermal infrared wavelength region, both types of common tropical beach sand materials exhibit absorption bands that cause the spectral indices of refraction and absorption to vary greatly with relatively small changes in wavelength. This is another way of saying that quartz and calcite are anomalously dispersive in the thermal infrared wavelength region.

In the 2–25 μm wavelength region, a number of absorption bands that give rise to strong reflectance maxima occur in silicates and carbonates as a result of primary stretching and bending of ligands or bonds between ions in the crystal lattice. These reflectance bands, all caused by interatomic motions, have been given the German name *reststrahlen,* which means residual rays. They are so called because at the turn of this century, crystals (especially of salts) were used as reflectance filters in the longer infrared wavelength regions.

Figure 2.5 shows spectral thermal infrared reflectance curves for fractured and cleaved surfaces of quartz and calcite, respectively (Salisbury et al. 1991), and Figure 2.6 shows $n(\lambda)$ and $k(\lambda)$ for the ordinary and extraordinary rays (two different polarizations) of both materials (Vincent 1972). It is clear from Figure 2.6 (for either polarization) and Equation 2.2 that quartz is extremely opaque, because $k(\lambda)$ is large, at wavelengths around 9 μm. Quartz opacity is caused by the vibrations of silicon when bounded by four oxygen atoms in an SiO_4 tetrahedron. Even the smallest of grains would gobble up all light of this 9 μm wavelength that was not reflected off the first surface. Likewise, it is clear from Figure 2.6 that calcite is extremely opaque to the ordinary ray at wavelengths near 7.1 μm; calcite opacity is caused by the internal vibrations between carbon and oxygen ions of the CO_3^{-2} anion in calcium carbonate ($CaCO_3$). In other words, for both ordinary and extraordinary rays in quartz near 9.0 μm or for the ordinary ray in calcite near 7.1 μm, there would be no volume reflectance, leaving specular reflectance in Equation 2.3 as the only component of total reflectance.

In summary, the tropical beach, whether composed of quartz or calcite, would exhibit mostly volume reflectance in the visible wavelength region, but would exhibit mostly specular reflectance for certain wavelengths in the thermal infrared wavelength region.

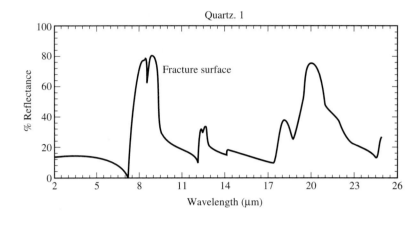

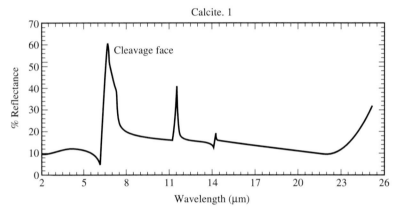

Figure 2.5 Thermal infrared reflectance spectra of relatively smooth surfaces of quartz and calcite crystals. (After Salisbury et al. 1991.) *(Reprinted by permission of the Johns Hopkins University Press.)*

THE EFFECT OF GRAIN SIZE ON THE REFLECTANCE PROPERTIES OF A BEACH

Why should we care whether the specular component or the volume component dominates the total reflectance of the beach? The answer is that a change in the average grain size of the beach has an opposite effect on the specular and volume reflectances. Therefore, for materials like quartz and calcite (and indeed for most materials), which exhibit volume reflectance domination of total reflectance at some wavelengths and specular reflectance domination at other wavelengths, a given change in grain size may cause the total reflectance of the beach to increase at some wavelengths and decrease at other wavelengths.

What would you expect to happen to visible light reflected off transparent sand grains if the average grain size of the beach (typically greater than 250 μm) were reduced by half? Would the amount of light reflected off the beach increase or

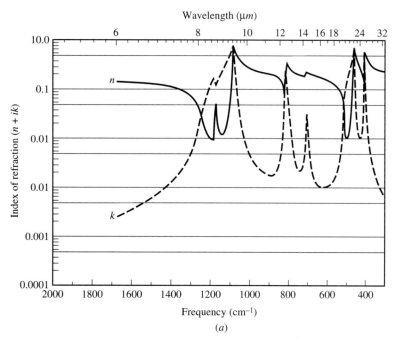

(a)

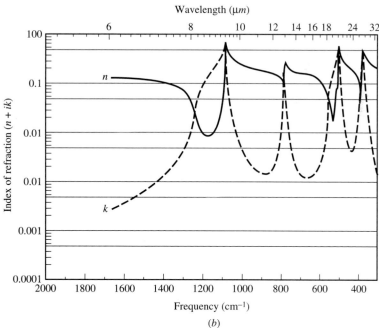

(b)

Figure 2.6 Complex index of refraction of (a) the ordinary index of quartz and (b) extraordinary index of quartz. (After Vincent 1972.) *(Courtesy of R. K. Vincent.)*

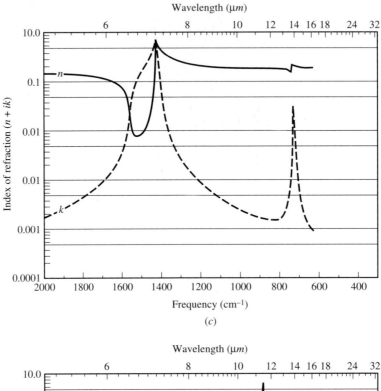

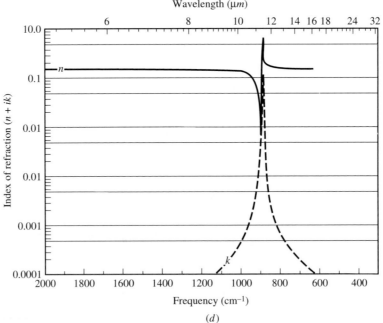

Figure 2.6 Complex index of refraction of (c) ordinary index of calcite and (d) extraordinary index of calcite. (After Vincent 1972.) *(Courtesy of R. K. Vincent.)*

decrease? Well, halving the grain size of unpacked sand would cause the light to encounter twice as many air–grain interfaces per unit of path-length traveled, thereby increasing the amount of light backscattered from the beach surface. Therefore, for wavelengths of light at which a material is fairly transparent (dominated by volume reflectance), decreased grain size will result in an increased brightness of a beach of such material. This is why finer-grained beaches, whether quartz or calcite in composition, reflect more visible sunlight than coarser-grained beaches of the same composition. Hence, they get brighter as the sand gets finer grained.

An example of this behavior, where smaller grain size yields higher reflectance, is shown in Figure 2.7 for calcite in the 0.4–2.5 μm wavelength region (Grove et al. 1992). The cross-over of reflectance spectra for the two largest grain sizes in the 0.6–0.9 μm wavelength is an artifact caused by the way the samples were handled. Investigators at the Jet Propulsion Laboratory smoothed the samples with a spatula, which preferentially packs the smaller size range, changing its reflectance compared to that of sifted samples.

An example of the opposite behavior, where smaller grain size yields lower reflectance, is shown in Figure 2.8 for pyrite (Grove et al. 1992). Pyrite evidently has large enough refractive and absorption indices in the 0.4–2.5 μm wavelength region to endow it with dominantly specular reflectance behavior throughout the region. Each pyrite grain encountered absorbs some of the light entirely; the more surfaces encountered, the greater the reduction in overall reflectance of the total sample surface. This effect can be more easily understood by realizing that even a

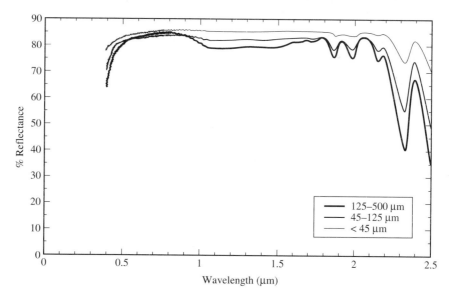

Figure 2.7   Directional reflectance in the 0.4–2.5 μm wavelength region for three calcite particulate sizes: (a) 125–250 μm, (b) 45–125 μm, and (c) less than 45 μm. Note how reflectance is generally lower for larger particle sizes because volume reflectance dominates throughout most of this wavelength region. *(Provided through the courtesy of the Jet Propulsion Laboratory, California Institute of Technology, Pasadena, California.)*

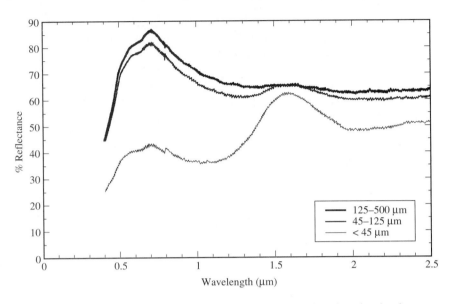

Figure 2.8   Directional reflectance in the 0.4–2.5 μm wavelength region for three pyrite particulate sizes: (a) 125–250 μm, (b) 45–125 μm, and (c) less than 45 μm. Note how reflectance is generally higher for larger particle sizes because specular reflectance dominates throughout most of this wavelength region. *(Provided through the courtesy of the Jet Propulsion Laboratory, California Institute of Technology, Pasadena, California.)*

near-perfect mirror has a reflectance less than 1.0 in value, and 10 mirrors will have a combined reflectance that is less than the reflectance of one mirror, as demonstrated in the previous section for the example of specular reflectance off three sand grains in succession.

At wavelengths near 9 μm, quartz acts only as a specular reflector, like a mirror. Decreasing the grain size increases the number of mirrors off which the light reflects before it leaves the beach surface, thereby decreasing specular reflectance. This relationship implies that if a quartz beach were observed in reflection by shining thermal infrared radiation of 9 μm wavelength on it, the beach would get darker as the grain size became smaller, just the opposite of what happens in the visible wavelength region for the same quartz sand. Therefore, for wavelengths of light at which a material is fairly opaque, decreased grain size will result in a decreased reflected brightness of a beach of such material. However, once the grain size becomes less than the optical depth of the grain, the behavior of the reflectance as a function of grain sizes changes (Salisbury and Wald 1992) because there is once again a volume reflectance contribution to the total reflectance.

A similar argument can be made for calcite near a wavelength of 7 μm. Figure 2.9 shows reflectance spectra of a polished slab (most calcite grains of sand will not be this smooth) and of several grain sizes of calcite, where both ordinary and extraordinary ray polarizations of reflected infrared radiation are present (Vincent and Hunt 1968). The strong absorption band around 7 μm in wavelength causes the reflectance to decrease

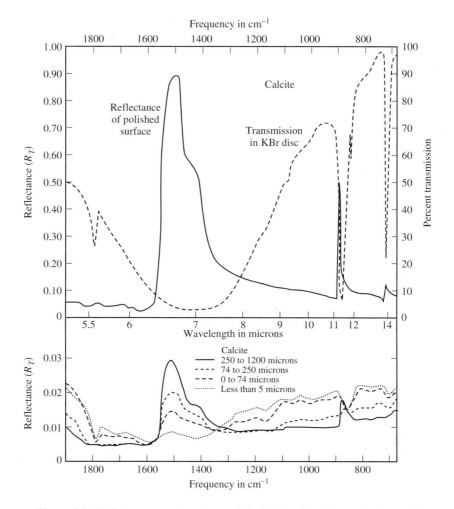

Figure 2.9 Reflectance spectra of a polished slab of calcite and of several particulate sizes of calcite (1–4), where both ordinary and extraordinary ray polarizations of reflected infrared radiation are present. (After Vincent and Hunt 1968.) *(Courtesy of "Applied Optics," Optical Society of America.)*

with decreasing grain size of the sample. As discussed in the previous paragraph, this decrease is the expected behavior for opaque grains, as long as grain size is greater than optical depth of the grain. It just so happens, however, that the atmosphere at this wavelength region is practically opaque (Figure 2.1) because of atmospheric water vapor. Consequently, the strongest absorption band of calcite cannot be well exploited in terrestrial remote sensing.

The calcium carbonate absorption band near 11.3 μm wavelength, however, is within an atmospheric window and can be a useful example of what happens to light backscattered from grains at wavelengths where there is a change from opaque to transparent behavior as grain size is reduced. As shown in Figure 2.9, the medium-strength absorption band around 11.3 μm produces a flip in spectral reflectance from a maximum to a minimum (centered at a slightly different wavelength) as grain size is reduced. Calcite goes from opaque behavior to transparent behavior in the vicinity of this 11.3 μm absorption band as grain size decreases, because $n(\lambda)$ and $k(\lambda)$ change moderately with changing wavelength, rather than strongly, as was the case of the calcite absorption band near 7 μm. Another way of describing this effect is that volume scattering produces absorption bands that are exhibited as reflectance minima because more photons are absorbed in absorption band centers, where $k(\lambda)$ is greatest, than in absorption band wings, where $k(\lambda)$ is lower. Conversely, surface scattering produces absorption bands that are exhibited as reflectance maxima because the surface reflectance is highest in absorption band centers, where $k(\lambda)$ is greatest, and is lower in absorption band wings, where $k(\lambda)$ is lower.

Three important observations can be summarized from the figures and discussion of this chapter thus far:

1. *The spectral locations of absorption bands depend on chemical composition of the material;* for example, quartz and calcite absorption bands in the thermal infrared wavelength region have different spectral locations because Si and O ions in quartz are connected by a "spring" with a different bond strength than that of the "spring" connecting C and O ions in calcite.

2. *The brightness, or magnitude, of the spectral reflectance depends primarily on the size of the reflecting grains.*

3. *Absorption bands appear as reflectance minima in transparent materials* (such as quartz and calcite in the visible wavelength region), *whereas absorption bands appear as reflectance maxima in opaque materials* (such as quartz at wavelengths near 9 μm or pyrite in the visible).

Note well that when we use the terms *transparent* or *opaque* to explain optical behavior, we must designate both a wavelength region and the material because the complex index of refraction of any material is generally not constant over large ranges of wavelength.

SEPARATION OF COMPOSITIONAL AND GRAIN-SIZE EFFECTS AND EXTRAPOLATION FROM BEACHES TO SOILS AND ROCKS

The total reflectance of our tropical beach depends on the grain size as well as the chemical composition of the beach materials. However, as the three summary observations of the previous section show, the effects of grain size and composition on beach reflectance are different. This distinction encourages us to separate the two effects.

When chemical composition is the physical parameter of greatest importance, it is helpful to suppress brightness differences between spectral reflectances of different locations on the beach at any given wavelength while enhancing relative brightnesses at different wavelengths for a given sample. The simplest and most straightforward method for achieving this goal is spectral ratioing, a detailed discussion of which is given in Chapter 5. Spectral ratioing involves the division of the reflectance recorded in one wavelength region by the reflectance recorded in another wavelength region for the same spot on the ground (in the case of remote sensing data) or on the sample (in the case of a lab sample). Instead of reflectance, a spectral parameter directly proportional to reflectance is often recorded; the spectral parameter recorded at one wavelength region is divided by the same parameter recorded at another wavelength to obtain a spectral ratio. There are several types of image processing algorithms that obtain this spectral ratio, some of which involve division of reflectance in one wavelength region by an average reflectance over a broader wavelength region (a variant of spectral ratioing) and some use first-derivative reflectance spectra (involving the change in reflectance as a function of wavelength, discussed in Chapter 3 as a method for mapping chemical properties of vegetative canopies).

Although this chapter has been devoted to spectral reflectance properties of a sandy beach, the reflectance properties of rocks are not so different. Rocks are assemblages of minerals that have interlocking grains or are bound together by various types of cement (usually silica or calcium carbonate cement). Particles in loose, unconsolidated media, like a sandy beach, are primarily separated by voids filled with air or water. Even when the particle size of a beach equals the grain size of a rock, there is a difference in what fills the interstices.

To consider the effect on reflectance of mixing several minerals together, we will take the simpler case of a particulate medium comprised of several mineral constituents, with air filling the interstices between particles. It is possible for us to estimate the spectral reflectance of a mixed-mineral particulate sample by using a linear combination of the reflectance spectra of its mineral constituents, weighted by the percentage of area on the sample's surface that is covered by each mineral constituent. The following equation describes this estimation for the total spectral reflectance of a mixed particulate sample at wavelength λ:

$$R^T(\lambda) = \sum_{i=1}^{n} f_i R_i^T(\lambda) \qquad \text{(Eqn. 2.4)}$$

where

f_i = fraction or percentage of the ith mineral constituent covering the sample surface, where $\sum_{i=1}^{n} f_i = 1.0$

n = total number of mineral constituents in the particulate sample

$R_i^T(\lambda)$ = spectral reflectance (the total of specular and volume reflectance) at wavelength λ for the ith mineral constituent alone

This linear mixing scheme becomes invalid when the particle sizes and spaces between particles become smaller than the typical distances over which scattering and interference of electromagnetic radiation occur. (As a rule of thumb, this distance is less than roughly five times the wavelength of incident electromagnetic radiation.) This scheme will also become invalid where optically thick and optically thin mineral particles are mixed (Hapke 1993).

Thus far we have talked about volume reflectance and specular reflectance according to whether electromagnetic rays did or did not penetrate one or more grains in a soil or rock surface. Now we need to define some reflectance terms that relate to the manner in which the soil or rock surface is illuminated, as well as how the reflected energy from its surface is measured. The most fundamental term for reflectance that will be used in this book is defined as *spectral hemispherical reflectance* or *diffuse reflectance* as follows:

$\rho^h(\lambda)$ = reflected electromagnetic radiation divided by the incident radiation (assuming that both the incident and reflected radiation are constant for any solid angle over the hemisphere above the plane of the surface of interest)

Two other reflectance terms must now be defined, because they relate to the laboratory spectra given in the next chapter. Figure 2.10 demonstrates the difference among these two and hemispherical reflectance (diffuse reflectance). In *directional hemispherical reflectance* measurements, the sample is illuminated by electromagnetic radiation that is incident from a small solid angle, described as directional, but the reflected light is collected from a solid angle of π steradians, described as hemispherical, such that the base of the collection hemisphere is centered on and parallel to the mean surface of the sample. In *biconical reflectance* measurements, both the il-

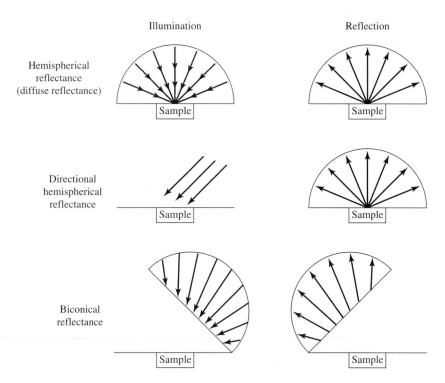

Figure 2.10 Diagrams of the illumination and reflection geometries of (a) hemispherical reflectance, (b) directional hemispherical reflectance, and (c) biconical reflectance. *(Courtesy of R. K. Vincent.)*

lumination and the collection of electromagnetic radiation is over π steradians, but the base of neither the illumination cone nor the collection cone is centered on the normal to the mean surface of the sample. Instead, the cones of illumination and collection, respectively, are centered on the angles of incidence and specular reflection, which are equal and on opposite sides of the sample surface normal. These two types of reflectances are different in that some forward scattering within the sample is permitted in the biconical case, because the collection hemisphere is centered about the angle of reflection and can extend below the sample surface, whereas no forward scattering within the sample is recorded in the directional hemispherical case, because the collection hemisphere is centered on the normal to the sample surface and does not extend below the sample surface.

The reflectance spectra for minerals in the following chapter are given in terms of *directional hemispherical reflectance* relative to a diffuse Halon standard in the 0.4–2.5 μm wavelength region, and of *biconical reflectance* relative to a mirror standard in the 2.0–25.0 μm wavelength region. The rock spectra in the following chapter are directional hemispherical reflectances for both the 0.4–3.0 μm and the 3–15 μm wavelength regions. Salisbury (1993) has shown, by measuring reflectance of rock samples with both directional hemispherical and biconical geometrical arrangements, that the two types of spectra are very similar, except for a multiplicative constant that is fairly independent of wavelength. The similarity holds especially if a "collar" is used on the biconical device to eliminate most forward scattering effects.

If $\rho(\lambda)$ is the directional hemispherical reflectance relative to a Halon standard that represents the ordinates of the 0.4–2.5 μm spectra in the following chapter, the diffuse reflectance (or hemispherical reflectance) for a rock or mineral surface can be given by:

$$\rho^h(\lambda) = b\rho(\lambda) \qquad \text{(Eqn. 2.5)}$$

Here b is a constant that is equal to $\pi/d\Omega_i \cos \theta_i$ for a Lambertian surface (a perfectly diffuse surface that reflects light equally into all solid angles of the hemisphere). Also, θ_i is the angle of incidence and $d\Omega_i$ the solid angle subtended by the incident radiation in the directional hemispherical measurement apparatus. If $\rho^c(\lambda)$ is the biconical reflectance that represents the ordinates of the 2.0–25.0 μm spectra in the previous chapter, the diffuse reflectance for a Lambertian surface can be given by:

$$\rho^h(\lambda) = b'\rho^c(\lambda) \qquad \text{(Eqn. 2.6)}$$

where b' is a constant that is different for every different biconical spectrometer.

What is important here is that both b and b' are independent of wavelength, under the assumptions that natural materials are Lambertian surfaces at all wavelengths and that the reflectance standards in both cases are spectrally flat (constant reflectance throughout the respective wavelength regions in which they are employed). These are acceptable assumptions, to first order, for the treatment of most terrain spectra, except for water and some types of vegetation that are very non-Lambertian, such as trees.

Two additional points should be made. First, although spectral reflectance was the primary optical parameter discussed above for all wavelengths, *the corollary optical parameter in the thermal infrared wavelength region is spectral emittance,* which is the supplement of spectral reflectance under conditions of thermal equilibrium. Kirchhoff's law states that under isothermal conditions (in which the surface in question is neither heating up nor cooling off), the following equation holds:

$$\epsilon(\lambda) = 1 - \rho^h(\lambda) \qquad \text{(Eqn. 2.7)}$$

where

$\epsilon(\lambda)$ = spectral emittance

Second, *not all minerals and rocks can be uniquely mapped in a practical manner,* even though their chemical compositions may be unique. Some minerals and rocks have more pronounced absorption bands than others and/or their absorption bands may occur at more favorable spectral locations, such as within an atmospheric window or where no other common materials exhibit absorption bands. Also, currently available remote sensing data collection systems do not collect multispectral data in some of the spectral bands that contain important compositional information, such as multiple thermal infrared spectral bands. Consequently, multispectral remote sensing methods can map some chemical compositions better than others. Analogously, it would be easier for you to pick up the correct airplane passenger on the sole basis of a verbal description if the passenger had purple hair, instead of white hair, especially if the arrival room was also white. If you were color blind, however, hair-color description, no matter how unique, would help only to the extent of predicting whether the passenger's hair was darker or lighter than the arrival-room walls.

REVIEW

The quest for new geological and environmental information and the reality of limited budgets make rapid mineralogically or chemically diagnostic mapping of Earth's surface highly desirable. The most efficient method for obtaining the data required for such mapping is with spaceborne and airborne multispectral sensors. This method requires those who employ remote sensing for compositional mapping to take a spectroscopist's view of the world, because the information obtained lies in interpreting rapid changes of spectral reflectance, absorption, and emittance as a function of wavelength. Unlike the laboratory spectroscopist, however, remote sensing scientists and engineers are constrained to work in more restrictive windows of atmospheric transparency because the atmospheric paths encountered during remote sensing applications are longer than those in the laboratory. Remote sensing scientists are motivated to image the multispectral data so that spatial relationships between different chemical compositions and where they occur are preserved.

The indices of refraction and absorption of single grains of soil or rock control the spectral locations of reflectance minima and maxima that contain the most important compositional information. Brightness variations across the scene contain information about the average grain size of the material under observation, as well as information about slope and aspect. As will be more fully discussed later, the separation of effects between brightness and absorption band spectral location will often be necessary for the creation of unambiguous information about chemical or mineralogical composition.

Although the next chapter will present laboratory reflectance spectra of minerals and rocks, the reflectances measured in the lab differ by multiplicative constants from hemispherical reflectance (diffuse reflectance), which is the same in all directions for a perfectly diffuse (Lambertian) surface. In the thermal infrared wavelength region, it is the spectral emittance (the complement of the hemispherical reflectance, under thermal equilibrium conditions) that is the optical parameter that carries most of the compositional information.

Not all rocks and minerals can be uniquely mapped with multispectral remote sensing data, given the limitations of current data collection systems.

Exercises

1. Electromagnetic radiation travels at the speed of light c (approximately 3×10^8 m/sec in a vacuum), and its wavelength is given by the equation $\lambda = c/v$, where v is the frequency in $\sec^{-1}$, or Hertz (Hz). What is the frequency (in Hz) of the following wavelengths of electromagnetic radiation: 0.45 μm, 1.0 μm, 12 μm, 3 cm, and 28 cm? What remote sensing wavelength regions do these fall into? *Note:* 1 μm $= 10^{-6}$m and 1 cycle/sec $= 1$ Hz.

2. Figure 2.6 shows plots of the real (n) and imaginary (k) parts of the index of refraction of quartz and calcite. (*Note:* The frequency in this plot is given the spectroscopist's units of cm^{-1}, and is defined as $10,000/\lambda$, where λ is the wavelength in μm.) Use Equation 2.1 (Fresnel's equation) with $g = 1.0$ to calculate the specular reflectance for the following values of n and k: $n = 1, k = .02; n = 0.1, k = 1.0; n = 8, k = 8; n = 2, k = .02$. For the ordinary ray of quartz (Figure 2.6), name at least one wavelength at which each of these four reflectances would occur. How do these calculated reflectances compare with polished quartz reflectances in Figure 2.5?

3. A diorite outcrop consisting of visibly dark minerals and visibly bright minerals weathers into clastic sediments of increasingly smaller grain size as they are carried farther away from the outcrop.

   (a) What will happen to the visible reflectance of the bright minerals and dark minerals, considered separately as two classes, as they move toward the sea?

   (b) Again considering the bright and dark minerals separately, what will happen to their respective reststrahlen band reflectances in the thermal infrared wavelength region?

Cited References

GROVE, C. I., S. J. HOOK, and E. D. PAYLOR II. 1992. *Laboratory Reflectance Spectra of 160 Minerals, 04. to 2.5 Micrometers.* JPL Publication 92-2. Pasadena, Calif.: Jet Propulsion Laboratory.

HAPKE, B. 1981. Bidirectional Reflectance Spectroscopy. 1. Theory. *Journal of Geophysical Research* 86:3039–3054.

———. 1984. Bidirectional Reflectance Spectroscopy. 3. Correction for Macroscopic Roughness. *Icarus* 59:41–59.

———. 1986. Bidirectional Reflectance Spectroscopy. 4. The Extinction Coefficient and the Opposition Effect. *Icarus* 264–280.

————. 1993. *Theory of Reflectance and Emittance Spectroscopy*. New York: Cambridge University Press.

HUGHES SANTA BARBARA RESEARCH CENTER, 1991. Spectrum of Electromagnetic Waves, Wall Chart.

JENKINS, F. A., and H. E. WHITE. 1957. Fundamentals of Optics. 3rd ed. New York: McGraw-Hill Book Co. 509–533.

SALISBURY, J. W. 1993. Chapter 4: Mid-Infrared Spectroscopy: Laboratory Data. In *Remote Geochemical Analysis: Elemental and Mineralogical Composition*, eds. C. M. Pieters and P. A. J. Englert, 79–98. Cambridge, England; Cambridge University Press.

SALISBURY, J. W., and A. WALD. 1992. The Role of Volume Scattering in Reducing Spectral Contrast of Reststrahlen Bands in Spectra of Powdered Minerals. *Icarus* 96:121–128.

SALISBURY, J. W., L. S. WALTER, N. VERGO, and D. M. D'ARIA. 1991. *Infrared (2.1–25 micrometer) Spectra of Minerals*. Baltimore, Md.: Johns Hopkins University Press.

VINCENT, R. K. 1972. Emission Polarization Study on Quartz and Calcite. *Applied Optics* 11:1942–1945.

VINCENT, R. K., and G. R. HUNT. 1968. Infrared Reflectance from Mat Surfaces. *Applied Optics* 7, no. 1: 53–59.

Additional References (Uncited)

HAPKE, B. 1993. Chapter 2: Combined Theory of Reflectance and Emittance Spectroscopy. In *Remote Geochemical Analysis: Elemental and Mineralogical Composition,* eds. C. M. Pieters and P. A. J. Englert, 31–42. Cambridge, England; Cambridge University Press.

3

Spectra of Earth Materials

▼ ▼ ▼ ▼ ▼ ▼ ▼ ▼ ▼ ▼ ▼ ▼ ▼ ▼ ▼

VEGETATION, WATER, AND SNOW

Imagine the same tropical beach that was invoked in the previous chapter, but lift your mind's eye from the sand to the trees, shrubs, and grasses growing on the shore. Even though there are many different varieties of vegetation to be found, all of them are principally composed of some or all of the same six constituents: water, cellulose (an amorphous carbohydrate polymer that is the main constituent of plant tissues and fibers), lignin (the chief noncarbohydrate constituent of wood; a polymer that functions as a natural binder and support for the cellulose fibers of woody plants), nitrogen, chlorophyll (the common pigments that impart shades of green color to vigorous plants), and anthocyanin (a class of water-soluble pigments that imparts colors ranging from blue to most shades of red to flowers and other plant parts, including autumn leaves). As an example of their relative amounts, Curran and Kupiec (1995) performed a foliar biochemical assay for 360 slash pine needles and found their mean content to be 57.90% water, 35.50% cellulose, 22.5% lignin, 0.98% nitrogen, and 1.71 milligrams/gram-dry-weight (0.171%) of chlorophyll. The sum of these numbers exceeds 100% because these percentages are means of each constituent, not averages. As will be discussed under the subsection on remote sensing applications to forests and wetlands in Chapter 9, Curran and Kupiec (1955) were able to find high correlations between reflectances recorded by a new type of airborne sensor package and most of these foliar bio-chemical concentrations.

As with most pigments, a little bit of chlorophyll goes a long way toward making vegetation green. There are two types, chlorophyll a ($C_{55}H_{72}MgN_4O_5$) and chlorophyll b ($C_{55}H_{70}MgN_4O_6$), which exhibit the absorption spectra shown in Figure 3.1 (Schanda 1986). Chlorophyll a (blue-green) has absorption bands at 0.43 μm and 0.66 μm, whereas chlorophyll b (yellow-green) has absorption bands at 0.45 μm and 0.64 μm. Chlorophyll a is three times more abundant in nature than chlorophyll b (Schanda 1986).

Figure 3.2 shows the fractions of total light intensity on the upper surface of a mature orange leaf that are due to reflectance, absorbance, and transmittance (Gausman et al. 1971). For example, at wavelengths of 0.5 μm, about 7% of the incident light is reflected and the remaining 93% is absorbed, while at 0.55 μm, about 9% is reflected, 1% is transmitted, and 90% is absorbed. This higher reflectance at 0.55 μm (within the visible green part of the spectrum) is the reason why vigorous vegetation is green. However, 9% reflectance is still relatively dark compared to most soils and rocks, so vigorous vegetation appears as dark green in the visible. Notice in Figure 3.2 that for visible wavelengths (less than 0.67 μm), absorbance dominates and is correlated with photosynthesis by chlorophyll in green plants. High reflectance and high transmittance dominate at wavelengths of 0.75–1.35 μm, a relationship that implies that a green leaf would appear very bright when viewed from either above or below the leaf. It is in this wavelength region that most reflected sunlight can be detected from lower canopies in a multiple canopy forest, or from grass obscured by overhead trees. The reflectance and transmittance minima and absorbance maxima of this orange leaf (Figure 3.2) in the water absorption bands at 1.45 μm and 1.95 μm suggest that water content of leaves could best be determined in the 1.4–2.0 μm wavelength region.

Figure 3.3 shows the spectral reflectance of green grass, dead grass, Virginia Pine, and Scarlet Oak (Abrams et al. 1984) and the spectral bands of an airborne multispectral scanner (NS-001 Thematic Mapper Simulator) that was designed to simulate the spectral bands of LANDSAT TM, as well as to collect data in some ad-

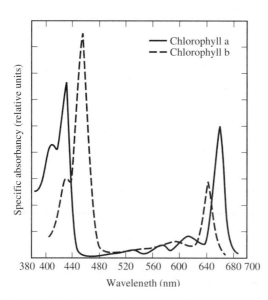

Figure 3.1   Absorption spectra of chlorophyll a (blue-green) and chlorophyll b (yellow-green). *(After Schanda 1986. Courtesy of Springer-Verlag GmbH and Co. KG.)*

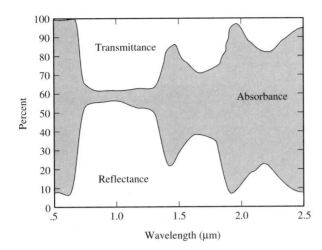

Figure 3.2 Fractions of the total light incident on the upper surface of a mature orange leaf that is reflected, absorbed, and transmitted. *(Gausman et al. 1971. Reprinted with permission of the Environmental Research Institute of Michigan.)*

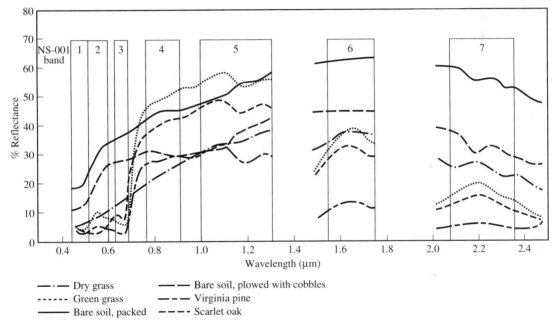

— · — Dry grass   — — Bare soil, plowed with cobbles
· · · · · Green grass   — — — Virginia pine
——— Bare soil, packed   – – – – Scarlet oak

Figure 3.3 Field reflectance spectra of green grass, dead grass, Virginia Pine, and Scarlet Oak. (After Abrams et al. 1984.) The spectral bands of a NASA-owned airborne multispectral scanner designated as NS-001 Thematic Mapper Simulator are given. NS-001 bands 1–4 and 7 are the same as LANDSAT TM bands 1–4 and 7. NS-001 band 6 is the same as LANDSAT TM band 5. *(Abrams et al. 1984. Reprinted by permission of the American Association of Petroleum Geologists.)*

ditional spectral bands. Multispectral scanners will be discussed in the next chapter. This figure demonstrates that in the visible blue and red wavelength regions (NS-001 bands 1 and 3, corresponding to LANDSAT TM bands 1 and 3), dry grass is brighter than green grass, whereas in NS-001 band 4 (which corresponds to LANDSAT TM band 4), green grass is brighter than dry grass. In NS-001 bands 2 and 6 (LANDSAT TM bands 2 and 5), reflectance of green grass and dry grass are approximately equal, and in NS-001 band 7 (LANDSAT TM band 7), dry grass is brighter than green grass. As green grass dries out, its reflectance in TM bands 2 and 5 remains unchanged, its reflectance in TM band 4 decreases, and its reflectance in TM bands 1, 3, and 7 rises.

Three other important observations can be made from Figure 3.3. First, all of the green vegetation spectra exhibit a great rise in reflectance over a short range of wavelengths from a reflectance low in the visible red (about 0.65 μm) to the top of its steep rise around 0.75 μm in the reflective infrared. As the nonvegetative spectra in this chapter show, this spectral characteristic is not shared (to this extent) by other natural materials. Second, reflectance spectra of deciduous vegetation (like Scarlet Oak) display the same general shape as coniferous vegetation (like Virginia Pine), but with higher reflectance (brighter) for wavelengths shorter than 1.3 μm. One reason for this is the greater shadowing that occurs on needles of coniferous vegetation than on more horizontally oriented leaves of deciduous vegetation. Third, as green vegetation dries out and becomes less vigorous, an unidentified absorption band, also exhibited by dried leaves and bark, can be found at wavelengths near 2.3 μm. In leaves that are not dried out, water absorption bands suppress this band. Spectral reflectance of vegetation is almost featureless and low at thermal infrared wavelengths, though there are subtle differences related to cuticular waxes that could be diagnostic of deciduous species, at least in laboratory spectra (Salisbury and Milton 1988).

Vegetation displays different thermal behavior than the land surface because leaves and needles of plants very efficiently exchange heat with the surrounding air, giving them the same temperature as the air at most times. Since the air temperature tends to be cooler than the land surface by day and warmer by night, so too are leaves and needles. Evapotranspiration is a second-order effect acting to also cool leaves by day. Water-stressed plants often appear warmer than well-watered plants because there is less evapotranspiration when there is less water available to the plants.

Returning one last time to our imaginary tropical beach, let your mind's eye wander away from land, toward the sea, which represents the most common surface material on the face of the Earth, namely, water. Figure 3.4 shows how the calculated spectral reflectance of deep ocean water varies with different amounts (in milligrams per cubic meter) of chlorophyll (Derr 1972). Disregarding specular observations associated with sun glint, where the angle of incidence equals the angle of observation and both angles are in the same plane, deep water is fairly dark at all wavelengths, as evidenced in Figure 3.4 by the spectral curve of ocean water with the least chlorophyll (1 mg/m^3).

Although water has low reflectance in the visible light region, its visible transmittance is high, especially in the blue/green region of the spectrum. Thus, in shallow water (less than 30 meters deep), a bright reflective bottom is visible and semiquantitative water depth mapping (bathymetry) is possible. Such bathymetric mapping is helped by the decreasing transmission of water with increasing wavelength in the

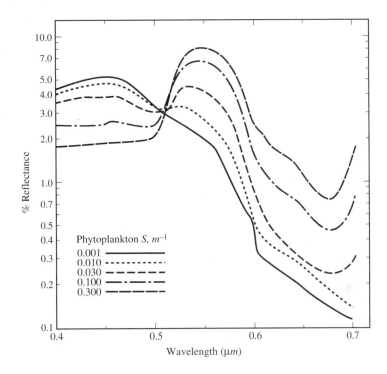

Figure 3.4 Calculated change in bulk reflectance of ocean water with increasing concentration of phytoplankton. *(After Derr, 1972. Reprinted with permission of the Environmental Research Institute of Michigan.)*

visible, and a comparison of long-to-short visible wavelengths can yield water depth, if the bottom materials are uniformly bright (Polcyn and Lyzenga 1973). Deep water is noteworthy for being darker than most other materials, especially in the reflective infrared wavelength region, where its reflectance is typically less than 1% and transmittance is so low that penetration of light is on the order of a few inches deep (discussed in Chapter 9 under forest and wetland remote sensing applications).

Large bodies of water, like deep lakes and the ocean, tend to maintain a more constant temperature than land during a diurnal cycle. During the day, sunlight is absorbed over a substantial column (about 70 meters) of clear water (less if phytoplankton or suspended solids are present), instead of over the first few millimeters of the surface of rocks and soils on land, and warmer water convects heat to the surface at night. Thus, large bodies of water are warmer than land at night and cooler than land in daytime, excluding times of the year when snow and ice are common. Color Plate CP10 is a nighttime thermal infrared image of the Great Lakes at about 2 A.M. on August 22, 1978, as recorded by the Heat Capacity Mapping Mission (no longer in orbit). Note that the lakes are redder (warmer) than the land, except for the cooler waters of Lake Superior and the upwelling cool bottom waters (from a convection cell) paralleling the Wisconsin Coast, due to the eastward push of Lake Michigan surface waters by West-to-East winds (Vincent et al. 1981).

Pure seawater reflects mostly in the blue wavelength region, around 0.45 μm, but the reflectance peak shifts to the green around a wavelength of 0.55 μm if phytoplankton are present in significant amounts, as Figure 3.4 indicates. This shift can be exploited for mapping the content of chlorophyll a, even in inland waters. Dekker et al. (1995) note in their summary of remote sensing for inland water quality that chlorophyll a content in the Vecht lakes region of the central Netherlands was highly correlated with a spectral ratio of narrow spectral bands located in the 0.698–0.714 μm and 0.671–0.684 μm wavelength regions, and that chlorophyll a content could be determined within an accuracy of approximately 9.5 mg/m^3. With airborne narrow-band scanners, they also found it possible to determine the cyano- phycocyanin (produced by cyanobacteria) content within an accuracy of 20 mg/m^3, and the Seston dry weight (suspended solids) within an accuracy of 4.9 mg/m^3. Relatively low spectral resolution satellite data, like LANDSAT TM and SPOT-1 sensors, do not allow analytical algorithm development for these features, however, and cannot discriminate between chlorophyll a and suspended matter (Dekker et al. 1995). The spectral bands on current satellites are just too broad for this task, indicating a need for narrow-band sensors in space for water-quality mapping.

After water and vegetation, snow is the next most common material covering Earth's surface. Bare ground is the next most common and is considered in the next two sections of this chapter. Snow is brighter than most Earth materials at wavelengths short of about 1.3 μm, as shown by computed reflectance spectra in Figure 3.5 for snow of three particle sizes (frost, fine, and coarse), or degrees of freshness (Salisbury et al. 1994). As snow ages, its particle size increases and its reflectance decreases (Salisbury et al. 1994), a change that indicates the dominance of volume reflectance over specular reflectance at wavelengths short of 3.0 micrometers. Snow exhibits a much lowered reflectance in the 1.5–2.5 μm wavelength region, relative to shorter wavelengths, and it decreases in reflectance between the wavelengths of 1.1 μm and 1.5 μm more than any other known substance commonly found at Earth's surface. The reflectance of snow is very low (less than 3%) in the thermal infrared wavelength region (Salisbury et al. 1994). Also, snow is typically colder than most Earth materials, especially in broad daylight.

MINERALS AND ROCKS

The purpose of this section is to introduce a diverse collection of mineral and rock spectra that display a variety of reflectance behaviors. Because minerals each have a different chemical composition or crystalline structure and because rocks consist of assemblages of minerals, there are obviously far too many reflectance spectra of minerals and rocks to include in any one book. A bibliography of papers and books

A.
Snow: Coarse (1), Fine (2), and Frost (3) (0.3-3.0 μm)

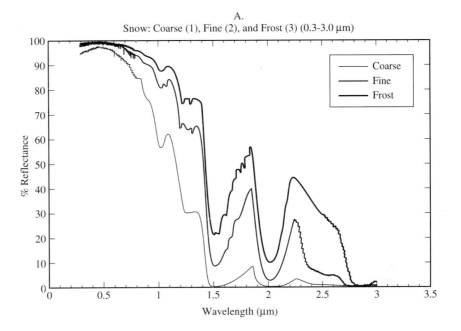

B.
Snow: Coarse (1), Fine (2), and Frost (3) (3-14 μm)

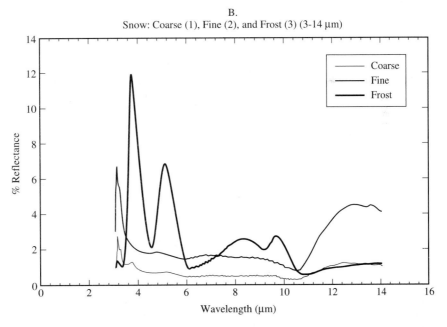

Figure 3.5 Computed reflectance spectra of three different textures of snow (coarse, fine, and frost) for (a) the 0.3–3.0 μm wavelength region. (b) the 3–14 μm wavelength region. *(Courtesy of Salisbury and D'Aria, personal communication.)*

with many rock and mineral reflectance spectra is included at the end of this chapter, in Additional References for Chapter 3. Unless otherwise noted, all of the spectra shown in the following figures are for samples that had a particle size range typical of a sandy beach (about 100–500 μm), which is equivalent to a rough natural rock surface. A coarse particulate sample and a rough rock surface display similar spectral behavior.

The 0.4–2.5 μm wavelength spectra of minerals in the following subsections all came from a spectral library of 160 minerals (Grove et al. 1992) produced by the Jet Propulsion Laboratory at the California Institute of Technology. The directional hemispherical reflectance measurements in that library were made relative to a polytetrafluoroethylene powder standard, which goes by the trade name of Halon. The spectra were then normalized to absolute directional hemispherical reflectance (effects of the standard powder reflectance were removed) in later drafts of the final report, which is the version referenced here. The 2–25 μm spectra of minerals in the following subsections came from a spectral library produced by Salisbury et al. (1991). The thermal infrared spectra were produced by biconical reflectance measurements (discussed in the previous chapter) relative to a mirror standard. Since the measurement methods and the mineral samples were not identical in the 0.4–2.5 μm and the 2–25 μm wavelength spectral libraries, spectral reflectance curves from the two libraries for a given mineral are not in perfect quantitative agreement in the 2–2.5 μm overlap region.

The particle sizes of most of the mineral samples cited below are 125–500 μm for the 0.4–2.5 μm wavelength spectra and 74–250 μm for the 2–25 μm wavelength spectra, or medium-sand-grain-sized. For some fine-grained samples that do not ordinarily or ever occur in sand-sized particle ranges, such as ferric oxides and clays, the particle size was 0–45 μm for the 0.4–2.5 μm wavelength spectra and either 0–2 μm or 0–75 μm for the 2–25 μm wavelength spectra, or clay-silt-grain-sized. Whenever spectra of the clay-silt-grain-sized samples are used, the graph title includes the words "Less than 45 μm," or something similar.

Several measured spectral reflectance curves of rock surfaces will be shown later in this chapter. It is possible to simulate the reflectance spectrum of a rock with a sand-grain-sized surface from the reflectance spectra of its mineral constituents, using Equation 2.4 and the assumption that the cement (silica or calcium carbonate) is an additional mineral constituent, as the problems at the end of this chapter demonstrate.

Examples of Silicate Minerals

Silicate rock-forming minerals can be divided into two broad classes, mafic (visibly dark) and felsic (visibly bright). Ferromagnesian silicates, which contain large amounts of iron and magnesium in the crystal lattice, are mafic minerals. Rock-forming silicates with little iron and magnesium, are felsic minerals. The reflectance spectra of silicate minerals in the 0.4–2.5 μm wavelength region are dominated by ferrous iron (Fe^{2+}), hydroxyl ions (OH^-), and water (H_2O) in the crystalline structure of the silicate. First, a few words are in order about absorption bands that are caused by these two ions and one compound, bearing in mind that *the exact spectral location of an absorption band can vary with the mineral because an ion's interaction with the surrounding crystal field determines the position of the absorption band.*

Ferrous iron, like other transition metal ions, can produce a number of different absorption bands, depending on the crystalline structure of the mineral involved and the levels of impurities present. The coordination number and site symmetry of the transition metal ion, the type of bond or ligand formed (with oxygen or some other element), the metal ligand interatomic distances, and the distortion of the metal ion site all help determine at what wavelength a particular absorption band will occur (Hunt and Salisbury 1970). Ferrous iron absorption bands occur in the wavelength vicinity of 1.0 μm.

Hydroxyl absorption bands in the reflective infrared wavelength region are caused by overtone and combination bands of primary hydroxyl ion molecular vibration bands located at wavelengths longer than 2.5 μm. Overtone bands occur at wavelengths where the frequency (proportional to the reciprocal of the wavelength) of a primary absorption band is doubled, tripled, and so on, and combinational bands occur where the frequencies of two or more primary absorption bands are added. The positions of the hydroxyl absorption bands depend on the position of the OH^- ion in the crystalline lattice, which can be at different sites with different potential fields (Hunt and Salisbury 1970). The most common hydroxyl overtone and combination bands are at 1.4 μm (overtone of the O—H stretch), 2.2 μm, and 2.3 μm. Water (H_2O) displays absorption bands at both 1.4 μm and 1.9 μm, but the presence of the 1.4 μm band without the 1.9 μm band indicates that hydroxyl groups other than those in water are present (Hunt and Salisbury 1970).

Figure 3.6 shows directional hemispherical reflectance spectra of eight rock-forming silicate minerals (fayalite, an olivine; enstatite, a pyroxene; tremolite, an amphibole; biotite, a mica; anorthite, a plagioclase feldspar; albite, a plagioclase feldspar; orthoclase, a potassium feldspar; and quartz) for the 0.4–2.5 μm wavelength region (Grove et al. 1992) and biconical reflectance spectra for the 2–25 μm region (Salisbury et al. 1991). The 0.4–2.5 μm region will be discussed for all eight minerals, with liberal help from the description of where the absorption bands are located from Hunt and Salisbury (1970), followed by a discussion of their spectral features in the 2–25 μm region.

The broad but shallow reflectance minimum centered near 1.0 μm in fayalite, the darkest silicate mineral in Figure 3.6, is caused by ferrous iron. Enstatite displays two strong reflectance minima caused by ferrous iron absorption at wavelengths of 0.9 and 1.84 μm. Tremolite reflectance minima at 1.39, 2.11, 2.30, 2.38, and 2.46 μm are caused by hydroxyl ion overtone and combination bands, respectively, from primary absorption bands located in the thermal infrared wavelength region. (*Note:* OH^- bands are also typically found in spectra of metamorphic minerals and rocks, as well as in clays.) At shorter wavelengths, tremolite displays a ferrous iron absorption band at a wavelength of 1.03 μm. Biotite displays very shallow reflectance minima near wavelengths of 0.74 and 1.21 μm that are caused by ferrous iron absorption. Biotite also displays very shallow reflectance minima near 2.33 and 2.39 μm that are caused by OH^- overtone and combination bands. Note that biotite

A.1
Fayalite (1) and Enstatite (2) (0.4-2.5 μm)

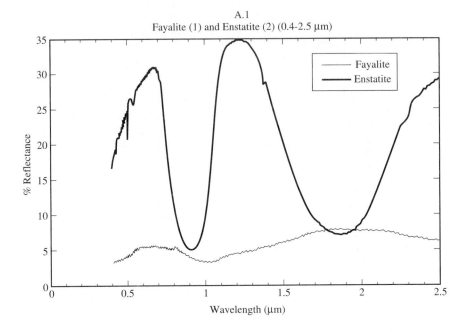

A.2
Fayalite (1) and Enstatite (2) (2-25 μm)

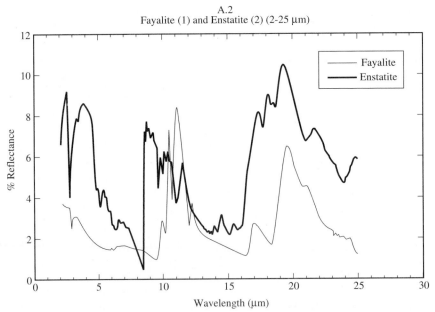

Figure 3.6 Directional hemispherical reflectance spectra in the 0.4–2.5 μm wavelength region and biconical reflectance spectra in the 2–25 μm wavelength region of eight rock-forming silicate minerals: (a) fayalite and enstatite. *(Figures 3.6 a–d provided through the courtesy of the Jet Propulsion Laboratory, California Institute of Technology, Pasadena, California, and Salisbury et al., 1991.)*

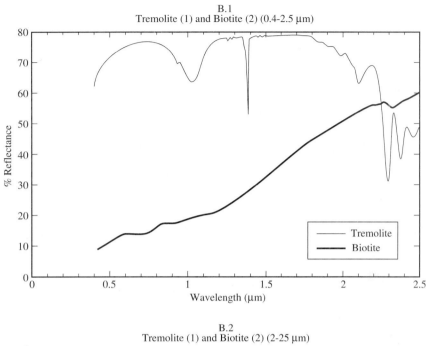

B.1
Tremolite (1) and Biotite (2) (0.4-2.5 μm)

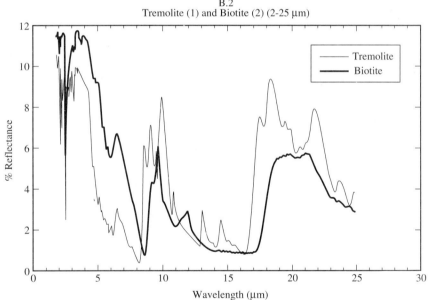

B.2
Tremolite (1) and Biotite (2) (2-25 μm)

Figure 3.6 *(continued)* (b) tremolite and biotite

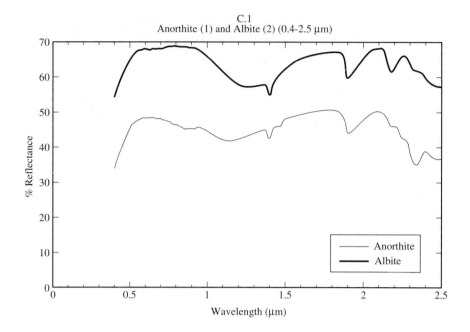

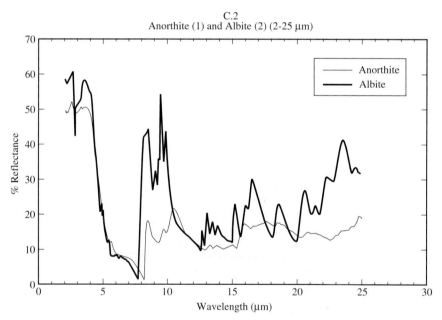

Figure 3.6 *(continued)*   (c) anorthite and albite

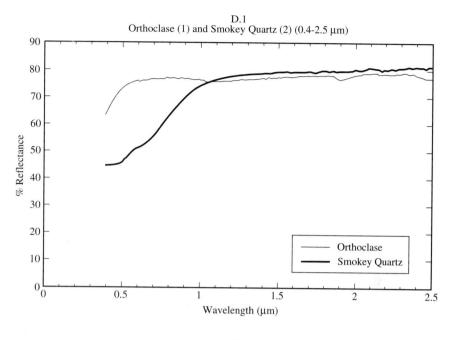

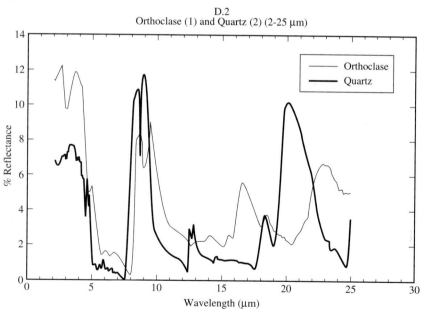

Figure 3.6 *(continued)* (d) orthoclase and smoky quartz

rises almost steadily in reflectance from near 10% at 0.4 μm to near 60% at 2.5 μm, which is an unusual spectral curve shape.

Anorthite and albite are moderately bright and display a few shallow spectral features. The relatively narrow absorption bands (reflectance minima) near 1.4 and 1.9 μm come from water in fluid inclusions. The broad reflectance minima in anorthite near 1.18 μm and in albite near 1.26 μm are likely caused by ferrous iron absorption. The 2.2 μm band in albite is likely caused by incipient alteration, presumably to sericite or kaolinite, both of which display a reflectance minimum near this wavelength. The two small reflectance minima near 2.34 μm and 2.48 μm may be caused by the hydroxyl ion, but the first of these occurs between two tremolite bands described above. The orthoclase and quartz are bright and almost featureless, except for the lowered reflectance of quartz at wavelengths short of 1.1 μm, which is caused by radiation damage (smoky quartz).

In the 2–25 μm wavelength region, a number of reststrahlen bands (manifested chiefly as strong reflectance maxima, as discussed in the previous chapter) occur in silicates, caused by vibrations of the SiO_4 tetrahedra, the building blocks of silicate minerals, within the crystalline lattice. Although overtone and combination bands of these vibrations (which are absent in glass because it has no crystalline lattice) occur in the 8.0–14.0 μm wavelength region, the largest quartz reststrahlen bands in this region are centered at wavelengths of 8.5 and 9.0 μm. Fayalite, the iron-rich version of olivine, displays its largest reststrahlen bands in the 8–14 μm atmospheric window centered at wavelengths of 10.7, 11.2, and 12.1 μm. Differences in the crystalline structure between quartz and fayalite that are related to how these minerals are formed from a silicate melt are responsible for the different locations of these reststrahlen bands.

The justification for this statement starts with a diagram of Bowen's reaction series, shown in Figure 3.7 (after Montgomery and Dathe 1994), which was devised over 65 years ago by geologist Norman L. Bowen. Bowen combined careful laboratory studies with wide-ranging field observations of igneous silicate rocks

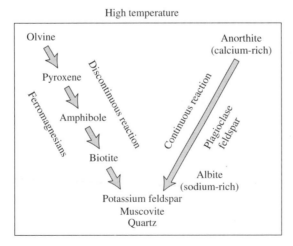

Figure 3.7 Bowen's reaction series. *(From Carla W. Montgomery and David Dathe, Earth Then and Now, 2nd ed. Copyright © 1994, Times Mirror Higher Education Group, Inc., Dubuque, Iowa. All rights reserved. Reprinted by permission.)*

to show the order in which silicate minerals solidify from a magma as the molten rock cools. The discontinuous reaction series, so-called because minerals with different crystalline structures form as the temperature of crystallization decreases, consists of ferromagnesian minerals that increasingly share more oxygen atoms among silica tetrahedra (SiO_4), the building blocks of silicate minerals, as the temperature of crystallization decreases. For example, the silicate minerals that crystallize at the highest temperatures are olivines. Olivines are orthosilicates that share no oxygen atoms between silica tetrahedra. Pyroxenes, which solidify at the next highest temperatures on the discontinuous reaction side, are single-chain silicates (inosilicates) that share two oxygen atoms between tetrahedra per silica tetrahedron. Amphiboles, next lowest, are double-chain silicates (also inosilicates) that intertetrahedronally share an average of 2.67 oxygen atoms per silica tetrahedron. Biotite mica, the discontinuous reaction series member with the lowest temperature of solidification, is a sheet silicate (phyllosilicate) that shares an average of three oxygen atoms per silica tetrahedron with other tetrahedra. At the end of the crystallization sequence (lowest temperatures) are potassium feldspar, muscovite mica, and quartz. Both potassium feldspar and quartz are framework silicate minerals (tectosilicates), which intertetrahedronally share four oxygen atoms per silica tetrahedron. Muscovite, like biotite, is a sheet silicate (phyllosilicate) mineral.

A comparison of Figure 3.7 with the thermal infrared spectra in the 8–14 μm atmospheric window of Figure 3.6 shows that as Bowen's reaction series proceeds along the discontinuous reaction series side from the mineral with the highest (fayalite olivine) to the mineral with the lowest (quartz) crystallization temperature, there is a systematic shift of the center of the reststrahlen band of the minerals to shorter wavelengths as the crystallization temperature decreases. Therefore, the shift is partly caused by changes in bond length and bond strength associated with the change in structure as more oxygen atoms are shared per silica tetrahedron with decreasing temperature of crystallization. These could also be called changes in the degree of polymerization of the SiO_4 tetrahedra within the crystalline lattice. In quartz (fully polymerized), all of the oxygen atoms in every SiO_4 tetrahedron share an electron with oxygen atoms of another SiO_4 tetrahedron, whereas in fayalite (no polymerization), none of the SiO_4 tetrahedra share electrons with other SiO_4 tetrahedra, but instead share electrons only with iron ions. The other igneous-rock-forming silicate minerals in Figure 3.6 display reststrahlen bands that fall somewhere between the reststrahlen bands of these two end members, because of lesser extremes of SiO_4 polymerization.

The four minerals at the top left of Bowen's reaction series are mafic minerals, and the three at the bottom are felsic minerals. Therefore, the felsic shift is the name we will give to the reststrahlen band shift to shorter wavelengths as mineral composition changes from mafic to felsic.

Although polymerization of the SiO_4 tetrahedra is a major part of the story, however, it cannot be all of it. The continuous reaction series consists of plagioclase feldspar that maintains the same crystalline structure of framework silicates (tectosilicates), which intertetrahedronally share four oxygen atoms per silica tetrahedron, throughout the reduction in temperature of crystallization. However, there are some substitutions of Ca, Al and Na, Al combinations for silicon in the tetrahedra. The plagioclase feldspars are a solid solution that changes only in relative calcium-to-sodium and aluminum-to-silicon content. Anorthite is the high-temperature end member of the continuous reaction series, a calcium plagioclase that contains no sodium, and albite is the low-temperature end member, a sodic plagioclase that contains no calcium. Examination of the 8–14 μm portion of the thermal infrared reflectance spectra of Figure 3.6 reveals that the reststrahlen band centers of anorthite, albite, and orthoclase (a potassium feldspar) occur at shorter wavelengths than the mafic minerals of the discontinuous reaction series, from olivine through biotite, and that quartz has the shortest wavelength occurrence of reststrahlen bands of any silicate mineral.

The following four spectra are clay minerals, all of which exhibit a strong absorption band near 2.7 μm, which is near the 2.5 μm upper limit of the reflectance spectra. Therefore, the 2–25 μm spectrum of each mineral will be discussed immediately following the 0.4–2.5 μm spectrum discussion for the following minerals, rather than in the order followed for the previous silicate minerals. Clays are phyllosilicates (sheet silicates), which share silicon atoms in their SiO_4 tetrahedra in two dimensions only.

Reflectance spectra of kaolinite, $Al_2Si_2O_5(OH)_4$, and montmorillonite, expressed by the general formula $(Al,Mg,Fe)_4(OH)_n(Si,Al,Fe)_8O_{20-n}(OH)_n \cdot 6H_2O$, are shown for samples of clay-sized particles in Figure 3.8 (Grove et al. 1992; Salisbury et al. 1991). Kaolinite, an alteration product of aluminum silicates, particularly feldspar, displays strong hydroxyl absorption bands (reflectance minima) with sharp, double-troughed minima near 1.39, 1.41, 2.17, and 2.20 μm wavelengths (Hunt and Salisbury 1970). The 2.20 μm absorption band is caused by Al—OH ligands. The 2.7 μm OH$^-$ band (reflectance minimum) in kaolinite is typical of a mineral that contains OH$^-$, but little water (no 1.9 μm band), as in this case. The reflectance maxima in the thermal infrared spectrum near 8.8 μm and 9.4 μm, which are barely evident in this sample with clay-sized particles, are caused by Si—O—Si and Si—O stretching vibrations. The slight, broad reflectance maximum at 11.9 μm is caused by a fundamental Al—O—H band, and the broad, very shallow reflectance maximum near 17.5 μm is caused by an Al—O—Si stretching mode (Salisbury et al. 1991).

Montmorillonite has strong bound water absorption bands (reflectance minima) at 1.4 and 1.9 μm (Hunt and Salisbury 1970). The slight, 0.95 μm reflectance minimum is caused by ferrous iron absorption. The 2.2 μm reflectance minimum is due to OH$^-$ absorption (overtone band). The 2.7 μm reflectance minimum is much broader than the one for kaolinite in this region because it is caused by water in the crystalline lattice. In the thermal infrared, the peaked reflectance maxima near 8.8 μm and 9.2 μm, the broad, shallow reflectance maximum in the 11–13 μm region, and the maximum in the 16–20 μm region are caused by the same features as for kaolinite, except that many more vibrations (less sharp peaks) are allowed in montmorillonite, due to its lack of lattice symmetry and the numerous exchangeable cations and water molecules present (Salisbury et al. 1991). These would be more

A.1
Kaolinite (1) and Montmorillonite (2) Less Than 45 μm (0.4-2.5 μm)

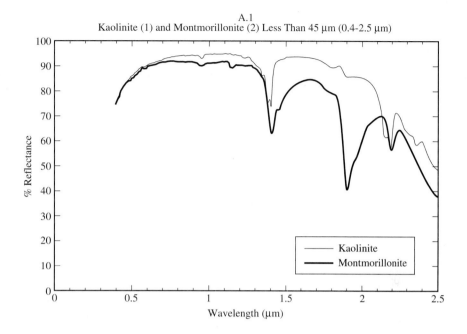

A.2
Kaolinite (1) and Montmorillonite (2) Less Than 2 μm (2-25 μm)

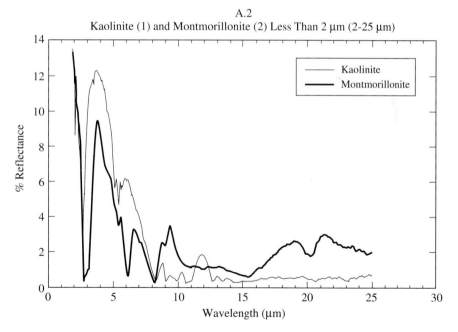

Figure 3.8 Directional hemispherical reflectance spectra in the 0.4–2.5 μm wavelength region and biconical reflectance spectra in the 2–25 μm wavelength region of four clay minerals: (a) kaolinite and montmorillonite. *(Figures 3.8a–b provided through the courtesy of the Jet Propulsion Laboratory, California Institute of Technology, Pasadena, California, and Salisbury et al., 1991.)*

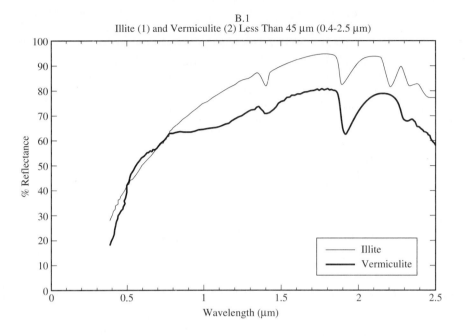

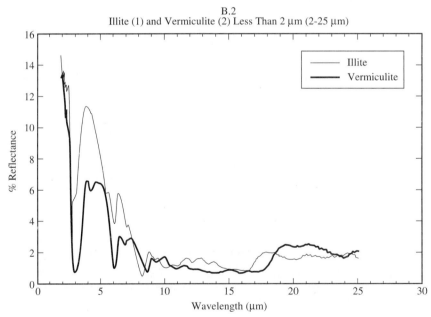

Figure 3.8 *(continued)*   (b) illite and vermiculite

evident in a packed-powder spectrum, but the clay-sized particle spectrum is more typical of field conditions.

Figure 3.8 also shows reflectance spectra of two other phyllosilicates named illite, $(K,H_3O)(Al,Mg,Fe)_2(Si,Al)_4O_{10}(OH)_2 \cdot H_2O$, and vermiculite, $(Mg,Fe^{+2},Al)_3$ $(Al,Si)_4O_{10}(OH)_2 \cdot 4H_2O$, for both the 0.4–2.5 (Grove et al. 1992) and the 2.0–25.0 μm wavelength regions (Salisbury et al. 1991). They both share the 1.4 μm and 1.9 μm water absorption bands with montmorillonite, but only illite of the two shares hydroxyl ion absorption bands at 2.2 μm (single-trough, not a doublet) and 2.31 μm with kaolinite. Vermiculite has an absorption band at 2.3 μm, likely due to the hydroxyl ion. Notice that illite and vermiculite are darker than kaolinite and montmorillonite at wavelengths short of 2.0 μm, especially in the visible wavelength region. There are spectral similarities with kaolinite and montmorillonite in the large reflectance maxima at wavelengths longward of 8 μm.

The reflectances for all four clays (except for montmorillonite at 8–10 μm) are less than 2% in the 8–14 μm atmospheric window because they have such small particle sizes. Partly for this reason, the most characteristic information about clays can be found by remote sensing sensors in the 1.4–2.5 μm wavelength region, since the 2.7 μm reflectance minimum shared by all four clays is not in an atmospheric window. For the other silicate minerals, however, the placement of reststrahlen bands in the 8–14 μm wavelength region is the most diagnostic of mineral composition. This begs for multispectral data in the 8–14 μm region, data that in the past has been available from aircraft, but not via satellite sensors.

Examples of Ferric Oxide and Ferric Hydroxide Minerals

Figure 3.9 shows reflectance spectra of two ferric oxides (hematite and magnetite) and a ferric hydroxide (goethite) in the 0.4–2.5 μm region (Grove et al. 1992) and in the 2.0–25.0 μm region (Salisbury et al. 1991), clay-to-silt-sized particles. The 0.48 μm and 0.85 μm absorption bands in hematite (Fe_2O_3), seen here as reflectance minima, are caused by electronic transitions in the outer shells of the Fe^{3+} ion. Magnetite is dark and almost featureless in the entire 0.4–2.5 μm region, like most metals. The goethite ($FeOH_2$) sample, although an iron hydroxide, shows only a hint of a hydroxyl absorption band at 2.2 μm and 2.45 μm, but shows ferric iron absorption at 0.48 μm and 0.9 μm. An important characteristic of the hematite spectrum is that it increases greatly in reflectance from 0.5–0.72 μm; thus hematite is nature's red pigment, giving rise to red, pink, orange, and yellow rocks and soils as it is mixed with other pigments. Goethite is a similar pigment, but lesser in the amount of reflectance increase with increasing wavelength in the visible wavelength region, i.e., it is less red than hematite. Hematite and goethite, common weathering products of iron-rich minerals, often coat particles of sand and clay.

It should be noted that none of the three has reflectance above about 4% in the 8–14 μm thermal infrared wavelength region, where the silicate minerals have their major reststrahlen bands. This fact explains why desert varnish tends to partially mask the quartz reststrahlen bands of long-exposed desert sand (Salisbury and D'Aria 1992; Vincent 1973). Desert varnish is a term given to the coating of

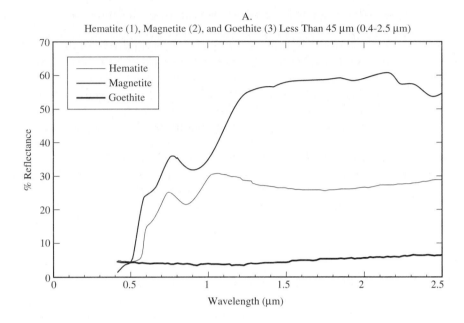

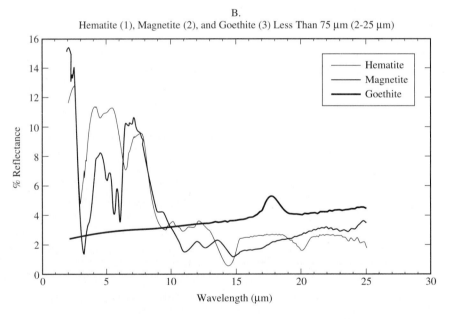

Figure 3.9 Directional hemispherical reflectance spectra in the 0.4–2.5 μm wavelength region and biconical reflectance spectra in the 2–25 μm wavelength region of ferric oxides hematite, magnetite, and the ferric hydroxide goethite. *(Provided through the courtesy of the Jet Propulsion Laboratory, California Institute of Technology, Pasadena, California, and Salisbury et al., 1991.)*

desert sand grains by much finer particles of manganese and iron oxihydrides mixed with montmorillonite clay (Potter and Rossman 1979). Since the manganese and iron oxihydrides are spectrally dark and somewhat featureless in this wavelength region, desert varnished rocks and soils take on the spectrum of montmorillonite between wavelengths of 8.0 μm and 14.0 μm. Magnetite (Fe_3O_4) displays a more metallic spectrum (opaque and featureless) than the other two at all wavelengths and is resistant to weathering.

In summary, hematite shows a generally increasing reflectance with increasing wavelength in the visible wavelength region, whereas magnetite is uniformly dark, and goethite is somewhere in between the other two, in the 0.4–2.5 μm region. In the 2.0–25.0 μm region, hematite and magnetite reflectances are dark and reasonably flat. Hematite-coated grains, including those that are desert varnished, can have their reststrahlen bands muted by the thermal infrared spectrum of hematite.

Examples of Carbonate Minerals

Carbonate minerals are characterized by the presence of the carbonate (CO_3) radical in their chemical formulae, and are most often found in sedimentary rocks. Four carbonate reflectance spectra are shown in Figure 3.10 for the 0.4–2.5 μm (Grove et al. 1992) region, and reflectance spectra for two of those (calcite and dolomite) are shown for the 2.0–25.0 μm (Salisbury et al. 1991) wavelength region, the other two carbonate spectra being unavailable. Calcite ($CaCO_3$) displays absorption bands (reflectance lows here) at 1.87, 1.99, 2.15, and 2.33 μm, all caused by combination and overtone bands of the C—O vibrations (internal to the CO_3^{-2} ion) in the 6–14 μm wavelength region (Hunt and Salisbury 1971). Dolomite, $CaMg(CO_3)_2$, exhibits the same bands, though slightly shifted to shorter wavelengths by Mg (1.86, 1.98, 2.13, and 2.30 μm). The reflectance minima near wavelengths of 0.52 and 0.93 μm in smithsonite, $ZnCO_3$, are probably caused by electronic transitions in Zn, but might possibly be caused by iron impurity. The reflectance minima at 2.0, 2.16, and 2.35 μm in smithsonite are carbonate combination and overtone bands, shifted slightly longward from those of calcite. Smithsonite is a zinc ore produced by ground water action in limestone rocks. Siderite, $FeCO_3$, displays a strong absorption band (reflectance low) near 1.15 micrometers, caused by Fe^{2+}, and carbonate bands at 1.9 and 2.32 μm. The visible reflectance of both smithsonite and siderite are far less distinctive than their reflective infrared absorption bands in the vicinity of 1.0 μm. This means that though these minerals are difficult to identify by eye in the field, a visible/reflective IR sensor would "see" them as fairly unique in nature for having such a low 1.0 μm reflectance, compared to their visible red (0.6–0.67 μm) reflectances.

The 2.0–25 μm reflectance spectra of calcite and dolomite display reststrahlen bands near 6.6 (6.5) μm, 11.4 (11.3) μm, and 14 (13.8) μm (the first two seen as reflectance maxima and the latter one as a reflectance minimum), except that the dolomite bands are shifted slightly, relative to the calcite bands (Salisbury et al. 1991). There are also sharp carbonate absorption bands (reflectance minima) in the

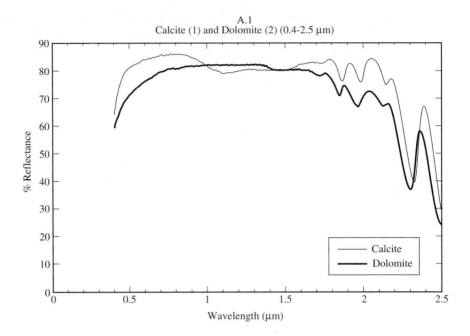

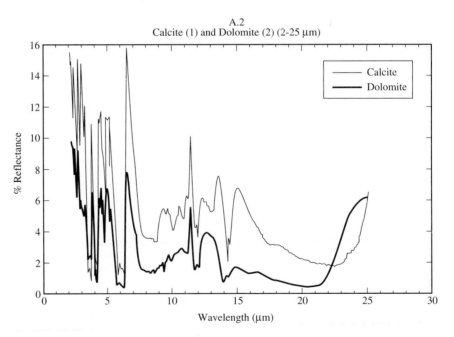

Figure 3.10 Directional hemispherical reflectance spectra in the 0.4–2.5 μm wavelength region and biconical reflectance spectra in the 2–25 μm wavelength region of four carbonate minerals: (a) calcite and dolomite *(Figures 3.10a–b provided through the courtesy of the Jet Propulsion Laboratory, California Institute of Technology, Pasadena, California, and Salisbury et al., 1991.)*

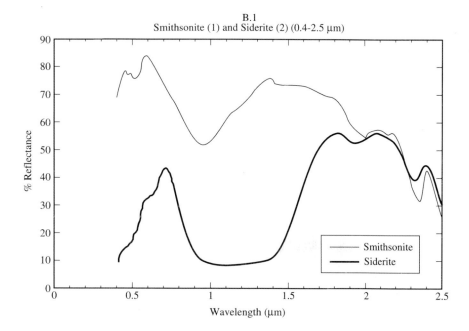

Figure 3.10 *(continued)*   (b) smithsonite and siderite (no 2–μm data were available)

3.3 μm, 4.0 μm, and 4.5 μm spectral regions that are inside an exploitable atmospheric window.

In summary, the chief carbonate bands are in the 1.92–2.55 μm, 3.3–4.5 μm, 11.3–11.5 μm, and 13.5–14.5 μm wavelength regions, but siderite and smithsonite (a zinc ore) also have deep reflectance minima near 1.0 μm.

Examples of Sulfate and Sulfide Minerals

The sulfates and sulfide minerals all share the element sulfur, but the sulfate spectra are dominated by the sulfate radical (SO_4), whereas the sulfides are not. Figure 3.11 shows reflectance spectra of two sulfates (alunite and anhydrite) and two sulfides (pyrite and pyrrhotite) for the 0.4–2.5 μm (Grove et al. 1992) and 2.0–25.0 μm (Salisbury et al. 1991) wavelength regions. The two sulfides are shown for clay-silt-sized particle samples.

Alunite, $KAl_3(SO_4)_2(OH)_6$, displays OH^- stretching mode overtones at 1.0 and 1.48 μm and a fundamental Al—O—H bending mode at 2.17 μm, with its first overtone at 1.75 μm (Hunt et al. 1972). The 1.27 μm absorption band of alunite is a combination of the Al—O—H band with an OH^{-1} stretching mode overtone, and the small band at 2.33 μm could be another such combinational absorption

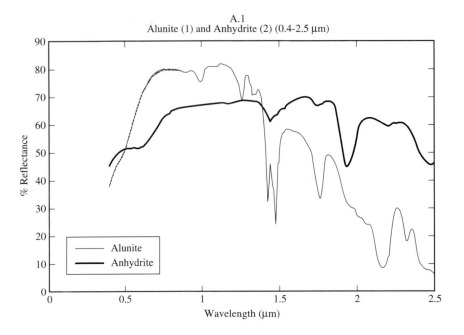

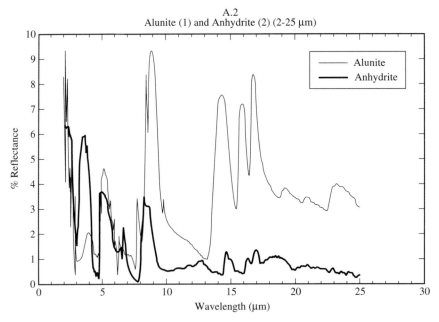

Figure 3.11 Directional hemispherical reflectance spectra in the 0.4–2.5 μm wavelength region and biconical reflectance spectra in the 2–25 μm wavelength region of two sulfate minerals and two sulfide minerals: (a) alunite and anhydrite *(Figures 3.11a–b provided through the courtesy of the Jet Propulsion Laboratory, California Institute of Technology, Pasadena, California, and Salisbury et al., 1991.)*

B.1
Pyrite (1) and Pyrrhotite (2) Less Than 45 µm (0.4-2.5 µm)

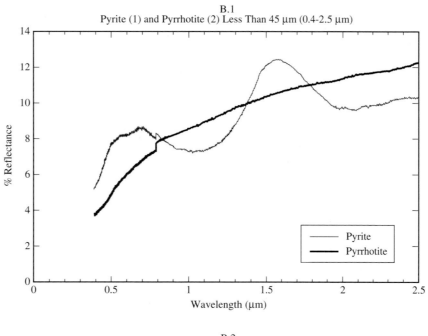

B.2
Pyrite (1) and Pyrrhotite (2) Less Than 75 µm (2-25 µm)

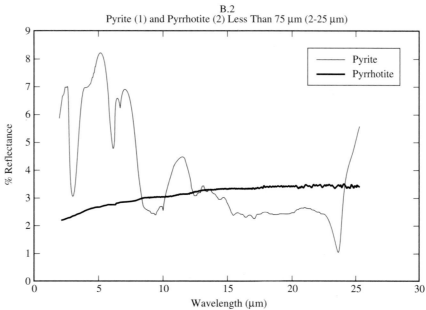

Figure 3.11 *(continued)*   (b) pyrite and pyrrhotite

band of Al—O—H and a different OH^{-1} overtone. The sharp, weak feature at 1.42 μm and the strong feature near 2.5 μm are also combinational bands of OH^{-1} bands (Hunt et al. 1972). The 8.9 μm reflectance maximum is caused by the fundamental sulfate ion stretching mode, and the three reflectance maxima near 14.2 μm, 16 μm, and 17 μm are probably caused by three bending modes of the sulfate ion. Alunite is only produced by hydrothermal alteration and is not a weathering product.

Anhydrite is an evaporite mineral. In the 0.4–2.5 μm region, anhydrite ($CaSO_4$) displays some water absorption bands at 1.44 and 1.94 μm (Hunt et al. 1972), even though the hydrated form of anhydrite has another name, gypsum. There are small reflectance minima at 2.2 μm and 2.48 μm, presumably from overtones of the sulfate ion reststrahlen bands at longer wavelengths. In the 2.0–25 μm region, the reflectance minima at 2.8 and 6.2 μm are caused by water, but the reflectance maxima at 7.9 μm and 8.4 μm are fundamental stretching modes of the sulfate ion. The reflectance minimum at 4.6 μm is probably a combination tone of the sulfate ion and water combination tones, though water is not supposed to be present in anhydrite. The reflectance of anhydrite is very dark (around 1%) at wavelengths longer than 10 μm. Fortunately, in both the reflective infrared and the thermal infrared wavelength regions, the common evaporite mineral anhydrite is spectrally distinct from alunite, an important hydrothermal alteration mineral associated with precious metal deposits (Chapter 7).

The sulfides pyrite, FeS_2, and pyrrhotite, $Fe_{1-x}S$, display fairly featureless, opaque behavior in the 0.4–2.5 μm for larger particle sizes, but this fine-particle pyrite sample displays a ferrous iron band centered about 1.1 μm, and another broad minimum near 2.05 μm that could an overtone of the Fe—S fundamental. In the 2.0–25.0 μm wavelength regions, pyrite displays absorption bands as reflectance minima at 3.0, 6.1, and 6.8 μm, and as reflectance maxima at 11.5 μm and 25 μm, caused by an Fe—S fundamental vibrations. The thermal infrared spectrum of pyrite is quite unlike any other mineral. Pyrrhotite is featureless throughout its spectrum, like a metal.

In summary, the sulfates exhibit typical absorption bands of the sulfate ion in the 8.7–8.9 μm wavelength region and a number of absorption minima in the 1.4–2.8 μm region. Alunite is particularly important for mineral exploration because it only occurs as the result of hydrothermal alteration and anhydrite is a common evaporite mineral. These two are spectrally distinct, especially in the reflective and thermal infrared wavelength regions. The sulfides display more featureless reflectance spectra, but pyrite rises greatly in reflectance at wavelengths longer than 23 μm, unlike any other mineral.

Examples of Igneous Rocks

Igneous rocks, with the exception of rare igneous carbonates called carbonatites, are composed of mixtures of silicate minerals that solidified from molten rock (underground magma or above-ground lava) as it cooled. In the previous discussion of silicate minerals, it was noted that the position of the silicate reststrahlen band shifted to shorter wavelengths as the degree of polymerization of the SiO_4 tetrahedra within the crystalline lattice increased. Mafic minerals, which contain more iron and magnesium than felsic minerals, have little SiO_4 tetrahedra polymerization and display longer wavelength reststrahlen bands, whereas felsic minerals have the most polymerization and display shorter wavelength reststrahlen bands. Igneous rocks composed primarily of mafic or felsic minerals are called mafic or felsic rocks, respectively. From Equation 2.4 and the foregoing discussion, we might expect the reflectance maxima of igneous rocks to shift to shorter wavelengths with increasing felsic composition. The following spectra show that this felsic shift does occur for igneous silicate rocks.

The spectral libraries for all of the mineral spectra shown in the previous figures of this chapter exclude spectra of rocks or soils (Grove et al. 1992; Salisbury et al. 1991). Library spectra of igneous rocks have been around for a few decades (Lyon 1964; Vincent et al. 1975). However, a more recent spectral library has become available in digital form. Salisbury et al. (1988) collected directional hemispherical reflectance spectra of natural, rough surfaces of igneous rocks in the 2.5–13.5 μm wavelength region. Digital data from that paper, plus reflectance spectra in the 0.4–2.5 μm region for the same igneous rock samples, are available on the Internet via FTP from the Johns Hopkins FTP site: rocky.eps.jhu.edu. There are many more spectra at that data site than are shown here.

Some of the igneous rock spectra from that library are displayed in Figure 3.12, arranged in order of increasingly mafic (decreasingly felsic) composition, except for the last spectrum of anorthosite, which has little quartz or mafic minerals. Descriptions of the rock samples represented are given in Table 3.1 on page 60 (Salisbury et al. 1988). The first three pairs of spectra represent extrusive (fine-grained) versus intrusive (coarse-grained) rocks with similar compositions, such as rhyolite and granite. Note how the strongest reflectance maximum in the 8.0–12.0 μm wavelength region generally occurs at longer wavelengths for more mafic (less felsic) rocks. For instance, the reststrahlen band (reflectance maxima) for dunite occurs about 11 μm, for basalt and gabbro occur about 10.0–10.5 μm, for andesite and diorite occur around 10 μm, and for rhyolite and granite occur around 9 μm. As will be discussed in Chapter 5, the felsic shift in igneous silicate rocks is one of the most important, exploitable relationships between remote sensing measurements and the composition of exposed rocks and soils.

Examples of Sedimentary Rocks

Sedimentary rocks, of course, contain different types of minerals, not just silicate minerals. All of the sedimentary rock specimens (Hunt and Salisbury 1976a) discussed in this section and metamorphic rock specimens (Hunt and Salisbury 1976b) discussed in the next section have directional hemispherical reflectance *of polished surfaces* of each rock specimen taken from the Johns Hopkins FTP site: rocky.eps.jhu.edu. Thus, *the absolute reflectance values of the spectra are not representative of the natural rock surfaces,* whereas the wavelength location of the reststrahlen bands most likely are. More spectra are available from that Internet FTP site than are shown here.

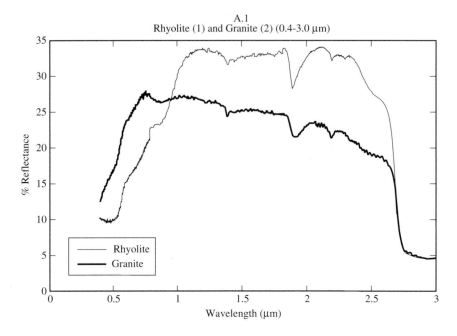

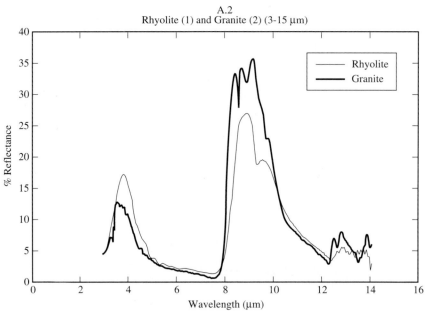

Figure 3.12 Directional hemispherical reflectance spectra in the 0.4–3.0 μm and the 3–15 μm wavelength region of natural surfaces of eight igneous silicate rocks: (a) rhyolite and granite. *(Figure 3.12a–b courtesy of Salisbury and D'Aria, personal communication and Salisbury et al., 1988, USGS Report.)*

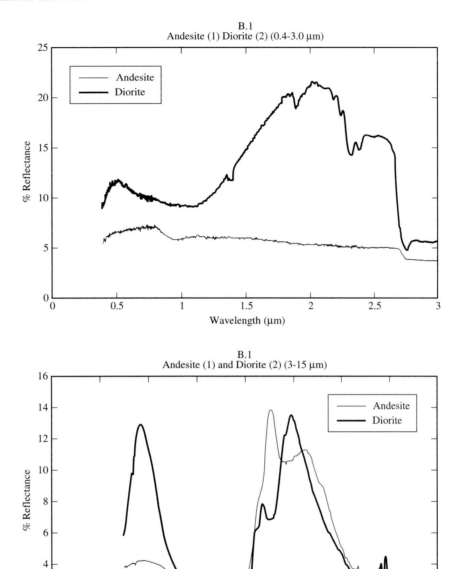

Figure 3.12 *(continued)* (b) andesite and diorite

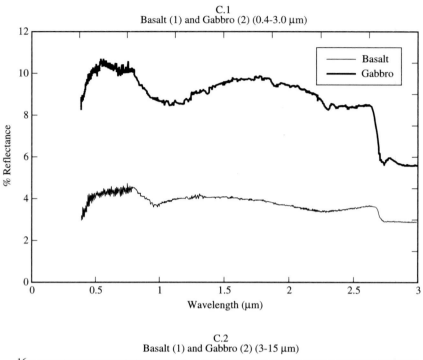

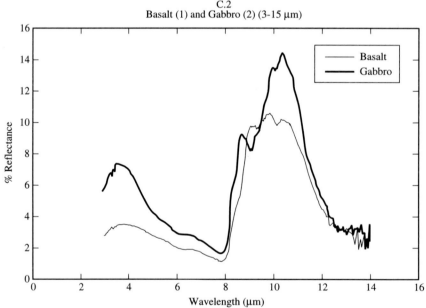

Figure 3.12 *(continued)*   (c) basalt and gabbro

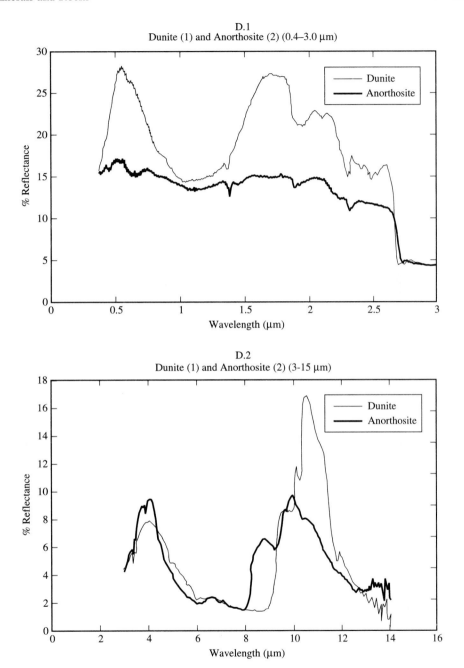

Figure 3.12 *(continued)*   (d) dunite and anorthosite

TABLE 3.1 DESCRIPTIONS OF IGNEOUS ROCKS (SALISBURY ET AL. 1988) FROM WHICH REFLECTANCE SPECTRA OF FIGURE 3.12 WERE MEASURED.

RHYOLITE

Designation

rhyolite.h1

Locality

Pennsylvania Hill near Rosita, Custer County, Colorado.

Hand Sample Description

A reddish, banded, porphyritic rock with the phenocrysts small (< 1 mm) and variable in size.

Petrographic Description

The groundmass is glassy, microlitic, and spherulitic. The phenocrysts consist of sanidine > plagioclase (oligoclase in part) > biotite > quartz. Accessories are apatite and magnetite.

GRANITE

Designation

granite.h2

Locality

Santa Rita Mountains, Arizona.

Hand Sample Description

A gray, medium-grained rock consisting of white feldspar grains, some of which contain red or pinkish cores, quartz, and minor biotite. Some of the feldspar appears to be perthitic.

Petrographic Description

Typically these rocks are hypidiomorphic with granular texture. Quartz, orthoclase, and plagioclase make up at least 93% of the rocks. Quartz anhedra possess moderate to strong undulatory extinction, the orthoclase anhedra are slightly kaolinitized. Perthite is common in the orthoclase with the amount of albite in the orthoclase 8%. The plagioclase is often sericitized, while the biotite is largely altered to chlorite. The modes for this sample were 40.4% orthoclase, 38.5% quartz, 19.7% plagioclase, 1.1% biotite, and 0.3% magnetite. There was no sign of carbonate in thin section, but spectrum and chemical analysis both indicate presence of a small amount.

Microprobe Analysis

The analysis revealed the plagioclase composition to be albite, with no more than 5% anorthite.

ANDESITE

Designation

andesite.h1

Locality

Saipan.

Hand Sample Description

The sample is about 4 × 3 cm, brown on the weathered surface and dark gray on fresh surfaces. It is porphyritic with the phenocrysts making up about 25–30% of the rock. The groundmass is gray and microcrystalline. The phenocrysts are < 1 mm and consist of plagioclase laths, pyroxene, and opaques, in that order of abundance, with pyroxenes nearly as abundant as the feldspars.

Petrographic Description

The modes for phenocrysts for this sample gave 26.75% plagioclase, 8.5% augite, 1.05% magnetite, 0.95% hypersthene, and 62.75% groundmass. The groundmass for these samples consists of labradorite, augite, hypersthene, magnetite, ilmenite, trydimite as isolated grains and aggregates of small wedge-shaped crystals closely associated with small patches of intergrown granular quartz and chalcedony, anorthoclase, and partly devitrified glass.

Microprobe Analysis

The composition of the plagioclase phenocrysts was found to range from An_{59} to An_{62} (labradorite), while the composition in the groundmass ranged from An_{50} to An_{55}. The augite phenocryst composition was Wo 34%, En 43%, Fs 23%, that of the hypersthene phenocrysts was En 67%, Fs 33%. Augite in the groundmass was Wo 24%, En 45%, Fs 31%.

DIORITE

Designation

diorite.h1

Locality

Near Azusa, Los Angeles County, California.

Hand Sample Description

A dark gray, medium-grained rock composed of plagioclase, a dark mafic phase, and minor biotite.

(continued)

TABLE 3.1 *(continued)*

Petrographic Description

A hypidiomorphic rock with moderately well-developed gneissoid foliation due to the subparallel alignment of subhedral green hornblende, which is partly replaced by biotite. The anhedral plagioclase is frequently zoned gradationally and often shows either Carlsbad or albite twinning. The cores of most of the feldspars are slightly sericitized and in a few regions of the sample are heavily sericitized. Epidote was a common accessory mineral, much less commonly apatite. The modal analyses gave: 51% plagioclase, 39% hornblende, 3.4% biotite, 0.8% epidote, with the remaining percentages composed of sericite and apatite.

Microprobe Analysis

The plagioclase composition was rather homogeneous, ranging from An_{40} to An_{48} (andesine). However, some grains were quite heterogeneous. One grain had a composition of Or_{62} at the border where Ab dropped to about 10%; another feldspar was heterogeneous as well, with a composition ranging from 0.5% to 12% CaO. The hornblende analysis revealed a homogeneous composition equivalent to pargasite.

BASALT

Designation

basalt.h1

Locality

Chimney Rock, Somerset County, New Jersey.

Hand Sample Description

A greenish-black aphanitic rock.

Petrographic Description

Microphenocrysts of augite (some glomeroporphyritic) are set in a matrix of thin laths of labradorite, granular clinopyroxene, and dark, essentially opaque glass, which is subordinate and interstitial. Some of the glass has been altered to a brown iron-rich chlorite. The modes gave 31% feldspar, 30% pyroxene, 31% groundmass, 0.2% opaque, and 7.8% miscellaneous (alteration products).

Microprobe Analysis

The feldspar composition ranged from An_{67} to An_{74} (labradorite-bytownite); pyroxene phases probed were augite and pigeonite.

GABBRO

Designation

gabbro.h1

Locality

Escondido, San Diego County, California.

Hand Sample Description

A dark gray medium-grained rock composed of plagioclase, a mafic mineral, and minor opaques.

Petrographic Description

An equigranular, hypidiomorphic rock composed of subhedral plagioclase laths showing albite twinning, sometimes Carlsbad twinning, and frequently gradational zonation. There are anhedra of green pyroxene and colorless pyroxene (as exsolution), minor green hornblende, and abundant opaque anhedra. Modes were: 66% plagioclase, 28% pyroxene, and 6% opaque.

Microprobe Analysis

Bytownite was the predominant feldspar composition, ranging from An_{80} to An_{84}. One plagioclase analyzed proved to be andesitic (An_{42}). The pyroxene phases proved to be green, subcalcic augite and exsolved (colorless) salite. Opaque phases were either magnetite or a mixture of ilmenite and magnetite.

DUNITE

Designation

dunite.h1

Locality

Near Balsam, Jackson County, North Carolina.

Hand Sample Description

A green, medium- to fine-grained granular rock composed almost entirely of olivine, with some scattered opaques (chromite?). There is a brown weathering stain on one surface.

Petrographic Description

A mosaically textured aggregate of anhedral olivine with serpentine alteration along some of the fractures. Chromite (?) occurs as scattered subhedra. The modal analysis gave: 89% olivine, 8% serpentine, and 3% opaque.

Microprobe Analysis

Microprobe analysis revealed an olivine composition of F091. The opaque phase showed Fe > Al > Mg with little titanium, and is probably

(continued)

TABLE 3.1 *(continued)*

chromite. Serpentine (antigorite) had 4% substitution of FeO for MgO. Minor chlorite approximated clinochlore in composition.

ANORTHOSITE

Designation
anorthos.h1

Locality
Near Elizabethtown, Essex County, New York.

Hand Sample Description
The hand sample consists of very dark gray, large tabular plagioclase with interstitial bluish-green granulated feldspar groundmass, with some opaque areas 1–3 mm in size scattered through the sample. The Marcy anorthosite from which this sample was taken is described in Buddington, A. F., 1938, Adirondack igneous rocks and their metamorphism: Geological Society of America Memoir No. 7, pp. 19–33 and 258–259.

Petrographic Description
Very coarse-grained sample consisting of subhedral plagioclase (96%) with some areas showing mortar texture. The plagioclase groundmass apparently is cryptocrystalline due to granulation. There are minute scattered amphibole anhedra (0.6%) and 0.7% opaques. There was no sign of calcite in thin section, nor was the rock analyzed for CO_2, but the spectrum clearly indicates the presence of a small amount of carbonate.

Microprobe Analysis
Microprobe analysis indicated a range of feldspar grains and groundmass composition from An_{10} to An_{50} (oligoclase-andesine); various cryptocrystalline amphiboles (either Tschermakite or hastingsite), opaques high in titanium but also high in silica and lime (a mixture), and possibly some zoisite. Like most "anorthosites" of the Adirondacks, average plagioclase composition is andesine. Hence, these rocks are sometimes classed with lime diorites instead of gabbros and referred to as "andesinites."

Figure 3.13 shows directional hemispherical reflectance spectra of polished surfaces of eight selected sedimentary rocks (dark gray limestone, dolomitic limestone, phosphatic shale, black shale, siltstone, dark gray siltstone, red sandstone, and arkosic sandstone spectra) in the 0.4–3.0 μm and 3–15 μm wavelength regions (Hunt and Salisbury 1976a). Table 3.2 on p. 67 provides a description of each of the sedimentary rock samples with spectra shown in Figure 3.13, as they are given in the rocky.eps.jhu.edu FTP site, with the exception of the dark gray siltstone title, where the original name was kept (Hunt and Salisbury 1976a).

The fossiliferous limestone and dolomitic limestone have similar spectra in the 3–13 μm region, except for the 8.5–9.5 μm region where the quartz and other silicate mineral content of the dolomitic limestone sample produces silicate reststrahlen bands. The dolomitic limestone also exhibits a marked ferrous iron absorption band near 1.1 μm. The phosphatic shale specimen, from Wyoming, is made up almost entirely of recrystallized phosphate nodules, with some presence of carbonate and quartz grains. The reststrahlen features between wavelengths of 9 and 10 μm are characteristic of the phosphate mineral apatite. The black shale specimen shows evidence of carbonate minerals near a wavelength of 7 μm and of silicate minerals (probably clay and feldspar) in the 10 μm region, suggesting more silicate than carbonate present, as was confirmed by digestion in HCl (Hunt and Salisbury 1976a).

The (regular) siltstone sample displays evidence of fine-grained feldspar (probably albite) and clay minerals, on the basis of its muted reststrahlen bands near 10 micrometers. The dark gray siltstone sample contains about 50% clay, 20%

A.1
Fossiliferous (1) and Dolomitic (2) Limestone (0.4–3.0 μm)

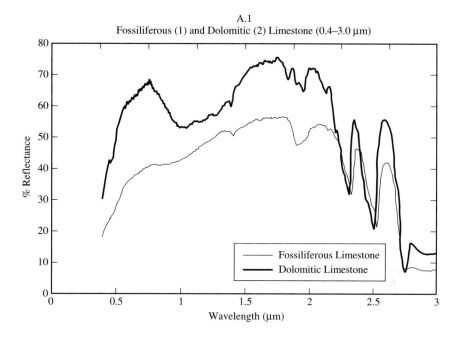

A.2
Fossiliferous (1) and Dolomitic (2) Limestone (3-15 μm)

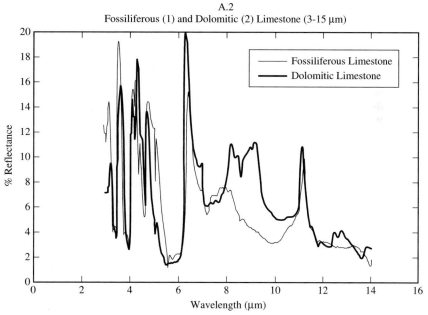

Figure 3.13 Directional hemispherical reflectance spectra in the 0.4–3.0 μm and the 3–15 μm wavelength region of polished surfaces of eight sedimentary rocks: (a) fossiliferous and dolomitic limestone. *(Figures 3.13a–d courtesy of Salisbury and D'Aria, personal communication.)*

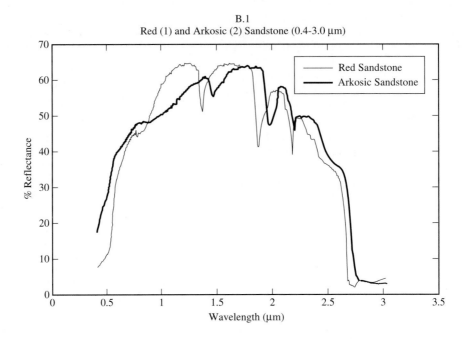

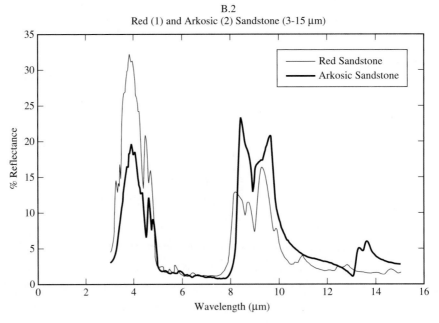

Figure 3.13 *(continued)*   (b) red and arkosic sandstone

C.1
Regular (1) and Dark Grey (2) Siltstone (3-15 μm)

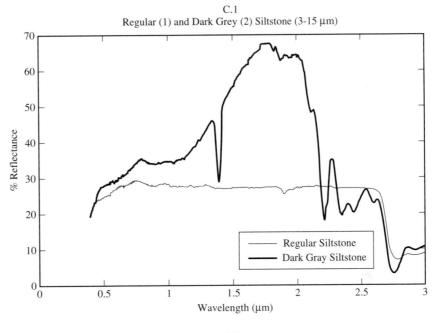

C.2
Regular (1) and Dark Gray (2) Siltstone (3-15 μm)

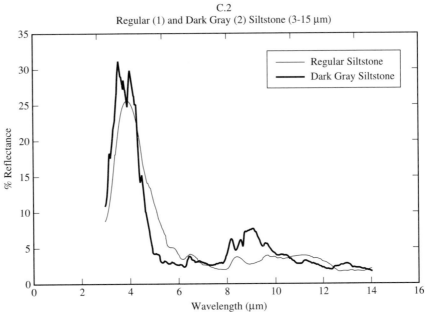

Figure 3.13 *(continued)* (c) regular and dark gray siltstone

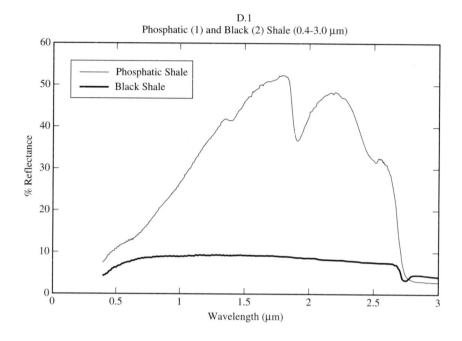

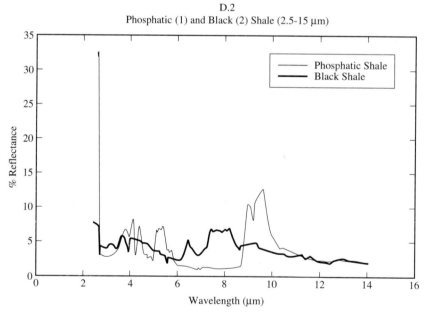

Figure 3.13 *(continued)*   (d) phosphatic and black shale

TABLE 3.2 DESCRIPTIONS OF SEDIMENTARY ROCKS FROM WHICH REFLECTANCE SPECTRA OF FIGURE 3.13 WERE MEASURED. (HUNT AND SALISBURY 1976A; FTP SITE ROCKY.EPS.JHU.EDU)

FOSSILIFEROUS LIMESTONE

Designation

Limeston.h1

Description

A sample consisting of a homogeneous microcrystalline carbonate matrix with scattered flecks of reddish-brown material and abundant fossils (primarily small bivalves) with recrystallized calcite; some bivalves still show laminar structure. Very minor quartz was also noted in this section.

DOLOMITIC LIMESTONE

Designation

Limeston.h2

Description

This somewhat heterogeneous sample consists of bands of pure fine-grained-to-microcrystalline dolomite alternating with bands of coarser dolomite crystals and subangular aligned quartz grains showing pressure solution texture. There is also a greenish-brown material (clay?) scattered throughout the sample in small amounts.

PHOSPHATIC SHALE

Designation

Shale.h2

Description

This sample consists of whitish to brown oolites of phosphatic material ~ 0.5–1.5 mm in size. Microscopically the material is collophane (amorphous calcium carbonophosphate, yellowish brown, and isotropic) in oolitic form and frequently zoned. Scattered fine-grained angular-to-subangular quartz grains occur throughout the sample, many in the cores of ooids; silica recrystallization has also occurred in some spots. There is a slight porosity to the rock, and carbonaceous material is present in many of the oolite cores and rims.

BLACK SHALE

Designation

Shale.h4

Description

This is very similar to #388 but with more clay in the matrix and fewer forams (mostly uniserial, some biserial and plani/trocho spiral types). There is possibly more quartz than carbonate in the groundmass, while again both opaques and carbonaceous material are present here.

SILTSTONE

Designation

Siltston.h1

Description

This is a very fine-grained lineated/laminated sample consisting of brownish clay and very fine quartz particles (about equal amounts) along with tiny slivers of another phase (feldspar?). There are abundant small flecks and discontinuous lenses of carbonaceous and opaque material as well.

DARK GRAY SILTSTONE

Designation

Siltston.h2

Description

This sample contains a microcrystalline groundmass or matrix of quartz, carbonate, and possibly some clay. There are irregular regions of larger anhedra of carbonate and quartz with some grains of muscovite, glauconite, and green biotite as well.

RED SANDSTONE

Designation

Sandston.h7

Description

This is a well-sorted fine-grained sandstone with slight porosity and good compaction. The well-rounded to subrounded quartz grains are all stained with ferruginous material, which, along with some sericite, makes up the scant matrix. The quartz is usually strain-free with moderate amounts of fluid inclusions, and there are minor amounts of microcline and a few areas of silica redeposition.

ARKOSIC SANDSTONE

Designation

Sandston.h1

(continued)

TABLE 3.2 *(continued)*

Description	
This sample is a moderately sorted fine medium-grained rock with slight porosity. The quartz grains are well-rounded to subrounded with some fluid inclusions and ferruginous staining. There is a trace of microcline and regions of concentrated car-	bonaceous granules and/or opaques on the quartz grains usually intimate with the ferruginous material. The quartz grains are cemented with secondary overgrowths of silica (chalcedony), and there are very minor secondary carbonate crystallites in some of the pores.

quartz, and 30% calcite, and the spectrum is dominated by kaolinite, quartz, and calcite. The red sandstone sample is interesting because its reststrahlen band in the 8.5–9.5 μm region is muted, presumably by ferric oxide staining of some of its quartz grains. The sample also exhibits a steep reflectance rise in the 0.4–0.7 μm wavelength region, giving it the red color (from ferric oxide absorption in the 0.4 μm vicinity). The arkosic sandstone (Colorado), which does not contain many small ferric oxide particles, displays the quartz reststrahlen band very prominently. The pink orthoclase fragments present in the specimen do not show up well, if at all, in this spectrum.

In summary, the sedimentary rocks show the presence of silicate, carbonate, and phosphate minerals in their spectra, and some evidence of ferric oxide staining in red sandstone is present in the lowered reflectance of the quartz reststrahlen band in the 8.5–9.5 μm region. The limestones, shales, siltstones, and sandstones appear to have significantly different reflectance spectra. This difference indicates that they can probably all be separated by multispectral sensors, if data from spectral bands in the visible, reflective infrared, and thermal infrared wavelength regions are available.

Examples of Metamorphic Rocks

Figure 3.14 shows the directional hemispherical reflectance of polished surfaces of eight different metamorphic rocks. The metamorphic rock descriptions (given in Table 3.3) and spectra discussed in this section all come from Hunt and Salisbury (1976b). Note the strong carbonate reflectance features (similar to the polished calcite spectrum of Figure 2.5) of both the white marble and serpentine marble. The carbonate absorption bands short of 6 μm are all evidenced as reflectance minima (volume reflectance dominates), even though these are polished samples. The serpentine marble (New York) specimen, however, is 20% serpentine, 5% phlogopite, and the remainder mostly calcite; this composition causes the spectrum to display spectral features of both calcite and serpentine (near 10 μm), which is a silicate mineral.

The pink quartzite specimen displays prominent quartz spectral features (near 9 and 13 μm) despite the fact that the recrystallized quartz grains (approximately 90% by mass content) contain ferric oxide minerals. Evidently, quartz covered the ferric oxide during recrystallization, and this process would account for the prominent quartz reststrahlen bands. The albite gneiss sample is composed of

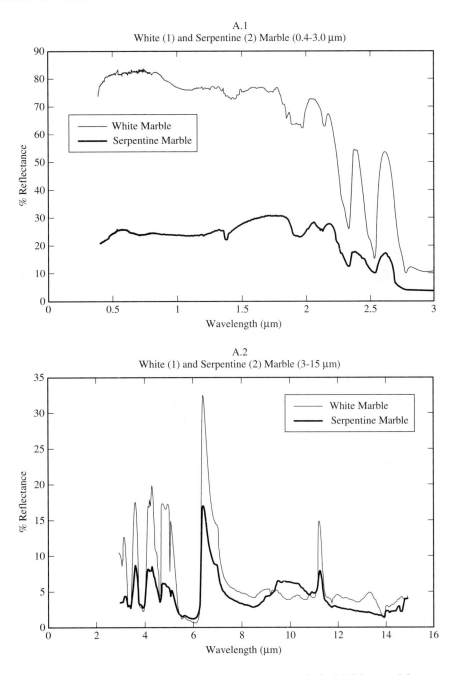

Figure 3.14 Directional hemispherical reflectance spectra in the 0.4–3.0 μm and the 3–15 μm wavelength region of polished surfaces of eight metamorphic rocks: (a) white and serpentine marble. *(Figures 3.14a–d courtesy of Salisbury and D'Aria, personal communication.)*

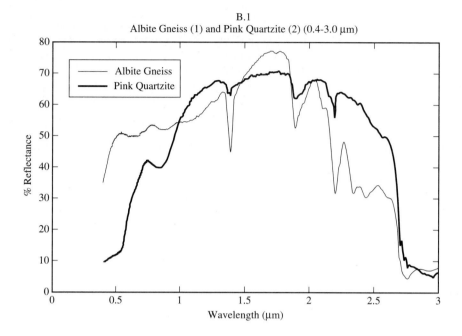

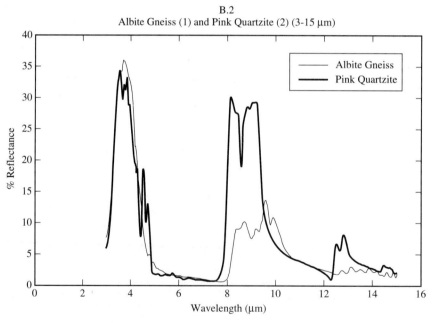

Figure 3.14 *(continued)*   (b) albite gneiss and pink quartzite

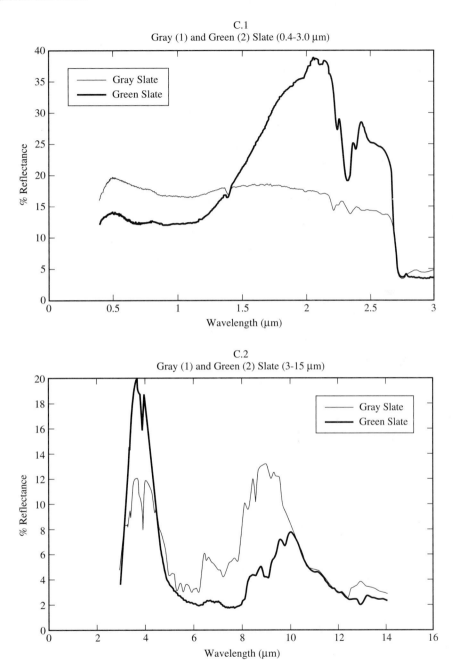

Figure 3.14 *(continued)*   (c) gray and green slate

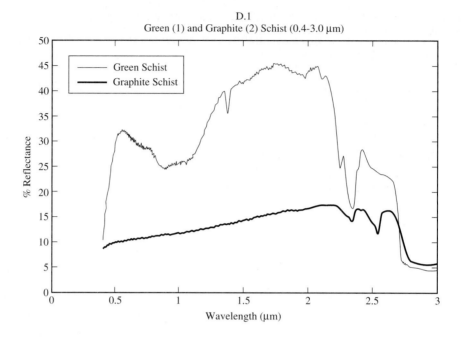

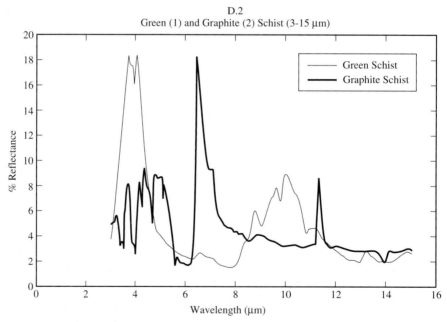

Figure 3.14 *(continued)*   (d) green and graphite schist

TABLE 3.3 DESCRIPTIONS OF METAMORPHIC ROCKS FROM WHICH REFLECTANCE SPECTRA OF FIGURE 3.14 WERE MEASURED.

WHITE MARBLE

Designation
> marble.h6

Description
> This rock contains coarse-grained carbonate with scattered, small, subrounded quartz grains and minor subhedral tremolite which has altered to talc in a few spots.

SERPENTINE MARBLE

Designation
> marble.h2

Description
> This sample consists of coarse-grained anhedral carbonate with numerous regions of serpentine (possibly brucite—$MgOH_2$, at least in part—probe) as an alteration/reaction of the carbonate. There are scattered subhedra of monticellite, tremolite, and phlogopite, with a few grains of idocrase (?) also noted in this section.

ALBITE GNEISS

Designation
> gneiss.h5

Description
> This coarse-grained rock has been sheared enough to give it mortar texture. The abundant albite subhedra (some with perthite) often evidence brittle deformation, while other areas have recrystallized. In addition to this phase there are muscovite euhedra (secondary to deformation), and larger primary muscovite heavily sheared to sericite in foliated masses or bands throughout the sample. Modes were 67% albite and 33% muscovite.

PINK QUARTZITE

Designation
> qtzite.h3

Description
> A medium-grained quartzite composed of subrounded equant quartz with some secondary silica (chalcedony) and quartz deposition; sometimes the chalcedony is pseudomorphic to quartz grains. These quartz grains contain numerous fluid inclusions, are often strained, and occasionally show pressure solution effects. Ferric material (plus clay?) is scattered throughout the sample giving this rock its reddish hue. There are trace amounts of zircon and lesser brown tourmaline (detrital), with a few scattered opaque/carbonaceous grains noted in this section. The modal analyses for this rock gave 90.4% quartz, 6.96% muscovite, 1.74% opaques, and 0.94% other.

GRAY SLATE

Designation
> slate.h1

Description
> This is a light gray, somewhat banded rock. It is fine-grained to microcrystalline and consists of carbonate, quartz, some feldspar, and clay. Abundant carbonaceous granules are scattered in discontinuous bands throughout the section.

GREEN SLATE

Designation
> slate.h2

Description
> This is a medium-grained sample dissected by thin veins. Microscopically, it consists of about equal amounts of actinolite and recrystallized albite lathes. The groundmass also consists of some chlorite, and carbonate, which is especially prominent in the veins, along with scattered pyrite. Modes were 59.3% amphibole, 25.7% albite, 10.8% chlorite, 3% calcite, and 1% pyrite.

GREEN SCHIST

Designation
> schist.h1

Description
> This is a medium to coarse-grained sample containing tremolite-actinolite crystals and talc. Microscopically, the amphibole appears to be tremolite as it does not have the typical green color of actinolite. This mineral is heavily altered to talc, and in addition to these phases there is quartz and some carbonate, as an alteration product, along with opaque grains. There may be heavily altered

(continued)

TABLE 3.3   *(continued)*

plagioclase present in this sample as well. Modes were 50.6% talc, 38% amphibole, 5.4% quartz, 3.6% calcite, 2.4% plagioclase remnants.	*Description* This rock contains coarse, irregular bands of brown, tan and gray material. Microscopically there are finer grains of calcite surrounded and enmeshed in a network of graphite bands and veins. There are also larger recrystallized calcite grains that are not associated with the graphite. Modes for this rock gave 62.5% calcite and 37.5% graphite.
GRAPHITE SCHIST *Designation* schist.h10	

mostly albite, which is apparent when its reflectance spectrum is compared to that of albite in Figure 3.6.

The gray slate has enough carbonate content to display absorption bands near 4 μm (reflectance minimum) and 7 μm (reflectance maximum). The green slate has a greenish color that suggests recrystallization of clay minerals to chlorite, with a very small amount of carbonate present. The features at 1.0 μm (ferrous iron absorption band) and 10 μm are similar to a spectral feature of chlorite. The green schist and graphite schist samples display distinctly different spectral features, owing to the greater carbonate content of the latter, giving rise to distinct carbonate bands near 4.0 μm, 7.0 μm, and 11.4 μm. The graphite schist specimen consists of graphite flakes (50%) in thin layers, interleaved with calcite (50%), which explains the predominantly calcite spectrum, since graphite has no significant spectral features in the thermal infrared wavelength region.

As with sedimentary rocks, metamorphic rocks contain silicates, carbonates, and other minerals, which gives them a variety of spectral features. With spectral bands in the visible, reflective infrared, and thermal infrared, multispectral remote sensing data should provide sufficient spectral contrast to separate these metamorphic rock samples from one another.

What should be made clear from the data and discussion in these three subsections, however, is that *it is easier to employ lab spectra and/or multispectral remote sensing data to separate rocks on the basis of whether they contain silicate or carbonate minerals (or both) than to differentiate rocks on the basis of whether they are igneous, sedimentary, or metamorphic.* In other words, it is the mineral composition that chiefly controls the spectral properties of rocks.

REVIEW

Vegetation, water, and snow have reflectance spectra that are unlike all rocks and minerals. The visible and reflective infrared spectra of vegetation are controlled mostly by chlorophyll *a* and chlorophyll *b,* which are the natural pigments that make grass green. However, the amounts of water, nitrogen, cellulose, lignin, and anthocyanin (another family of pigments) also contribute to the spectral behavior of plants, but to a lesser extent than do the two chlorophylls. The thermal infrared reflectance spectra of vegetation are relatively dark and featureless, and the temperature of vegetation tends to stay near the surrounding air temperature.

Water is transparent in the visible wavelength region, but changes color, depending on its suspended sediment load and its living organism content, with chlorophyll *a* the most important pigment for phytoplankton. In the reflective infrared wavelength region, water becomes increasingly opaque and is very opaque in the thermal infrared region. However, water is usually cool when land is warm (and vice versa), yielding good contrast between water and land in the thermal infrared wavelength region at most times of the day.

Snow is very bright in the visible wavelength region, but is dark at wavelengths longer than about 1.2 μm. Its spectrum is sensitive to the age and/or the degree of consolidation of the snow. There are many dissimilarities between snow and other natural materials.

The most common absorption bands in minerals in the 0.4–2.5 μm wavelength region are caused by ferric (Fe^{3+}) and ferrous (Fe^{2+}) iron ions in iron oxides and mafic silicates, the hydroxyl (OH^{-1}) ion in hydroxides and clays, H_2O in hydrated minerals, the carbonate CO_3^{-2} ion in carbonate minerals, and the sulfate SO_4^{-2} ion in sulfate minerals. Absorption bands in the visible, reflective infrared, and lower thermal infrared wavelength regions are manifested by reflectance minima.

In the 7.0–25.0 μm wavelengths, the primary reststrahlen bands of the silicate, carbonate, and sulfate minerals cause strong reflectance maxima that can be diagnostic of the bulk mineralogy of rocks. Exceptions to this are minerals with very small, clay-sized particles, which tend to greatly reduce the reststrahlen band reflectance maxima in the thermal infrared wavelength region.

There are dissimilarities of spectra among silicate, carbonate, sulfate, and sulfide minerals in the 0.4–2.5 μm, 3–5 μm, and 8–14 μm wavelength regions, which multispectral remote sensing can exploit for compositional mapping. Some minerals are more distinct than others, but if enough spectral bands were available in those wavelength regions, all or nearly all of them could be uniquely discriminated.

Rocks are aggregates of minerals, and the reflectance spectrum of a rock can be approximated by producing a linear combination of reflection spectra of the mineral constituents, each weighted by the mass percentage of that mineral in the rock. When rocks are almost monomineralic, their reflectance spectrum in the 2.0–25.0 μm wavelength region closely resembles the principal mineral constituent. An example of this nearly monomineralic condition is fossiliferous limestone, a sedimentary rock that has a reflectance spectrum very similar to the reflectance spectrum of the carbonate mineral, calcite. When silicate and carbonate minerals are both important constituents, the influence of the two types of minerals creates a rock reflectance spectrum that shows both silicate and carbonate reststrahlen bands. An example is serpentine marble, a metamorphic rock that displays a reflectance spectrum that is almost equal parts of calcite and serpentine reststrahlen bands. The most important wavelength region for separating rocks from one another by spectral means is the 8–14 μm thermal infrared atmospheric window region, where many of the reststrahlen bands diagnostic of mineral composition are located.

Ferric oxide coatings of quartz grains can moderately suppress quartz or other reststrahlen bands in a red sandstone, as a result of the low, somewhat featureless spectral reflectance of hematite in the 2–25 μm wavelength region. However, red quartzites display strong quartz reststrahlen bands, even with ferric oxide present, possibly because the ferric oxide coating in the parent sandstone was covered by quartz during the metamorphic recrystallization process.

It is clear from the spectra shown in this chapter that the compositional information we seek is imbedded in the spectral reflectance of Earth materials. The next chapter will deal with how spectral reflectance information related to composition can be separated from reflected sunlight and emitted thermal infrared radiation that emanates from the surface of Earth.

Exercises

1. What are the spectral bands for LANDSAT MSS, LANDSAT TM, and SPOT satellite sensors? (*Hint:* See Appendix B.) What are the ground dimensions of one frame of each of these sensors?

2. On day 1, an early-season snow falls on a still-green forest and melts off by the end of day 3. On day 2, LANDSAT IV collects data over the site, and on day 4, LANDSAT V collects data over the site. Compare what will happen to the brightness detected in each of the 7 Thematic Mapper spectral bands from day 2 (first overpass) to day 4 (second overpass). Your answer should be U for up (increase) on day 4, D for down (decrease) on day 4, or S (for staying about the same). Use the band numbering and wavelength limits of each spectral band given in Appendix B for LANDSAT TM.

3. How wide (taken at the half-height of the absorption band) would the band be and where would the band center be located for a spectral band to match the spectral widths of the following absorption bands: plankton chlorophyll, the ferrous absorption band in tremolite, the alunite absorption band with lowest reflectance in the 0.4–2.3 μm region, and the major reststrahlen band of quartz? Give both band width and band center location in μm. Also estimate band width as a percentage of band center (divide band width by center of band location and multiply by 100).

4. What are the approximate band center locations (in μm) of the major reststrahlen bands in the 8–14 μm region for the following silicate rocks: granite, diorite, gabbro, and dunite?

5. Using Equation 2.4, Figure 3.6 for the reflectance of smoky quartz, and Figure 3.10 for the reflectance of dolomite, calculate the reflectance of a mixture of 30% quartz mixed with 70% at the following wavelengths: 8.5 μm (the lowest reflectance peak of the quartz reststrahlen bands), 10 μm, 11.3 μm (the dolomite reflectance maximum), and 11.5 μm. Compare your results with the thermal infrared spectrum of dolomitic limestone in Figure 3.13 by normalizing your calculated reflectances (with a multiplicative constant) to the dolomitic limestone reflectance at the first quartz peak (8.5 μm wavlength), then using this constant to normalize your calculated reflectances at the other wavelengths.

Cited References

ABRAMS, M. J., J. E. CONEL, and H. R. LANG, eds. 1984. *The Joint NASA/Geosat Test Case Project Final Report,* vol. 2, part 2. Tulsa, Okla.: American Association of Petroleum Geologists.

CURRAN, P. J., and J. A. KUPIEC. 1995. Imaging Spectrometry: A New Tool for Ecology. In *Advances in Environmental Remote Sensing,* eds. F. M. Danson and S. E. Plummer, 71–88. New York: John Wiley & Sons.

DEKKER, A. G., T. J. MALTHUS, and H. J. HOOGENBOOM. 1995. The Remote Sensing of Inland Water Quality. In *Advances in Environmental Remote Sensing,* eds. F. M. Danson and S. E. Plummer, 123–142. New York: John Wiley & Sons.

DERR, V. E. 1972. *Remote Sensing of the Troposphere,* U. S. Dept. of Commerce, NOAA, Catalog No. C55. 602 T75, Stock No. C323-0011. (Also found in W. L. Wolfe and G. J. Zissis, eds., 1978. *The Infrared Handbook,* 3–113. Ann Arbor: Environmental Research Institute of Michigan.)

GAUSMAN, H. W., W. A. ALLEN, C. L. WIEGAND, D. E. ESCOBAR, and R. R. RODRIQUEZ. 1971. Leaf Light Reflectance, Transmittance, Absorption, and Optical and Geometrical Parameters for Eleven Plant Genera with Different Leaf Mesophyll Arrangements. In *Proceedings of the Seventh Symposium on Remote Sensing of the Environment,* 1599. Ann Arbor: Environmental Research Institute of Michigan.

GROVE, C. I., S. J. HOOK, and E. D. PAYLOR II. 1992. *Laboratory Reflectance Spectra of 160 Minerals, 0.4 to 2.5 Micrometers.* Jet Propulsion Laboratory Publication 92-2, 406 pp. Pasadena: California Institute of Technology.

HUNT, G. R., and J. W. SALISBURY. 1970. Visible and Near-Infrared Spectra of Minerals and Rocks: I. Silicate Minerals. *Modern Geology* 1:283–300.

HUNT, G. R., and J. W. SALISBURY. 1971. Visible and Near-Infrared Spectra of Minerals and Rocks: II. Carbonates. *Modern Geology* 2:23–30.

HUNT, G. R., J. W. SALISBURY, and C. J. LENHOFF. 1971. Visible and Near-Infrared Spectra of Minerals and Rocks: IV. Sulfides and Sulfates. *Modern Geology* 3:1–14.

HUNT, G. R., and J. W. SALISBURY. 1976a. Visible and Near-Infrared Spectra of Minerals and Rocks: XI. Sedimentary Rocks. *Modern Geology* 5:211–217.

HUNT, G. R., and J. W. SALISBURY. 1976b. Visible and Near-Infrared Spectra of Minerals and Rocks: XII. Metamorphic Rocks. *Modern Geology* 5:219–228.

LYON, R. J. P. 1964. Evaluation of Infrared Spectrophotometry for Compositional Analysis of Lunar and Planetary Soils: Rough and Powdered Surfaces, Final Report, Part 2. NASA Report CR-100.

MONTGOMERY, C. W., and D. DATHE. 1994. *Earth Then and Now.* Dubuque, Iowa: Wm. C. Brown Publishers.

POTTER, R. M., and G. R. ROSSMAN. 1979. The Manganese and Iron Oxide Mineralogy of Desert Varnish. *Chemical Geology* 25, no. 1–2:79–94.

POLCYN, F. C., and D. R. LYZENGA. 1973. Calculations of Water Depth from ERTS MSS Data. In *Proceedings of the Symposium on Significant Results Obtained from the Earth Resources Technology Satellite,* vol. 1, sec. B. 1433–1441. New Carrollton, Md.: NASA Goddard Space Flight Center.

SALISBURY, J. W., and D. M. D'ARIA. 1992. Emissivity of Terrestrial Materials in the 8–14 μm Atmospheric Window. *Remote Sensing of Environment* 42, no. 2:83–106.

SALISBURY, J. W., D. M. D'ARIA, and A. WALD. 1994. Measurements of Thermal Infrared Spectral Reflectance of Frost, Snow, and Ice. *Journal of Geophysical Research* 99, no. B12:24235–24240.

SALISBURY, J. W., and N. M. MILTON. 1988. Thermal Infrared (2.5–13.5μm) Directional Hemispherical Reflectance of Leaves. *Photogrammetric Engineering and Remote Sensing* 54:1301–1304.

SALISBURY, J. W., L. S. WALTER, and D. M. D'ARIA. 1988. *Mid-Infrared (2.5 to 13.5 Micrometers) Spectra of Igneous Rock.* U.S. Geol. Surv. Open File Report 88–686, Reston, Va.

SALISBURY, J. W., L. S. WALTER, N. VERGO, and D. M. D'ARIA. 1991. *Infrared (2.1–25 Micrometer) Spectra of Minerals.* Baltimore, Md: Johns Hopkins University Press.

SCHANDA, E., 1986. *Physical Fundamentals of Remote Sensing.* Berlin: Springer-Verlag.

VINCENT, R. K. 1973. *A Thermal Infrared Ratio Imaging Method for Mapping Compositional Variations Among Silicate Rock Types.* Ph.D. Dissertation, Department of Geology and Mineralogy, Ann Arbor. University of Michigan

VINCENT, R. K., L. C. ROWAN, R. E. GILLESPIE, and C. KNAPP. 1975. Thermal-Infrared Spectra and Chemical Analyses of Twenty-Six Igneous Rock Samples. *Remote Sensing of Environment* 4:199–209.

VINCENT, R. K., D. H. COUPLAND, and J. B. PARRISH. 1981. HCMM Night-Time Thermal IR Imaging Experiment in Michigan. In *Proceedings of the Fifteenth Symposium on Remote Sensing of Environment,* vol. 2, 975–984. Ann Arbor: Environmental Research Institute of Michigan.

Additional References (Uncited)

HOVIS, W., JR., and W. R. CALLAHAN. 1966. Infrared Reflectance Spectra of Igneous Rocks, Tufts, and Red Sandstone from 0.5 to 22 Micrometers. *Journal of the Optical Society of America* 56, no. 5:639–643.

HUNT, G. R., L. M. LOGAN, and J. W. SALISBURY. 1973. Mars: Components of Infrared Spectra and the Composition of the Dust Cloud. *Icarus* 18:459–469.

HUNT, G. R., J. W. SALISBURY, and C. J. LENHOFF. 1971. Visible and Near-Infrared Spectra of Minerals and Rocks: III. Oxides and Hydroxides. *Modern Geology* 2:195–205.

HUNT, G. R., J. W. SALISBURY, and C. J. LENHOFF. 1973. Visible and Near-Infrared Spectra of Minerals and Rocks: VI. Additional Silicates. *Modern Geology* 4:85–106.

HUNT, G. R., J. W. SALISBURY, and C. J. LENHOFF. 1973. Visible and Near-Infrared Spectra of Minerals and Rocks: VII. Acidic Igneous Rocks. *Modern Geology* 4:217–224.

HUNT, G. R., J. W. SALISBURY, and C. J. LENHOFF. 1974. Visible and Near-Infrared Spectra of Minerals and Rocks: VIII. Intermediate Igneous Rocks. *Modern Geology* 4:237–244.

HUNT, G. R., J. W. SALISBURY, and C. J. LENHOFF. 1974. Visible and Near-Infrared Spectra of Minerals and Rocks: IX. Basic and Ultrabasic Igneous Rocks. *Modern Geology* 5:15–22.

HUNT, G. R., J. W. SALISBURY, and C. J. LENHOFF. 1975. Visible and Near-Infrared Spectra of Minerals and Rocks: X. Stony Meteorites. *Modern Geology* 5:115–126.

HUNT, G. R. 1977. Spectral Signatures of Particulate Minerals in the Visible and Near-Infrared. *Geophysics* 42, no. 3:501–513.

HUNT, G. R. 1979. Near-Infrared (1.3–2.4 Micrometers) Spectra of Alteration Minerals—Potential for Use in Remote Sensing. *Geophysics* 44, no. 12:1974–1986.

LANG, H. R., M. J. BARTHOLOMEW, C. I. GROVE, and E. D. PAYLOR. 1990. Spectral Reflectance Characterization (0.4–2.5 and 8.0–12.0 Micrometers) of Phanerozoic Strata, Wind River Basin and Southern Bighorn Basin Area, Wyoming. *Journal of Sedimentary Petrologists* 60, no. 4:504–524.

LOGAN, L. M., G. R. HUNT, J. W. SALISBURY, and S. R. BALSAMO. 1973. Compositional Implications of Christiansen Frequency Maximums for Infrared Remote Sensing Applications. *Journal of Geophysical Research* 78, no. 23:4983–5003.

LYON, R. J. P. 1963. Evaluation of Infrared Spectrophotometry for Compositional Analysis of Lunar and Planetary Soils. NASA Report NASA-TN-D-1871.

ROSS, H. P., J. E. M. ADLER, and G. R. HUNT. 1969. A Statistical Analysis of the Reflectance of Igneous Rocks from 0.2 to 2.65 Microns. *Icarus* 11:46–54.

VANE, G., R. O. GREEN, T. G. CHRIEN, H. T. ENMARK, E. G. HANSEN, and W. M. PORTER. 1993. The Airborne Visible/Infrared Imaging Spectrometer (AVIRIS). *Remote Sensing of Environment* 44:127–143.

VINCENT, R. K. 1972. An ERTS Multispectral Scanner Experiment for Mapping Iron Compounds. *Proceedings of the Eighth International Symposium on Remote Sensing of Environment,* 1239–1247. Ann Arbor: Environmental Research Institute of Michigan.

VINCENT, R. K., and W. W. PILLARS. 1974. Skylab S-192 Ratio Codes of Soil, Mineral, and Rock Spectra for Ratio Image Selection and Interpretation. *Proceedings of the Ninth International Symposium on Remote Sensing of Environment,* 875–895. Ann Arbor: Environmental Research Institute of Michigan.

VINCENT, R. K. 1975. The Potential Role of Thermal Infrared Multispectral Scanners in Geological Remote Sensing. *Proceedings of the IEEE* 63, no. 1:137–147.

4 Spectral Radiance from Earth's Surface and Its Measurement by Electro-Optical Multispectral Sensors

▼ ▼ ▼ ▼ ▼ ▼ ▼ ▼ ▼ ▼ ▼ ▼ ▼ ▼ ▼ ▼

REFLECTED AND EMITTED RADIANCE AVAILABLE FOR DETECTION

Thus far we have been discussing directional hemispherical reflectance, $\rho(\lambda)$, and biconical reflectance, $\rho^c(\lambda)$, of natural materials, as measured in the laboratory. For simplicity, it has been assumed that all natural materials are Lambertian surfaces, described in Chapter 2 as a perfectly diffuse surface that reflects light equally into all solid angles of the hemisphere above the surface. This is a more reasonable assumption for rocks and soils than for trees and water, which depart greatly from Lambertian behavior. This assumption led to Equations 2.5 and 2.6, which state that both of these types of reflectances are directly proportional to hemispherical reflectance, $\rho^h(\lambda)$. As Chapter 3 has shown, there is important information in these spectral reflectances concerning the chemical and/or mineralogical composition of the material under examination, which we will call the target.

However, remote sensing instruments cannot measure spectral reflectances directly, but instead measure the power of electromagnetic radiation that is impinging upon the detector, as it is pointed at the target. That power, P_i, has units of watts and is given by the following equation (Silva 1978):

$$P_i = L_\lambda \, \Delta\omega \, \Delta\lambda \qquad \text{(Eqn. 4.1)}$$

where

L_λ = the total spectral radiance available to a multispectral sensor at some distance above Earth, in units of energy per unit time (given in watts), per unit area instantaneously being observed on the ground (given in

square meters and called a pixel), per unit solid angle (given in steradians) subtended by the sensor at some distance above Earth, per unit
wavelength (given in μm)

$\Delta\omega$ = the solid angle subtended by the sensor (given in steradians)

$\Delta\lambda$ = the wavelength range (given in μm) to which the sensor is filtered

We must next consider L_λ, which describes how sunlight illuminates terrain features on Earth's surface, how these terrain features radiate heat, and how the atmosphere absorbs and scatters light for both of these types of electromagnetic radiation. For simplicity, integrals have been removed from equations in the following chapters. The more complete integral equations are given in Appendix A. The total spectral radiance, L_λ, available to a multispectral sensor at some distance above Earth is given by:

$$L_\lambda = L_\lambda^r \tau(\lambda) + L_\lambda^r(Path) + L_\lambda^e(T)\tau(\lambda) + L_\lambda^e(Path)$$

$$= (1/\pi)sE_\lambda b\rho(\lambda)\tau(\lambda) + L_\lambda^r(Path) + L_\lambda^e(T)\tau(\lambda) + L_\lambda^e(Path) \qquad \text{(Eqn. 4.2)}$$

where

L_λ^r = detected spectral radiance that was reflected off the target in the direction of the detector at wavelength λ

s = a unitless "shadow/slope" factor that varies between 0 and 1.0 according to the percentage of each picture element (pixel) on the target that is in shadow ($s = 0$ for 100% shadow, $s = 0.6$ for 40% shadow, and so on)

E_λ = direct solar spectral irradiance impinging on the target at wavelength λ with units of watts/m^2-μm

b = constant from Equation 2.5

$\tau(\lambda)$ = atmospheric spectral transmittance at wavelength λ, a number between 0 and 1.0, along the path between target and sensor

$\rho(\lambda)$ = spectral reflectance (see Equation 2.5) of target at wavelength λ, which contains the information that we seek on target or background chemical composition

$L_\lambda^r(Path)$ = atmospheric spectral path radiance at wavelength λ, which accounts for light scattered into the detector that never encountered the target

$L_\lambda^e(T)$ = spectral radiance leaving the target at wavelength λ, which contains $\rho^c(\lambda)$ in a manner to be defined later

$L_\lambda^e(Path)$ = spectral radiance emitted by the atmosphere in the path between the target and the sensor and scattered at infrared wavelengths

Figure 4.1 demonstrates that solar spectral radiance, assumed to be emitted by a blackbody of the temperature of the sun and reflected off Earth's surface, would equal the amount of thermal spectral radiance (or heat from the ground) emitted by a blackbody of 300 K at a wavelength of approximately 4 μm. This result implies that the first two terms of Equation 4.2 dominate at wavelengths short of approximately 4.0 μm, and the second two terms dominate at wavelengths longward of approximately 4.0 μm, with the assumption of a ground temperature at about 27°C (300 K,

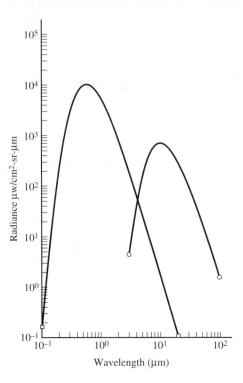

Figure 4.1 Terrain radiance for a terrain with visible/near infrared reflectance of 0.2 and thermal infrared emittance of 0.8 for sunlight reflected off the terrain (sun at 6000 K) and heat emitted by the terrain at 300 K. *(After Maxwell 1994. Reprinted with permission of the Environmental Research Institute of Michigan.)*

or 80.6°F). If the ground is warmer, the cross-over point (where the first two terms equal the second two terms) shifts to a wavelength below 4.0 μm, and if the ground is colder, it shifts to a wavelength longer than 4.0 μm, under the assumption of the same solar illumination for both conditions. For the remainder of this text, we will make the simplifying assumption that the first two terms will be applicable to remote sensing performed by sensors operating at wavelengths short of 4 μm (the visible and reflective infrared regions) and that the last two terms will be applicable to remote sensing performed by sensors operating at wavelengths longer than 4 μm (the thermal infrared region). This is a common, but often unstated, assumption in remote sensing literature.

Camera film and solid-state detectors operating in the visible and reflective infrared wavelength regions detect the first two terms in Equation 4.2 only, except for unusually hot targets, such as gas well flares, steel mills, volcanoes, and the like. Disallowing those exceptions, the following equation *is applicable for the visible and reflective infrared wavelength regions:*

$$L_\lambda = (1/\pi)sE_\lambda b\rho(\lambda)\tau(\lambda) + L_\lambda^r(Path) \qquad \text{(Eqn. 4.3.)}$$

The atmospheric path radiance term, $L_\lambda^r(Path)$, is caused by two types of scattering (Silva 1978). Rayleigh scattering, caused chiefly by the air molecules, has a λ^{-4} wavelength dependence that primarily affects the visible blue wavelength region and is responsible for the blue sky. The second type of scattering is by aerosol particulates, with a wavelength dependence ranging from $\lambda^{-1.2}$ to λ^{-2} at visible wavelengths. In most of the reflective infrared region, atmospheric path radiance is small enough to be ignored for most remote sensing tasks. In the visible wavelength region, where it cannot be ignored, the atmospheric path radiance term should be eliminated, because it contains no information about the target. A relatively simple empirical method (Vincent 1972) for accomplishing this goal will be discussed in the next chapter, which discusses image processing methods. During nighttime conditions, Equation 4.3 is approximately equal to zero.

Detectors operating in the thermal infrared region face a different condition. Unlike the visible and reflective infrared wavelength regions, the thermal infrared region offers detectable radiance during both night and day, since the radiance comes primarily from thermal emissions (heat) of the target itself. Equation 4.2 can be more completely described for the thermal infrared wavelength region as follows (Vincent 1975):

$$L_\lambda = L_\lambda^e(T) + L_\lambda^e(Path)$$

$$= \left\{ \epsilon(\lambda) \left[L_\lambda^e(bb, T) - \frac{E_\lambda(Sky)}{\pi} \right] + \frac{E_\lambda(Sky)}{\pi} \right\} \tau(\lambda) + L_\lambda^e(Path) \quad \text{(Eqn. 4.4)}$$

where

$\epsilon(\lambda)$ = target spectral emittance

$\tau(\lambda)$ = atmospheric spectral transmittance

$L_\lambda^e(bb, T)$ = the spectral radiance of a black body (with spectral emittance of 1.0) at the same temperature T in degrees Kelvin as the target

$E_\lambda(Sky)$ = the spectral irradiance from the sky incident from all directions on the diffuse target surface

The atmospheric path radiance is usually small enough to be ignored when observations are confined to the 3.0–5.0 μm, 8.0–14.0 μm, and 16.0–23.0 μm atmospheric windows because of three reasons: it only comes from the narrow column of air between the target and the sensor; most of the atmosphere between a high-altitude sensor and the target is considerably colder than the ground; and atmospheric emissivity is very low in those atmospheric windows. The spectral irradiance term, $E_\lambda(Sky)$, is larger than the path radiance term because irradiance comes from all directions in the sky, not just in the path between target and sensor. However, it will also be ignored for the sake of simplicity (perhaps an oversimplification) because under clear sky conditions, most of the atmosphere is colder than the ground and atmospheric emissivity is very low in atmospheric windows. With these assumptions, Equation 4.7 reduces to the following for the thermal infrared wavelength region:

$$L_\lambda = \epsilon(\lambda)\tau(\lambda)L_\lambda^e(bb, T) \quad \text{(Eqn. 4.5)}$$

A more rigorous version of Equation 4.5 appears in Appendix A.

Substitution of Equation 2.7 into Equation 4.5 yields the final equation for *the spectral radiance available for detection in the thermal infrared wavelength region:*

$$L_\lambda = [1 - b'\rho^c(\lambda)]\tau(\lambda)L_\lambda^e(bb, T) \quad \text{(Eqn. 4.6)}$$

A more rigorous form of Equation 4.6 is given in Appendix A.

Therefore, Equation 4.3 is the operable equation for relating spectral radiance to directional hemispherical reflectance for spectra in the 0.4–2.5 μm wavelength region. Equation 4.6 relates spectral radiance to biconical reflectance for the spectra given in the previous section for the 2.0–25.0 μm wavelength region, under the reasonable assumptions previously given. In the case of thermal infrared directional hemispherical spectra of rocks in the previous chapter, Equation 4.6 holds with a replacement of b for b' and $\rho(\lambda)$ for $\rho^c(\lambda)$. However, the sedimentary and metamorphic rock spectra were of polished surfaces, unlike natural surfaces.

Within the spectral radiance available for measurement by remote sensing sensors, it is the target spectral reflectance that contains the compositional information that we seek. What must be considered in the last section of this chapter is how these available spectral radiance equations are translated to measured spectral radiance by various kinds of remote sensing sensors. First, however, the kinds of sensors that are used for multispectral remote sensing will be described.

MULTISPECTRAL ELECTRO-OPTICAL SENSORS FOR REMOTE SENSING

There are two fundamental types of passive (employing natural sources of electromagnetic radiation) electro-optical imaging sensors, diagrammed in Figure 4.2, that have been used for commercial (civilian) multispectral data collection:

a. Multispectral scanners operate like a whisk-broom that scans side-to-side in a direction perpendicular to the flight direction, building up each line by sweeping along the scan line. The forward motion of the sensor platform creates new lines in the image. NASA's LANDSAT MSS and TM sensors and NOAA's GOES and AVHRR sensors are examples of multispectral scanners in orbit, and Daedalus Enterprises aerial scanners are examples of airborne multispectral scanners.

b. Linear arrays operate like a push broom, collecting an entire line at once, and using the forward motion of the platform to create new lines in the image. There are no moving parts required for producing an image with linear arrays, except for pointing mirrors, if off-nadir (something other than straight down) images are desired. The French SPOT and German MOMS sensors are examples of linear arrays in orbit, and the Canadian MIES sensor is an example of an airborne linear array.

c. Framing cameras take an instantaneous two-dimensional image on film, on an electronically scanned photoconductive surface, or on a two-dimensional solid-state array. With film, every grain of emulsion acts as a detector. Electronic framing cameras are like film cameras, except that electronically scanned photoconductive surfaces or two-dimensional solid-state arrays replace the film. The return-beam vidicon (RBV) cameras aboard the early LANDSAT satellites are examples of framing cameras with electronically scanned photoconductive surfaces, but there are yet no examples of two-dimensional, solid-state array, framing cameras on commercial satellites.

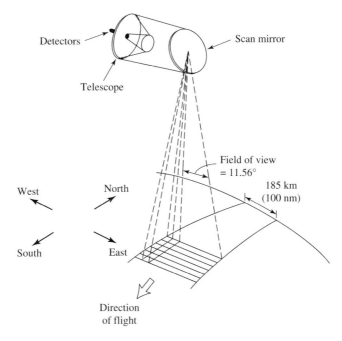

Figure 4.2 *(After Chen, 1985. Reprinted with permission from Academic Press.)* (a) Satellite whisk-broom scanner.

For all of these types of electro-optical imaging devices, the signals are normally amplified and digitized into binary numbers on the satellite platform, for digital transmission to ground stations.

The next two subsections explore differences among these two basic types of multispectral sensors in the manner that the incoming radiation is split up into distinct multispectral bands and in the radiometric and geometric corrections that must be applied to each. Then follows a subsection on hyperspectral imaging devices, which combine elements of two of the above three types into one sensor package that collects multispectral data in many narrow spectral bands. Finally, a list of present and upcoming commercial satellite sensors is given in Appendix B.

Splitting the Light

Multispectral scanners typically divide the light up into different wavelength regions with prisms, transmission filters, and dichroic mirrors, as sketched in Figure 4.3. Dichroic mirrors pass wavelengths longer than a certain wavelength and reflect shorter wavelengths, or vice versa. In airborne multispectral scanners, a single detector (though it can be several) is ordinarily used for each spectral band, such that all detectors are instantaneously co-registered, i.e., they look at the same instantaneous field of view on the ground (called a pixel, or picture element) at the same time. The instantaneous field of view for a multispectral scanner is one pixel. Each line, indeed the entire image, is built up one pixel at a time, and the image is an accumulation of

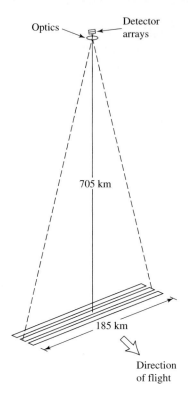

Direction
of flight

Figure 4.2 *(continued)*   (b) Push-broom
linear array.

lines. There is usually no data processing effort required to maintain co-registration of spectral bands, because the detectors are always co-registered in a well-built multi-spectral scanner.

Different solid-state materials are required for different wavelength regions of coverage. For instance, silicon (Si) detectors are usually employed in the 0.4–1.0 μm region, indium antinomide (InSb) detectors in the 1.0–5.0 μm region, and mercury-cadmium-telluride (HgCdTe) detectors in the 8–14 μm region. The latter two require cooling to liquid nitrogen temperatures (about 77 K). This presents little difficulty for multispectral scanners, since typically only one detector is employed per spectral band, meaning there are fewer detectors to cool for multispectral scanners than for linear or two-dimensional array devices. Some scanners have more than one detector per spectral band. LANDSAT TM, for example, scans six detectors simultaneously for each spectral band, building the image six lines at a time.

Linear array devices consist of a row of detectors, sometimes several thousand, that collect data for an entire line at a time. In other words, the field of regard (what's being imaged in one "gulp" of data collection) is an entire line of pixels. However, a different linear array, each filtered to a different narrow-band wavelength region, is used for each spectral channel. Filtering separate linear arrays to different wavelength regions can be accomplished in two fundamental ways: by time-delayed linear arrays

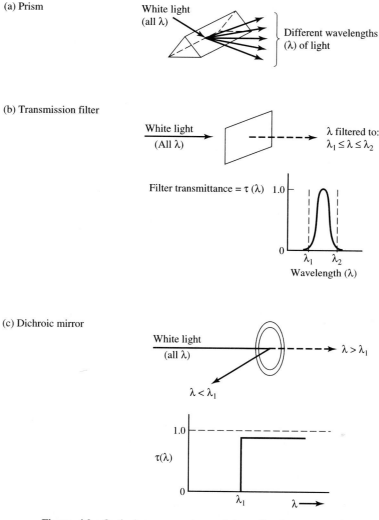

Figure 4.3 Optical components used to split electromagnetic radiation into different wavelength regions: (a) prism (b) transmission filter (c) dichroic mirror *(Figure 4.3a–c courtesy of R. K. Vincent.)*

that each have their own individual filter and by simultaneous-view linear arrays that are filtered by prisms, gratings, or dichroic mirrors. Figure 4.4 shows a diagram of these two types of linear array filtering. The first of these, involving several individually filtered linear arrays, can be lined in parallel such that they are all perpendicular to the direction of flight. This arrangement means that each array collects data from the same area on the ground at slightly different times. Thus, a translation of lines (like a time delay) is required before the resulting multispectral data set is in co-registration; the SPOT satellite works this way. This approach is suitable for a few spectral bands, but if

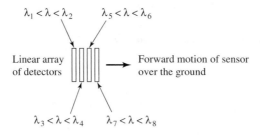

$\lambda_1 < \lambda < \lambda_2$ $\lambda_5 < \lambda < \lambda_6$

Linear array
of detectors

Forward motion of sensor
over the ground

$\lambda_3 < \lambda < \lambda_4$ $\lambda_7 < \lambda < \lambda_8$

a) Individually filtered linear array (viewed from above)

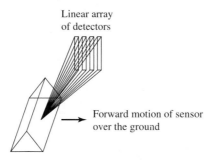

Linear array
of detectors

Forward motion of sensor
over the ground

b) Linear array filtered by an optical component, here a prism,
   (viewed from above)

Figure 4.4   Two methods for spectral filtering of linear array devices: (a) each linear array individually filtered. (b) splitting the light with an optical component (here a prism) in front of the linear array *(Figure 4.3a–b courtesy of R. K. Vincent.)*

many spectral bands are employed in this manner, misregistration among spectral bands can occur due to parallax effects.

The second way of filtering linear arrays divides the spectrum up for each line of detectors via a prism, grating, or dichroic mirror. This method requires a less efficient optical train, thus yielding less energy for detection than in the first example. However, spectral bands are instantaneously co-registered with one another in the second example. In the future, linear array sensors will have to utilize a mixture of both ways of splitting up the spectrum of incoming light. Dichroic mirrors will most likely be used to divide the spectrum into at least three broad wavelength regions, one each for the Si, InSb, and HgCdTe linear array sensitivity ranges. Multiple linear arrays of each type of detector will be used to further break up the spectrum into smaller wavelength intervals. Unlike multispectral scanners, which use the same detector element for all the pixels collected in an image for a particular spectral band, linear arrays require correction for slight differences of electronic gain and offset from detector to detector on the linear array. Electronic gain is a multiplicative factor and electronic offset is an additive (or subtractive) factor that are applied to the signal measured by each detector, as will be discussed in the next section. SPOT satellite data, when purchased, are already spectrally co-registered and normalized for pixel-to-pixel electronic gain and offset differences before the data are delivered.

Electro-optical framing cameras, which have only been deployed commercially in the form of the return-beam vidicon (RBV) camera aboard earlier LANDSAT satellites, have a field of regard that corresponds to the entire collected image, as do film cameras. Splitting up the spectrum of incoming light for electro-optical framing cameras can likewise be performed in a number of ways that require too much optical engineering to be included here.

It is interesting to compare the number of detectors and dwell times (the amount of time over which a detector collects and integrates energy for an individual pixel in a given spectral band) required for each of the three types of multispectral sensors for n spectral channels, q columns of pixels, and r rows of pixels. Dwell time is important because it is proportional to the amount of energy collected per measurement. The signal-to-noise ratio (the input-signal power divided by the noise-equivalent power) is higher for greater amounts of energy collected (Silva 1978). Therefore, the longer the dwell time, the less noisy the instrument.

A multispectral scanner requires only n detectors, but the dwell time allowed on each pixel would be roughly $1/qr$ times the length of time it takes for a satellite to move r lines forward. This is typically one-millionth of a second (a microsecond) for multispectral scanners.

For a linear array system that was individually filtered, it would take n linear arrays, each q pixels long, to collect the same amount of multispectral data. However, the dwell time per pixel can be $1/r$ times the length of time it takes for a satellite to move r lines forward, or something on the order of one-thousandth of a second (a millisecond).

A framing camera or two-dimensional array camera would require a number of detectors equal to the product qr for each array; these detectors could record the whole image within the exposure time allowed by the camera's shutter speed. The dwell times for n co-registered, individually filtered, two-dimensional arrays are typically longer (up to 10 times longer) than dwell times for a linear array. Current two-dimensional array technology limits the number of detectors per array to approximately 25,000,000 for Si detectors (covering the 0.4–1.0 μm wavelength range), 250,000 for InSb detectors (covering approximately the 1.0–6.0 μm wavelength range), and 62,500 for HgCdTe detectors (covering approximately the 5.0–14.0 μm wavelength range).

Longer dwell times allow for greater energy collection; hence, narrower spectral bands can be achieved with acceptable signal-to-noise characteristics. This relationship implies that multispectral scanners are more suitable for medium-width spectral bands, where the spectral band-width interval amounts to approximately 5–10% of the band-center wavelength. An example of medium-width spectral bands would be thermal infrared bands centered at a wavelength of 10 μm, with band widths of 0.5 or 1.0 μm, which would imply spectral bands in the wavelength ranges of 9.75–10.25 μm (5%) or 9.5–10.5 μm (10%). For narrow spectral bands on the order of 1% band-width-to-band-center ratio, which is in the realm of hyper-spectral imaging devices, two-dimensional array framing cameras are favored over multispectral scanners because of the longer dwell times of the former.

Dwell time is not the only consideration, however, as will be discussed later in the chapter in the section on geometric and radiometric corrections. Linear array and two-dimensional array devices do not cover as wide an area as multispectral scanners, which can scan to wider angles on both sides of nadir (directly down). This favors multispectral scanners for medium and low spatial resolution sensors that are designed to cover large areas in one image, such as the AVHRR sensor aboard the U.S. weather satellites. AVHRR has a 1.1 km pixel size (at nadir), but has a 2400 km swath width, i.e., it covers regions of 2400 km by 2400 km in one image. However, as will be addressed in a later section of this chapter, atmospheric path is longer at the edges of an image than at nadir, and increasing the swath width introduces appreciably greater atmospheric attenuation at the edges, relative to the center, of an image.

Hyperspectral Imaging Devices

The main objective of multispectral remote sensing is the acquisition of information about the chemical composition and location of a target. As the spectra of natural terrain elements in Chapter 3 demonstrate, many of the materials represented by those spectra are unique, if the entire spectrum in the atmospheric windows of 0.4–2.5, 3.0–5.0, and 8.0–14.0 μm are available. Until the last decade, technical limitations on the narrowness of spectral width and the number of spectral bands that could be collected by one sensor package forced applications of multispectral remote sensing to resemble the old radio game called "Name That Tune," whereby listeners were asked to name a mystery tune with as few notes as possible. In essence, spectra of each pixel on the terrain were averaged over a few relatively wide spectral bands (wider than 20% of the spectral band's center wavelength), and the identity of each pixel was sought from this information alone. For some terrain materials, this was adequate, as later chapters will attest, because the materials were sufficiently unique to be able to "name that tune" with a few notes. For many other materials, no identification was possible with such crude estimations of the reflectance or emittance spectrum of a pixel, because the available spectral bands were either too few or too spectrally broad to render identification possible.

Many minerals and practically all gases (such as methane, the most common constituent of natural gas) display critical spectral features that are on the order of 1% in spectral band-width-to-band-center (BW/BC) ratio. Identification of such features in their natural settings requires imaging of the ground with spectral bands that are about 20 times narrower, from a BW/BC standpoint, than a normal multispectral scanner. Devices that can perform such a task are called hyperspectral imaging devices because they collect multispectral image data in far greater numbers of spectral bands than do conventional multispectral scanners.

Figure 4.5 shows diagrams of two generic types of hyperspectral imaging devices. The type of hyperspectral device shown in Figure 4.5a is a two-dimensional array that is used in one dimension as a spectrometer and in the other dimension as

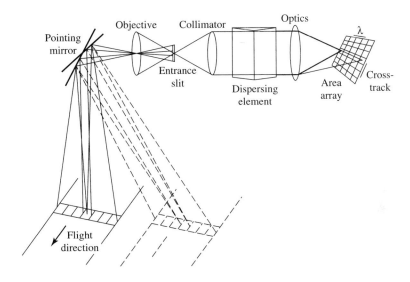

Figure 4.5 Generic types of hyperspectral imaging devices. (a) Two-dimensional array is used in one dimension as a spectrometer and in the other dimension as a push-broom linear array. *(Vane et al. 1993. Reprinted with the permission of Cambridge University Press.)*

a linear array. In a push-broom-like manner, the image is built up one line at a time, with no moving parts. The type in Figure 4.5b is a grating behind a slit that splits up entering radiation into vertical segments that line up with individual detectors of a vertical linear array. That array is then scanned whisk-broom-like from left to right across the scene. This makes each detector of the linear array act as a different, very narrow spectral band, and the scanning motion of the vertical linear array from side to side builds up an image one line at a time, similar to a multispectral scanner.

The first published hyperspectral imaging device was produced by Vane et al. (1984), although the term *hyperspectral* came a year later (Goetz et al. 1985). The instrument was called an airborne imaging spectrometer, or AIS-1. The AIS-1 employed a two-dimensional array in a push-broom mode behind a dispersing grating, similar to the type shown in Figure 4.5a. It acted the same as if a push-broom linear array, discussed earlier, were replaced by a prism and a two-dimensional array, with each row in the array covering a different spectral interval of the instantaneous field of view on the ground, in this case a single line of the resulting image. The AIS-1 employed a 32×32 element HgCdTe array, which produced an image 32 pixels wide, with 32 spectral bands between wavelengths of 1.2 μm and 2.4 μm. The AIS-2 employed a 64×64 element array and covered the 0.4–2.4 μm wavelength region (Vane and Goetz 1988).

The principal limitation of this type of hyperspectral imager design is that the width of the image in number of pixels is limited to the width of a two-dimensional solid-state array of detectors, which was commercially limited in 1984 and 1988. Whereas satellite and airborne multispectral scanners typically have approximately

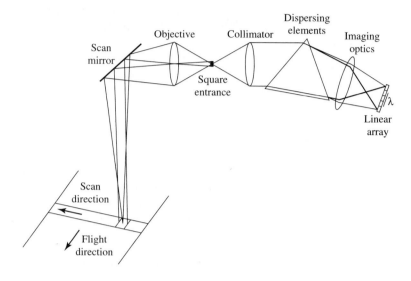

Figure 4.5 *(continued)*   (b) A grating is placed behind a slit that splits up entering radiation into vertical segments that line up with individual detectors of a vertical linear array. The array is then scanned whisk-broom fashion across the scene in a direction perpendicular to the flight path. *(Vane et al. 1993. Reprinted with the permission of Cambridge University Press.)*

6000 and 750 (in the case of scanners built by Daedalus Enterprises) pixels per line, respectively, the AIS-1 and AIS-2 had only 32 and 64 pixels per line, respectively. Even though two-dimensional arrays are available in larger sizes now, the limitation of image line length to the number of pixels on a single row of a two-dimensional sensor is a severe one for either spatial resolution or for swath-width of the image. Area arrays could perhaps be strung together to achieve a more acceptable image width.

A second generation of hyperspectral imaging, called the Airborne Visible/ Infrared Imaging Spectrometer (AVIRIS), employs linear arrays behind four gratings and a whisk-broom-type scan mirror, similar to the type diagrammed in Figure 4.5b, except that fiber optics were used in AVIRIS between the condensing mirror and four spectrometers, each covering a different wavelength region (Vane et al. 1993). AVIRIS was able to employ 614 cross-track pixels and 224 spectral bands between the wavelengths of 0.4 μm and 2.5 μm. Spectral radiance from four adjacent pixels in the image focal plane was passed through four optical fibers to four gratings, which dispersed the light up and down four different linear arrays, one of Si detectors and the other three of InSb detectors. Thus, each detector on each linear array behind a given grating spectrometer recorded a different spectral band for the same pixel on the ground, simultaneously with one another, thereby guaranteeing co-registration of the spectral bands for that grating. Postprocessing of the image took out the one-pixel offset (across-track, perpendicular to the flight direction) between adjacent optical fibers that fed the grating spectrometers.

As noted in Chapter 3 and discussed in Chapter 9, a recent remote sensing experiment with AVIRIS hyperspectral scanner data was conducted that mapped folial biochemical concentrations in a slash pine forest (Curran and Kupiec 1995).

Hyperspectral imaging devices of either type create huge amounts of data because there are so many spectral bands per pixel collected in the image. For the same pixel size, the amount of data collected by an imaging device is directly proportional to the number of spectral bands. For the same number of spectral bands, the amount of data collected by an imaging device is proportional to the reciprocal of the pixel area (or the square of the pixel linear dimension). Although it is difficult to argue against the fact that more spectral information permits more complete identification of a target, the number of pixels multiplied by the number of spectral bands to be recorded, transmitted, and processed is a practical limiting factor in current technology. Usually, this trade-off is between spectral resolution and either spatial resolution or amount of area covered on the ground (proportional to swath width).

The spectra in Chapter 3 show that some compositionally diagnostic absorption bands are broad, and others are narrow. The placement of the spectral bands is often more important than the width of the spectral band. Some of the most important information about mineral and rock identities can be found with spectral bands that have a BW/BC ratio greater than 10%. There are, however, some important minerals for which narrow spectral bands with a BW/BC ratio of 1% are crucial, as will be discussed in Chapter 7 (particularly in the subsection about gold exploration). The narrow band width of absorption bands in gases and the lesser need for fine spatial resolution in imaging gases make hyperspectral imaging greatly desirable for atmospheric environmental monitoring and natural gas exploration. There is a discussion of gas imaging in general in Chapter 5 and of methane imaging for solid waste landfill monitoring in Chapter 9.

One caveat about hyperspectral sensors must be heeded. Equation 4.1 shows that the power available for detection is directly proportional to the solid angle subtended by the sensor (which is, in turn, proportional to pixel area) times the band width of the spectral band. There is always less power (energy per unit time) in a narrow spectral band than in a wider one, and signal-to-noise decreases as the amount of detectable energy decreases. The most common way that this loss of energy associated with narrowing of the spectral bands has been countered in the past is with an increase in the pixel size, such that energy is gathered over a larger area at any one instant in time. Another approach, which is more costly, is to increase the collector size (the telescope in front of the sensor). Unless a breakthrough in increased sensor sensitivity occurs, hyperspectral sensors will either have coarser spatial resolution or require larger, more expensive collection optics than multispectral scanners flown on the same platform.

Radiometric and Geometric Corrections

Each of the fundamental types of imaging devices requires different types of collection-device-dependent geometric corrections and radiation balancing, to ensure equal radiance from nontarget sources (such as atmospheric radiance) throughout the scene being imaged. Because multispectral scanners sweep from side to side in a wider field of view than linear arrays, the pixels (or picture elements) at the left and right extremes along one scan line look through a longer atmospheric path. This situation is a greater problem with airborne scanners and coarse resolution satellite scanners, like AVHRR (Appendix B), which typically look 45 degrees to the left and right of nadir, than with higher resolution satellite scanners, which typically look about 6 degrees or less to the left and right of nadir. The longer atmospheric path length at higher scan angles causes higher radiances to be measured at the ends of a scan line than at the center, even if the ground were of homogenous reflectance everywhere. If such radiance imbalances remain uncorrected, the effects can be severe for the user. For instance, a target in the center of a multispectral scanner image will not have the same spectral characteristics as the same target at the edge of the image, even in perfectly flat terrain.

Geometric distortions also occur because of the scanning motion. Because a multispectral scanner looks side to side, the pixel sizes at the left and right extremes of a scan line are elongated in the scan direction compared with the pixel size at nadir. However, this problem is automatically corrected in recent LANDSAT data (including almost all LANDSAT TM data) and can be ignored for all but the oldest LANDSAT MSS data. Commercial software packages are now available that can automatically correct for both the radiation balancing and pixel geometry effects for airborne multispectral scanner data.

Linear arrays usually have even smaller fields of view than multispectral scanners. Pixels on the same line of a linear array are usually pointed down. Sometimes the array is pointed at oblique angles by a pointing mirror, as often occurs with the SPOT satellite. Thus, side-to-side radiance balancing and regularization of pixel size along the scan line are not needed for linear arrays. Neither of these two corrections is required for electronic framing cameras, which have front-end optics similar to photographic cameras.

There are two other kinds of geometric distortions, however, that can occur for either multispectral scanners or linear arrays, because the image is built up one pixel or one line at a time. The first is the motion of the sensor platform due to attitude changes while the image is being built up. The second, which applies only to space platforms, is the rotational motion of Earth beneath the satellite while the image is being built up. The latter is the more easily dealt with because the speed with which the rotation moves Earth's surface across the scene of observation is very predictable, even though that speed is a function of latitude on Earth's surface. Commercial satellite data sellers usually correct for Earth rotation in multispectral scanner data (like LANDSAT) or linear array data (like SPOT). It is because of this correction for Earth rotation during image build-up that the resulting satellite images are canted (slanted into a parallelogram, instead of a rectangle), such that the lower-left corner of the corrected image is farther left than the upper-left corner for a South-bound, polar-orbiting satellite, such as LANDSAT or SPOT.

The sensor platform attitude changes are unpredictable, except in heavy, gyro-stabilized space platforms, which can be physically constrained to maintain a very narrow attitude variation while an image is being built up. However, the attitude changes expected in airplanes or in inexpensive satellites require a different solution. Recently, a U.S. patent (Pleitner and Vincent 1989 and 1992) for a *Sensor Platform Attitude* (SPLAT) detector was awarded which, instead of physically constraining the platform, records changes in the attitude plane of the sensor platform and corrects the collected image for those changes on the ground after the image is collected. SPLAT works by adding a two-dimensional array, single-band camera with a powerful telescope to the sensor platform and collecting successive images such that the overlap between them is about 90%. The neighborhoods of at least 5 pixels in the master (older) image are correlated with the expected positions of those pixels in the slave (younger) image, and the x, y position (row and column numbers) of each of the five pixels is recorded. The correlation is done automatically, and approximately 10 bytes of information are saved for each successive image. The SPLAT sensor images themselves are not saved. Afterwards on the ground, the x, y positions of the five pixels are used to calculate the changes in the sensor platform attitude plane during successive images collected by the SPLAT sensor, and the line is automatically corrected for the attitude change. This procedure provides a practical method that uses no moving parts and yet can produce finer sensor platform attitude information than starpointers, inertial navigation systems, or multiple Global Positioning Satellite (GPS) receivers. It can make image data from aircraft or lightweight satellites as free of attitudinal distortions as image data from expensive, physically constrained satellite platforms.

There are some other geometric and radiometric corrections required that vary only with the altitude of the sensor platform above the terrain being imaged, the environmental conditions during data collection, and the map projection desired for the final image; these corrections are similar for all types of multispectral sensors. Included in these are corrections for Earth's curvature, atmospheric refraction, atmospheric transmittance, atmospheric path radiance, sun position, and selected map projection. Geometric corrections for terrain elevation variations (parallax effects) will be discussed in the following subsection and in Chapter 6, Spatial Image Processing Methods.

MEASURED SPECTRAL RADIANCE

Once the spectral radiance available for detection has passed through a multispectral sensor, it is converted to a detected spectral radiance for each spectral band. The two principal changes that occur in this conversion process are the creation of electronic gain and offset, both of which are constant over a single image for a single spectral band, once the corrections discussed above have been made.

For the sake of simplicity, it will be assumed that the spectral bands of the multispectral sensors are sufficiently narrow that the wavelength-dependent functions $E_\lambda, \rho(\lambda)$, $\tau(\lambda)$, $L_\lambda^r(Path)$, and $\rho^c(\lambda)$ are constant over the wavelength region λ_l^i to λ_u^i. These functions can then be replaced by the average value of each function over that wavelength interval, which will be designated by a change in variable from λ to i for the ith band and removed from the integral in the integral versions of the following two equations that are given in Appendix A. Further, the filter function $f(i)$ will be taken to be the value of the filter spectral transmittance at the wavelength center of the ith band. The following two equations represent the transformation of Equation 4.3 and Equation 4.6 from available radiation to radiation measured by the multispectral sensor in a specific spectral band for their respective wavelength regions:

$$L(i) = \frac{bsg(i)f(i)E(i)\tau(i)\rho(i)\Delta\lambda_i}{\pi} + g(i)f(i)L^r(i)(Path)\Delta\lambda(i) + a(i) \qquad \text{(Eqn. 4.7)}$$

for the *0.4–4.0 μm wavelength region,* and

$$L(i) = 2hc^2g(i)\tau(i)f(i)[1 - b'\rho^c(i)]L_{bb}(i, T) + a(i) \qquad \text{(Eqn. 4.8)}$$

for the *greater than 4 μm wavelength region,* where

$$\Delta\lambda(i) = \lambda_u^i - \lambda_l^i = \text{band width of the } i\text{th spectral band}$$
$$\lambda_l^i = \text{lower wavelength limit of the } i\text{th band}$$
$$\lambda_u^i = \text{upper wavelength limit of the } i\text{th band}$$
$$g(i) = \text{electronic gain for the } i\text{th band}$$
$$a(i) = \text{electronic offset for the } i\text{th band}$$
$$f(i) = \text{filter function for the } i\text{th band (filter transmittance)}$$
$$\tau(i) = \text{average atmospheric transmittance in the } i\text{th band}$$
$$\rho(i) = \text{average directional hemispherical reflectance of target in the } i\text{th band}$$
$$\rho^c(i) = \text{average biconical reflectance of target in the } i\text{th band}$$
$$L_{bb}(i, T) = \text{integrated radiance in the } i\text{th band for a black body at temperature } T \text{ (K)}.$$

These are the equations for radiance in the ith spectral band measured by a multispectral sensor. All of the compositional information is located in $\rho(i)$ and $\rho^c(i)$, for the reflective and thermal wavelength regions, respectively. However, if a digital electro-optical sensor has collected the data, the radiance for each spectral band, as given by either of the last two equations, must be digitized before being recorded. This means that the ith band radiance, which is initially a continuous number, must be broken up into discrete radiance ranges or intervals, and each of these intervals must be assigned an integer number, called a digital number (DN) value. The DN for each spectral band is limited to values of 0–255 for 8-bit data. The designation of 8-bit comes from the definition of a bit as 2^1, and from the fact that $2^8 = 256$ levels. Although old LANDSAT MSS data are only 7 bits (128 values of DN, corresponding to 128 intervals of radiance for each spectral band), LANDSAT TM and SPOT produce 8-bit data, as do most Daedalus Enterprises airborne multispectral scanners. Because there are many, equally spaced intervals of radiance (128 for a 7-bit sensor and 256 for an 8-bit sensor) represented by DN values, the DN value of the ith spectral band for a given pixel can be well approximated by

$$DN(i) = q(i)L(i) \qquad \text{(Eqn. 4.9)}$$

where $L(i)$ is the spectral radiance from either Equation 4.7 or 4.8, depending on the wavelength region of the ith band, and $q(i)$ is merely a proportionality constant with units of cm^2-steradians/watt, such that $DN(i)$ values are dimensionless integers. If the sensor is calibrated, the $q(i)$ are known for all the spectral bands. However, even if the sensor is uncalibrated, i.e., if the $q(i)$ are unknown, relative radiance levels are still recorded, and image processing can still be performed in the manner described in the next chapter. If the experimenter needs to calculate the actual radiance values for some reason, a calibrated sensor must be employed.

When a digital, electro-optical sensor collects data for one of its spectral bands or channels, it produces a digital image that consists of a $DN(i)$ value for each pixel (picture element) of the image. The resulting $DN(i)$ values are stored in a raster format data file, i.e., the $DN(i)$ values for all the pixels in the image are organized into a series of rows and columns. A digital multispectral image data set typically consists of a series of raster files of $DN(i)$ values, with a separate file for each spectral band or channel. The next chapter deals with the digital computer manipulation of multi-spectral image data files to produce enhanced images and classification maps of Earth's surface.

REVIEW

Equations for the spectral radiance available for detection were developed for the reflective (0.4–4.0 μm) and thermal (4–25 μm) wavelength regions. Electro-optical sensors available for commercial remote sensing include whisk-broom multispectral scanners, push-broom solid-state linear arrays, and framing cameras (including vidi-con tubes and two-dimensional solid-state arrays). Each requires its own brand of geometric and radiometric corrections, many of which are performed by the sellers of commercial remote sensing data.

Hyperspectral imaging devices of two different types have been built and tested, one that combines push-broom and frame camera technology with dispersive optics and the other that combines whisk-broom scanning and frame camera tech-nology with dispersive optics. Hyperspectral imaging is especially helpful for map-ping biochemical content of tree canopies, for identifying minerals that have narrow spectral absorption bands, and for imaging gases.

After systematic corrections have been made within each image of each spectral band, there remains an electronic gain and offset alteration of the avail-able spectral radiance, which transforms the latter into measured spectral radi-ance. The equations for measured spectral radiance have been developed for both the reflective and thermal wavelength regions. Mining these equations for compo-sitional information that is completely contained within the spectral reflectance of the target is one of the primary functions of image processing, the subject of the next chapter.

Exercises

1. What clue does Figure 4.1 give you about why plants on this planet have evolved with chlorophyll instead of some other chemical as their primary pigment?

2. An airplane flying 600 miles per hour (966 km/hr) is in the region being imaged by the LANDSAT TM scanner. How far will the airplane move between the collection of two adjacent lines of 6000 pixel/line, assuming it takes no time for reversing the scanner mirror between lines? For the same airplane in a SPOT image, how far will the airplane move if the time between adjacent lines is equivalent to the dwell time on one pixel? Compare both your answers to the pixel dimension of each of the two sensors.

Cited References

CHEN, H.S., 1985. *Space Remote Sensing Systems, An Introduction.* New York: Academic Press.

CURRAN, P. J., and J. A. KUPIEC. 1995. Imaging Spectrometry: A New Tool for Ecology. *Advances in Environmental Remote Sensing,* eds. F. M. Danson and S. E. Plummer, 71–88. New York: John Wiley & Sons.

GOETZ, A. F. H., G. VANE, J. E. SOLOMON, and B. N. ROCK. 1985. Imaging Spectrometry for Earth Remote Sensing. *Science* 228, no. 4704:1147–1153.

MAXWELL, J. R. 1994. Sensible Characteristics of Targets and Backgrounds. *Advanced Infrared Technology,* vol. 1, Engineering Continuing Education Conference 9412. Ann Arbor: University of Michigan.

PLEITNER, P. K., and R. K. VINCENT. 1989. System for Determining the Attitude of a Moving Imaging Sensor Platform or the Like. United States Patent No. 4,802,757.

PLEITNER, P. K., and R. K. VINCENT. 1992. System for Determining and Controlling the Attitude of a Moving Airborne or Spaceborne Platform or the Like. United States Patent No. 5,104,217.

SILVA, L. F. 1978. Chapter 2: Radiation and Instrumentation in Remote Sensing. In *Remote Sensing, The Quantitative Approach,* eds. P. H. Swain and S. M. Davis. New York: McGraw-Hill.

VANE, G., and A. F. H GOETZ. 1988. Terrestrial Imaging Spectroscopy. *Remote Sensing of Environment* 24:1–29.

VANE, G., J. E. DUVAL, and J. B. WELLMAN. 1993. Chapter 6: Imaging Spectroscopy of the Earth and Other Solar System Bodies. In *Remote Geochemical Analysis: Elemental and Mineralogical Composition,* eds. C. M. Pieters and P. A. J. Englert, 121–141. Cambridge, U.K. Cambridge University Press.

VANE, G., A. F. H. GOETZ, and J. B. WELLMAN. 1984. Airborne Imaging Spectrometer: A New Tool for Remote Sensing. *IEEE Trans. Geosci. Remote Sensing* 6:546–549.

VINCENT, R. K. 1972. An ERTS Multispectral Scanner Experiment for Mapping Iron Compounds. *Proceedings of the Eighth International Symposium on Remote Sensing of Environment,* 1239–1247. Ann Arbor: Environmental Research Institute of Michigan.

VINCENT, R. K. 1975. The Potential Role of Thermal Infrared Multispectral Scanners in Geological Remote Sensing. *Proceedings of the IEEE* 63, no. 1:137–147.

Additional References (Uncited)

VANE, G., T. G. CHRIEN, E. A. MILLER, and J. H. REIMER. 1987. Spectral and Radiometric Calibration of the Airborne Visible/Infrared Imaging Spectrometer (AVIRIS). In *Imaging Spectroscopy II,* vol. 834, ed. G. Vane, 91–106. Bellingham, Wash.: SPIE.

VANE, G., ed. 1990. *Imaging Spectroscopy of the Terrestrial Environment.* Bellingham, Wash.: SPIE 1298.

WOLFE, W. L., and G. J. ZISSIS, eds. 1978. *The Infrared Handbook.* Washington, D.C.: U.S. Government Printing Office. Prepared for the Office of Naval Research, Dept. of the Navy by IRIA Center of the Environmental Research Institute of Michigan, Ann Arbor.

5

Multispectral Image Processing Methods

▼ ▼ ▼ ▼ ▼ ▼ ▼ ▼ ▼ ▼ ▼ ▼ ▼ ▼ ▼

INTRODUCTION

By the end of the last chapter, we had developed expressions for measured spectral radiance in both the reflective wavelength region (short of 4.0 μm) and the thermal infrared region (wavelengths longer than 4.0 μm). These expressions, given in Equations 4.7 and 4.8, account for everything of importance that a sensor detects in the spectral radiance impinging on the detectors, for one pixel (picture element) at a time. Before returning to these expressions, we will take a brief detour while the following paragraph summarizes how the data are recorded for all of the pixels and all of the spectral bands in an image data set.

A raster array (r lines, each with a line length of q pixels) of these measured spectral radiances is recorded for each spectral band of the sensor as the multispectral image data set. In some cases, all of the pixels in the image for each spectral band of data are recorded as an independent file, and there are n (the number of spectral bands) separate files in the image data set. This is called a *band sequential* (or BSQ) raster array format, e.g., the whole image in band 1, the whole image in band 2, and so on, up to the whole image in band n. If each file contains one line of data for all n spectral bands, the raster array format is called *band interleaved by line* (or BIL), e.g., line 1 band 1, line 1 band 2, and so on up to line 1 band n; line 2 band 1, and so on up to line 2 band n, etc. If the n spectral bands are recorded for each pixel at a time, they are called a *band interleaved by pixel* (or BIP) raster array format, whether the file lengths are one line long or not.

Additionally, there is usually header information available for each raster array format that describes the characteristics of the image, such as the number of pixels per line (q), the number of lines (r), and the number of bits recorded per pixel. An image format implies that a certain header format will accompany a certain raster array format. For instance, RAW image format is the simplest because it always contains no header information at all, and usually implies that the raster array data format is BSQ (but not always). TIFF (for Tagged Image File Format) is a common image data format, but it is very complex (hundreds of pages of specifications) and has no standard for the allowable combination of "tags" for the image. There are many other image data formats, some of general use and some that are tied to specific image processing software packages. For example, the ER Mapper image processing commercial software package (Earth Resource Mapping Proprietary Ltd. 1995) employs an ER Mapper format that consists of a separate text header format and a binary BIL raster array format for the data. However, most commercial software packages for image processing are able to import LANDSAT, SPOT, and other types of commercial satellite data formats directly into the format of the specific software package with no special knowledge required of the user.

The purpose of this chapter is to examine image processing methods that are designed to transform multispectral data from one of these image data formats into an image display that either increases contrast between interesting targets and the background or yields information about the composition of certain pixels in the image. The following four sections of this chapter will each discuss an aspect of this process.

The first section deals with the problem of eliminating the additive terms in Equations 4.7 and 4.8 that contain no information about the target's chemical or mineral composition. For the reflective wavelength region short of 4.0 μm, this elimination involves the use of histogramming and dark-object subtraction, whereas the thermal wavelength region beyond 4.0 μm involves thermal infrared calibration from calibration sources on the sensor platform.

The second section is a discussion about the electronic display of color and human perception of color. Even though multispectral image data can be supplied for a sensor of many spectral bands, the image processing results can only be displayed in the three primary visible colors that the human eye can perceive. All other visible colors are combinations of those primary colors, so that color vision of the human eye effectively has three spectral bands—blue, green, and red. How do we display many spectral bands of information in only three spectral bands of display?

The subject of the third section is spectral and temporal ratioing, and its use as both as an image enhancement tool and as a preprocessing step for higher-order image processing methods. Spectral ratio imaging is an especially important tool for compositional mapping because it enhances information related to the location of absorption bands in the reflectance spectra of Earth materials. The fourth section deals with higher-order multispectral data processing schemes, such as principal components transformation (another image enhancement tool) and supervised and unsupervised multispectral classification.

The objective of this chapter is to elucidate relationships between a limited set of generic image processing algorithms and the underlying physical parameters that are being measured. Many image processing software packages are now commercially available, each with their own user documentation and some with their own unique image processing algorithms. Therefore, we will leave specific information about computer programs that implement the algorithms discussed below, as well as information about unique algorithms, to the relevant user manual of the commercial software package of your choice.

ELIMINATION OF ATMOSPHERIC PATH RADIANCE AND ADDITIVE ELECTRONIC OFFSET

The last two additive terms in Equation 4.7 and the last term in Equation 4.8 contain no information about the target and they may be different from image to image, as times of data collection change. For processing of data from one single pass of data collection, any image processing algorithm that is merely a linear transformation of these equations will not be affected by these terms, as they just change the intercept term of the linear transformation. However, for processing of image data from different passes of the sensor platform by linear transformations or for image processing by nonlinear transformations of even single-pass data, such as spectral ratioing, it is desirable to remove these nonessential terms that change with environmental conditions, but not with target composition.

The next two subsections deal with this problem for wavelengths short of 4.0 μm and for wavelengths beyond 4.0 μm (the thermal infrared), respectively.

Histogramming and Dark-Object Subtraction

Let us consider the raster array of data from the ith spectral band of an image data set. There is something useful to be learned from a simple investigation of how the spectral radiances in that band are distributed among the pixels of the data set. For instructive purposes, Figure 5.1a shows a small, 25-pixel raster array image data set with digital numbers $DN(i)$ of the ith spectral band shown for each pixel of the data set. For reasons that will become obvious later in this subsection, the first step in image processing is usually histogramming—a method of sampling the image (for example, pixels on every 10th column of every 10th row) for each spectral band. A histogram is a plot of the number of pixels versus $DN(i)$ value, showing how many pixels are represented by each $DN(i)$. Thus, a histogram of 8-bit data for one spectral band plots $DN(i)$ values from 0 to 255 as an abscissa versus number of pixels (population density) that have each of those $DN(i)$ values as an ordinate. Figure 5.1b shows a histogram of the ith spectral band for the small image data set of Figure 5.1a, where all of the pixels in the data set have been sampled. Note that the sum total of all the pixels in all the cells of the histogram is equal to the total number of pixels in the image data set (25 in this case), because every pixel has been sampled.

The following discussion applies only to a spectral band in the reflective wavelength region short of 4.0 μm. The lowest $DN(i)$ value of the ith spectral band with any pixels in that "bin" (the lowest nonempty bin) may be something other than zero, as it is in the Figure 5.1 example. If the image is of heterogeneous terrain that contains deep shadows or other dark objects (Crane 1971; Vincent 1972), defined here as an object from which no detectable reflected radiance is available to the sensor, the highest $DN(i)$ value with no pixels in its bin or any lower $DN(i)$ bin represents the sum of the

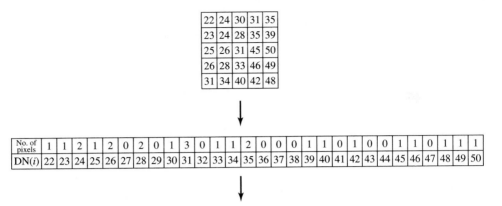

No. of pixels	1	1	2	1	2	0	2	0	1	3	0	1	1	2	0	0	0	1	1	0	1	0	0	1	1	0	1	1	1
DN(i)	22	23	24	25	26	27	28	29	30	31	32	33	34	35	36	37	38	39	40	41	42	43	44	45	46	47	48	49	50

(a) DN(i) values for ith spectral band of a raster array of a 25-pixel image data set

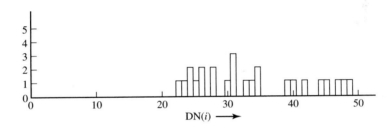

(b) DN(i) values for ith spectral band of a raster array of a 25-pixel image data set

Figure 5.1   A hypothetical 25-pixel image data set and its histogram for the ith spectral band. (a) Digital number values for the ith spectral band, DN(i), of a raster array of a 25-pixel image data set. (b) Histogram of the ith spectral band for the image data set in (a). The histogram shows the number of pixels in the image that have each of the digital number values represented in the image. *(Courtesy of R. K. Vincent.)*

last two terms of Equation 4.7 (called the additive terms) since either $s = 0$ or $\rho(i) = 0$ for a dark object. This $DN(i)$ value, which is one less than the smallest nonzero-bin $DN(i)$, is called the *dark object, DO(i)*, and represents the sum of the atmospheric path radiance and the electronic offset for the ith spectral band. The equation below results; it comes from the substitution of Equation 4.7 into Equation 4.9 for the case of $s = 0$:

$$DO(i) = q(i)[g(i)f(i)L'(i)(Path)\Delta\lambda(i) - a(i)] \qquad \text{(Eqn. 5.1)}$$

If this dark object is then subtracted from the $DN(i)$ of all other pixels in the image for that spectral band, the resulting $DN(i)'$ is given by the substitution of Equation 4.7 into Equation 4.9, as follows:

$$DN(i)' = DN(i) - DO(i)$$

$$= q(i)L(i)' = q(i)L(i) - q(i)[g(i)f(i)L'(i)(Path)\Delta\lambda(i) - a(i)]$$

$$= \frac{q(i)bsg(i)f(i)E(i)\tau(i)\rho(i)\Delta\lambda(i)}{\pi} \qquad \text{(Eqn. 5.2)}$$

where $L(i)'$ and $DN(i)'$ are the dark-object-corrected versions of the measured spectral radiance and digital number, respectively, for the ith spectral band at reflective wavelengths short of 4.0 μm. In the Figure 5.1 example, $DO(i) = 21$, though this data set is so small that there is no great probability that a pixel either in a shadow or with reflectance of zero exists within the data set. (We will return to the problems encountered with small data sets in just a few paragraphs.)

Each spectral band has its own dark object, and the $DO(i)$ vary from band to band. For spectral bands of wavelengths shorter than approximately 1.0 μm, the $DO(i)$ are usually nonzero and become progressively larger with decreasing wavelength because of scattering by the atmosphere. Atmospheric path radiance, except for smoky, smoggy, or foggy areas, is usually not detectable at wavelengths longer than about 1.0 μm, but the electronic offset term may be nonzero and can vary from band to band. Except for images of areas that have no significant shadows or dark-objects, it is usually safer to apply dark-object subtraction to all bands short of 4.0 μm in wavelength, such that *all of the terms left in Equation 5.2 are simply multiplicative factors of* $\rho(i)$, *which contains all of the target compositional information.*

Figures 5.2a, b, and c give examples of histograms of a LANDSAT TM sub-scene (about 21 km × 19 km) of a region around Santo Domingo, Dominican Republic (shown as Color Plate CP7) for the three visible TM bands (bands 1, 2, and 3 displayed as blue, green, and red), after dark-object subtraction. The image has

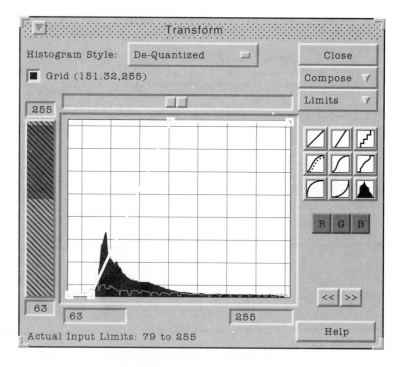

Figure 5.2 Histograms of TM bands 1 (a), 2 (b), and 3 (c) of a subframe area (approximately 21 km × 19 km) around Santo Domingo, Dominican Republic, shown in Color Plate 7. (a) Band 1. *(Courtesy of R. K. Vincent.)*

been linearly stretched from the lowest to one of the higher values of $DN(i)$ recorded for the input subscene image data to conform with the output display values of 0 to 255 in TM bands 1, 2, and 3. Note that $DO(1) = 79$ is highest, followed in descending order by $DO(2) = 21$, and $DO(3) = 14$, in rough agreement with expected scattering by atmospheric molecules and particles, a scattering that is a function of λ^{-2}, approximately. Unless dust, aerosol, or cirrus cloud cover is present, the TM bands (other than the thermal infrared band 6) at longer wavelengths than TM band 3 usually have $DO(i)$ of zero, unless the electronic offset is greater than zero.

There is a simple image processing procedure called linear contrast stretching that is similar, but not equivalent, to dark-object subtraction. When an image is linearly contrast-stretched, the $DN(i)$ values that represent typically the 3rd and 97th percentiles (3% and 97%) of the cumulative pixel population are set at 0 and 255 values of the display digital numbers, respectively, and all the original $DN(i)$ values are linearly stretched between those values. This removes more than the two additive terms in Equation 4.7. Linear contrast stretching is not a good substitute for dark-object subtraction, though it does usually produce a better-looking color-composite image of three spectral bands than if no stretching were performed. If dark-object subtraction is performed prior to contrast stretching, it will have no effect on the linearly contrast-stretched color-composite image of the same three spectral bands because linear contrast-stretching and dark-object subtraction are both examples of linear transformations.

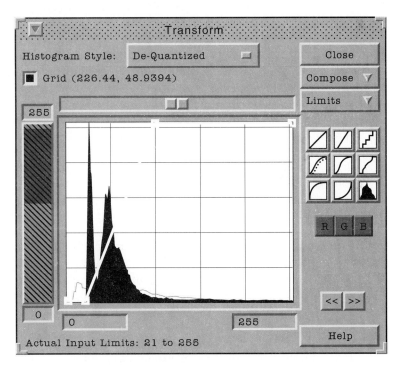

Figure 5.2 *(continued)* (b) Band 2.

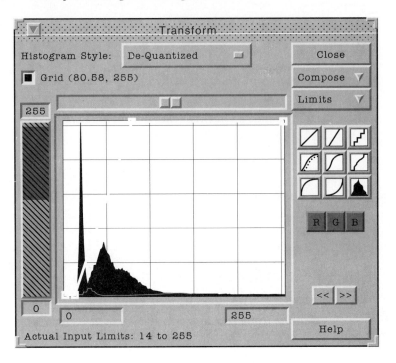

Figure 5.2 *(continued)* (c) Band 3.

The elimination of additive terms that have no target information is especially important for types of nonlinear image processing, some of which are discussed later. Therefore, it is prudent to perform dark-object subtraction as one of the first steps in image processing for spectral bands below 4.0 μm in wavelength coverage.

However, not all scenes will contain acceptably dark objects, especially if the scenes are small. Although empirical methods that can be automatically implemented, like the dark-object subtraction procedure, are most attractive to the user, there are always limitations to their use. For instance, Color Plate CP8 shows the whole LANDSAT TM frame of the Dominican Republic, from which the Santo Domingo subscene (outlined) was extracted. Figure 5.3a–c shows that the whole-frame dark object values are $DO(1) = 63, DO(2) = 20$, and $DO(3) = 13$. Comparison of histograms in Figure 5.3a–c and the subscene histograms in Figure 5.2a–c show that the $DO(i)$ of the whole-frame image are lower than the ones for the subscene image. This result occurs because the probability of sampling deep shadows or very dark objects that cover an entire pixel increases with the amount of area searched. The probability of finding good dark-object values also increases with the amount of water, cloud shadows, and topographic relief in the scene. In large, low-relief deserts, it is possible that there will be no accurate dark objects found in the entire frame.

There are two important points to be remembered from this example. First, the $DO(i)$, if wrong, are always overestimated, not underestimated, because nothing can be darker than the dark object, as defined above. A dark object that covers only part

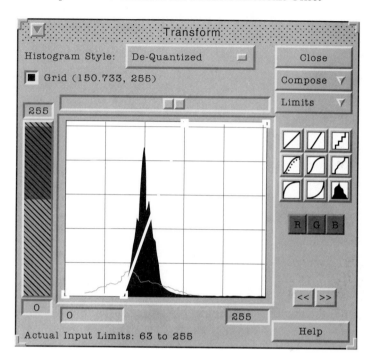

Figure 5.3 Histograms of TM bands 1 (a), 2 (b), and 3 (c) of a whole-frame area (approximately 185 km $\times$ 185 km) in the Dominican Republic, shown in Color Plate 8. The rectangle shows the area covered by the histograms in Figure 5.2 and the image in Color Plate 7. (a) Band 1. *(Courtesy of R. K. Vincent.)*

of a pixel, however, will lead to an overestimation of $DO(i)$. Second, the more pixels that are included in the histogram, the more likely it is that acceptable dark objects will be found. Of course, a dark object for one spectral band may be different from a dark object for another spectral band, because light scatters into shadowed areas in the visible spectral bands and the shadowed ground can reflect different amounts of light in different spectral bands.

A true dark object requires that the material on the shadowed ground have a sufficiently low reflectance in the ith spectral band so that the light reflected toward the sensor from the ground will be less than the detectable limit of the sensor for that band. An acceptable $DO(i)$ is one that is only slightly overestimated, such that the overestimation is small, say 10% or less of the empirically measured $DO(i)$ value. Therefore, the best protection against unacceptable dark objects for satellite data is the inclusion of quarter-frame or whole-frame areas in the histogramming phase, even though the area of interest may be a small subscene. Since the $DO(i)$ change with sun elevation and atmospheric conditions, it is necessary that the $DO(i)$ come from an image collected at the same time as the subscene of interest.

With aircraft data, which cover much smaller portions of the ground than satellite data, this precaution is more difficult to heed. However, it should be noted that the additive terms contain the effects of variations in both atmospheric path radiance

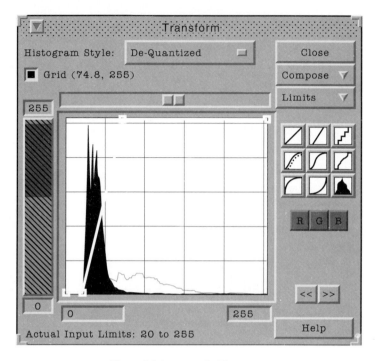

Figure 5.3 *(continued)* (b) Band 2.

and an electronic offset from image to image. Since aircraft fly much closer to the ground than satellites, the electronic offset may dominate atmospheric path radiance. In addition, electronic offset is far more variable for aircraft data than for satellite data, since it is common practice with aircraft multispectral scanner data to adjust electronic gain and offset from flight to flight. Satellite multispectral scanner electronic gains and offsets are adjusted much less frequently. Therefore, the temptation to eliminate correction for the additive terms with aircraft multispectral data on the basis of relatively short atmospheric paths should be overcome because correction for the additive electronic offset is even more important for aircraft data than for satellite data.

It is rare for an entire LANDSAT frame, which covers about 34,000 square km, not to contain an acceptable dark object. For cases where acceptable dark objects cannot be found, there are other methods for estimating the additive terms (Crippen 1986) of Equations 4.7 and 5.2, but they are more computationally complex and sometimes require more information than the dark object subtraction method.

Calibration of Thermal Infrared Spectral Bands

The thermal wavelength region is different from the reflective wavelength region in two important aspects regarding elimination of additive terms. First, there are no true dark objects, because nothing in the scene is at or even near 0 K in temperature.

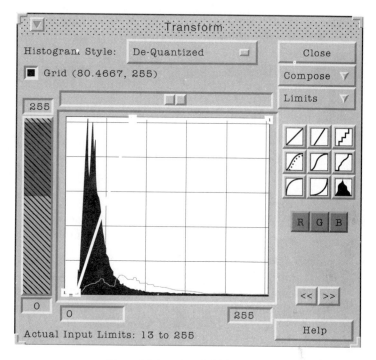

Figure 5.3 *(continued)* (c) Band 3.

Second, since atmospheric path radiance is being ignored in atmospheric windows, the only additive term is the electronic offset in Equation 4.8. Substitution of Equation 4.8 into Equation 4.9 and slight re-arrangement yields

$$DN(i) = q(i)L(i) = q(i)g(i)2hc^2\tau(i)f(i)[1 - b'\rho^c(i)]L_{bb}(iT) + q(i)a(i) \quad \text{(Eqn. 5.3)}$$

where

$$
\begin{aligned}
g(i) &= \text{electronic gain for the } i\text{th band} \\
a(i) &= \text{electronic offset for the } i\text{th band} \\
f(i) &= \text{filter function for the } i\text{th band (filter transmittance)} \\
\tau(i) &= \text{average atmospheric transmittance in the } i\text{th band} \\
\rho(i) &= \text{average hemispherical reflectance of target in the } i\text{th band} \\
\rho^c(i) &= \text{average biconical reflectance of target in the } i\text{th band} \\
L_{bb}(i, T) &= \text{integrated radiance in the } i\text{th band for a black body at tempera-} \\
&\quad \text{ture } T \text{ (K)}
\end{aligned}
$$

This equation gives the digital number for wavelength regions longer than 4.0 μm, the integral version of which is given in Appendix A as Equation 5.3a.

The additive term is simple enough to eliminate by using a calibrated thermal infrared sensor that has a hot plate and a cold plate at the edges of each line of data, usually maintained at temperatures somewhere near the lowest and highest temperatures expected for the ground. When viewing either the hot plate or cold plate, everything in Equation 5.3 between the $q(i)g(i)$ and $L_{bb}(i, T)$ is known because $\tau(i) = 1$ (for very short paths in the atmospheric windows), the term in brackets is equal to the known emissivity of the hot and cold plates, and the temperatures of the hot and cold plates are known. If $q(i)g(i)$ is considered one unknown and $q(i)a(i)$ is considered a second unknown, they can both be solved for because there are two plates, hot and cold, of known temperatures and emissivities. With $q(i)a(i)$ known, it is possible to subtract off the additive term, as in the following equation, where the $q(i)g(i)$ term is retained, though known, and integration is completed:

$$DN(i)' = q(i)L(i)' = q(i)L(i) - q(i)a(i)$$
$$= q(i)g(i)2hc^2\tau(i)f(i)[1 - b'\rho^c(i)]L_{bb}(i, T) \qquad \text{(Eqn. 5.4)}$$

where

$DN(i)'$ = the electronic-offset-corrected $DN(i)$ for the ith spectral band for wavelengths greater than 4.0 μm

Equation 5.4 is the operative equation for additive-offset-corrected $DN(i)'$ for spectral bands longer than 4.0 μm in wavelength.

It is worth noting that atmospheric path radiance has been not been eliminated by this procedure, since there is little or no atmospheric path radiance between the sensor and the hot or cold plates. Atmospheric path radiance was ignored earlier as negligible for wavelengths longer than 4.0 μm. If a user wishes to estimate path radiance and atmospheric emission reflected off the ground with atmospheric models, these terms would be included in Equation 5.4 in a complex manner. Besides the additive term created by atmospheric path radiance, atmospheric emission reflected off the ground will partially cancel out reststrahlen bands in infrared radiation emitted by the ground. However, within the atmospheric windows during clear weather and with ground temperatures near 300 K, we will ignore atmospheric emission and atmospheric path radiance.

THE ELECTRONIC DISPLAY AND HUMAN PERCEPTION OF COLOR

Multispectral image processing is the manipulation of an n-dimensional data set (n is number of spectral bands), such that it can be displayed either in three-dimensional (full color) or one-dimensional (black and white) form as an image that can be recognized by the human eye–brain perception system. Two of the most important characteristics of image processing are that it adds spatial context to even the most elemental detection schemes and that it can convey the highest amount of information to human beings in the shortest time; this latter characteristic of image processing leads to the cliché that a single picture is worth a thousand words.

Image processing of multispectral remote sensing data is a particularly important technology for geologists because the spatial configuration of surface compositional patterns are important clues to the geological history of a particular rock outcrop or residual soil exposure. For example, a geologist exploring for a particular type of ore body looks for the presence of minerals associated with the geochemical event that concentrated the ore and for the spatial patterns in which those associated minerals occur. Sometimes the geometric shape of the spatial occurrence of an associated mineral is the only aspect that separates an economic discovery from an occurrence of the same mineral not associated with ore.

Thus far, the word *color* has been used to refer to the change in spectral radiance leaving the target, as a function of wavelength. The color of the object of observation is not restricted to the visible wavelength region, as has already been discussed. The multispectral sensor detects spectral radiance coming from the object of observation in several spectral bands and the multispectral data are recorded. The recorded data are then input to multispectral image processing algorithms. The image processing results, however, are presented to the viewer as a visible image, because that is what a human can perceive. The color in a visible image that is displayed as an output product of a multispectral image processing system is called display color. It is restricted to the visible wavelength region. A subtle point in multispectral image processing is that whereas the color of an object is not restricted to the visible wavelength region, the display color is restricted to that region.

There are two basic types of color display, one called color additive and the other, its complement, called color subtractive. Color-additive displays are most common, because that is the way our eyes see visible colors reflected off observed objects. In color-additive displays, which include color television and color cathode ray tubes, white light can be produced (Avery and Berlin 1992) from the addition of three additive primary colors, red (0.6–0.7 μm), green (0.5–0.6 μm), and blue (0.4–0.5 μm). Color-subtractive displays are transmissive in nature and can produce all colors from the subtractive primary colors of yellow, magenta, and cyan. For instance, a yellow transmission filter passes both green and red, yet absorbs blue, its complementary color. Magenta passes both red and blue, absorbing green, its complementary color. Cyan passes both blue and green, absorbing red, its complementary color. For the remainder of this book, additive color displays will be assumed, with blue, green, and red acting as the three primary colors from which all other visible colors can be produced.

Although many multispectral sensors, including LANDSAT TM, have more than three spectral bands, there are still only three primary colors that can be displayed for human perception. How would you solve this dilemma? Fortunately, there are several solution paths for this dilemma. One approach you could take is to select three appropriate spectral bands for the remote sensing task at hand, contrast-stretch each of them, and display each one in a different primary color. This is somewhat like holding several actors in the wings of a small, three-actor stage, and calling them out, one trio at a time, for their appropriate scenes. A natural color image results only in the case where three input bands happen to cover the blue, green, and red portions of the visible wavelength region (like LANDSAT TM bands 1, 2, and 3) and those three bands are displayed as blue, green, and red, respectively. Color plates CP7 and CP8 are examples of natural color images. For all other combinations of input bands and display colors, the colors in the image will not be the same that a human would observe in the field, i.e., the colors in the image will be false. The most well-known false color image is one for which spectral bands of visible green, visible red, and reflective infrared (such as LANDSAT MSS bands 4, 5, and 7 from 0.5–0.6 μm, 0.6–0.7 μm, and 0.8–1.1 μm wavelength regions, respectively) are displayed in the primary colors of blue, green, and red. In such a false color image, green vegetation is displayed as very red, owing to its high reflectance in the reflective infrared wavelength region, relative to any visible wavelength.

Another approach you could take toward handling the problem of having more spectral bands than the number of primary colors in which to display them is to combine several of the spectral bands into three continuous-valued (256 brightness levels for 8-bit data) spectral parameters, which can then be displayed as combinations of blue, green, and red. This approach could generally be described as decreasing the dimensionality of the data. Spectral ratio imaging and principal components imaging are two such methods, discussed later in this chapter, that result in virtually continuous-toned images with $(2^8)^3 = 16,777,216$ possible colors (combinations of blue, green, and red), assuming that each spectral parameter consists of 8-bit data. Because they result in virtually continuous-toned images, both of those image processing methods are said to be forms of image enhancement.

Another image enhancement method that is especially important for the combination of multispectral data with nonmultispectral raster data deals with the display of three raster files in an enhanced color space called Intensity-Hue-Saturation (IHS) space. The intensity corresponds to the total brightness of a display color, hue corresponds to the average wavelength of the display color, and saturation corresponds to the deepness (the opposite of "pastelness") of the display color relative to gray; e.g., red has high saturation and pink has low saturation. In an IHS image, the intensity, hue, and saturation can be controlled independently. The three raster input files can be spectral bands from a multispectral scanner, digital elevation data, or digitized map data, as long as the input files of data are co-registered, or overlay one another. This makes an IHS image well suited for combining different spatial resolution multispectral images, combining digitized map data with multispectral images, or imaging geophysical data, as will be demonstrated in a later chapter. An excellent description of IHS transformation is given by Lillesand and Kiefer (1994).

Two other methods for decreasing the dimensionality of the data are supervised and unsupervised classification schemes, which result in classification images that display a relatively few discrete colors that correspond to discrete classes (usually numbering less than 50) of terrain features. These two classification methods, which are not forms of image enhancement, will also be discussed later in this chapter.

SPECTRAL AND TEMPORAL RATIOING

Spectral ratioing is a multispectral image processing method that involves the division of one spectral band by another, usually after some preliminary corrections have been made for atmospheric path radiance and/or additive offset introduced by the multispectral sensor. The spectral bands employed for a spectral ratio image (the image manifestation of a spectral ratio) are usually selected such that one spectral band is inside and the other is outside a wavelength region of spectral reflectance minimum or maximum of a particular target. Spectral ratios are attractive because they enhance compositional information, while suppressing other types of information about Earth's surface, such as terrain slope and grain size differences.

Temporal ratioing is a way of comparing changes in the area of interest that occur for a chosen spectral parameter, between two different dates of data collection. The chosen spectral parameter could be any spectral band or combination of spectral bands. Data from the two collection dates are co-registered and the data from one date is divided, pixel by pixel, by the data from the other date, for the selected spectral parameter. Temporal ratios of spectral ratios will be used in this section to experimentally verify the robustness, or high degree of invariance to changes in solar illumination and atmospheric conditions, of corrected spectral ratios.

As indicated by the preceding equations of this chapter, the correction of spectral ratios is different in the reflective wavelength region below 4.0 μm and in the thermal infrared wavelength region above 4.0 μm. For this reason, each of these wavelength regions will be dealt with separately. The three subsections below deal with spectral ratioing in the reflective wavelength region, temporal ratioing and vegetation indices in the same reflective wavelength region, and spectral ratioing in the thermal infrared wavelength region, respectively.

Spectral Ratioing in the Reflective Wavelength Region

A thoughtful inspection of Equation 5.2 for the reflective wavelength region below 4.0 μm reveals an important fact: all of the terms on the right side of this equation are approximately equal throughout the scene being imaged, except for s (the shadow-slope factor that accounts for the amount of shadow within a pixel), b (the constant that relates directional hemispherical reflectance of a target's lab spectrum to hemispherical reflectance of the target in the field), and $\rho(i)$, the directional hemispherical reflectance of the target in the ith spectral band. Since $\rho(i)$ is the term that contains all compositional information about the target, it varies from pixel to pixel throughout the image as the composition of the Earth's surface changes from place to place on the ground. The shadow-slope factor, s, also changes from pixel to pixel, depending on the percentage of shadow present in each pixel. Topographic slopes that face toward the sun are brighter than slopes facing away from the sun. This brightness difference, which is independent of wavelength to first-order approximation, changes from pixel to pixel as Earth's surface rises and falls from place to place on the ground. For the parts of Earth's surface that are non-Lambertian, the bidirectional constant, b, also changes with different angles of observation (from pixel to pixel), but it likewise is independent of wavelength, to first-order approximation. If there were some way to remove the s and b factors, all of the remaining factors in Equation 5.2 besides $\rho(i)$ would reduce to a single multiplicative factor that is approximately constant for the whole image. This reduction would hold as long as the image came from satellite sensors, which view the imaged scene within a narrow range of viewing angles. There is such a method, and it is called spectral ratioing, first performed in 1970 (Vincent and Thomson 1971) and explained in greater detail in 1977 (Vincent 1977).

A spectral ratio for a single pixel is simply the ratio of the spectral radiance measured in the one spectral band to the spectral radiance measured in another spectral band, after dark-object subtraction has been performed on both spectral bands. The spectral ratio of the ith and jth bands can be represented with the help of Equation 5.2 as follows for spectral bands below 4.0 μm in wavelength:

$$R(i,j)' = \frac{DN(i)'}{DN(j)'} = \frac{q(i)g(i)f(i)E(i)\tau(i)\rho(i)\Delta\lambda(i)}{q(j)g(j)f(j)E(j)\tau(j)\rho(j)\Delta\lambda(j)} \frac{\rho(i)}{\rho(j)} = C(i,j)\frac{\rho(i)}{\rho(j)} \quad \text{(Eqn. 5.5)}$$

where

> $R(i,j)'$ = the spectral ratio of the ith and jth spectral bands that have been dark object corrected

Since b and s are independent of wavelength, to a first-order approximation, they cancel out, which leaves the $C(i,j)$ term of Equation 5.5 as a multiplicative constant for the entire image. This is a great simplification because $C(i,j)$ does not vary from pixel to pixel. Only the ratio of spectral reflectances in the ith and jth spectral bands, which contains all of the compositional information about Earth's surface, varies from pixel to pixel throughout the image. The brightness information about each pixel has been selectively suppressed, by the dividing out of b and s in Equation 5.5, in favor of the color or hue of each pixel, which is represented by the ratio of spectral reflectances in the two bands. The terms *color* and *hue* also are meant to apply to invisible colors and hues for wavelength regions beyond the visible region. A physicist's way of describing what spectral ratios do for the remote sensing scientist is that they separate the effects of brightness information and hue information, suppressing the former in favor of the latter.

Once $R(i,j)'$ is calculated and quantized to digital values of 0–255, the following equation becomes true:

$$DN(i,j)' = q(i,j)R(i,j)' = q(i,j)C(i,j)\frac{\rho(i)}{\rho(j)} \quad \text{(Eqn. 5.6)}$$

where

$DN(i,j)'$ = the digital number for the $R(i,j)'$ spectral ratio

If an image is produced for which the image brightness is proportional to $DN(i,j)'$, the result is called a spectral ratio image. A spectral ratio image is bright for pixels that have a high $\rho(i)/\rho(j)$ and low for pixels that have a low $\rho(i)/\rho(j)$.

There are several desirable traits of spectral ratios, as defined for wavelength regions less than 4.0 μm in Equation 5.5, and of spectral ratio images. First, spectral ratios measured from multispectral remote sensing image data are easily related to reflectance spectra of terrain features, as measured by laboratory and field spectrometers. The only difference between $\rho(i)/\rho(j)$, a term that can be calculated from lab or field spectrometer data, and $R(i,j)'$, a term that can be calculated from the measured $DN(i)'$ and $DN(j)'$ of the ith and jth channels (after dark object subtraction) of a multispectral imaging sensor, is the multiplicative factor $C(i,j)$, which can be determined if there is at least one pixel in the image that lies on a target for which the remote sensing scientist has collected spectral reflectance data. Once the multiplicative factor $q(i,j)C(i,j)$ is known anywhere in a spectral ratio image, it is known everywhere in the image, under clear-weather conditions.

From a geological standpoint, this fact implies that spectral ratio images can be calibrated to ratios of reflectances from laboratory or field spectra of rocks and soils. The inhomogeneity of a single pixel limits this calibration more than do the approximations that have been used in the development of the above equations. However, for a given rock there is more direct connection between $R(i, j)'$ as measured by a remote sensing sensor and $\rho(i)/\rho(j)$ as measured by a laboratory or field spectrometer than there is between $L(i)'$ measured by a remote sensing sensor and $\rho(i)$ measured with a laboratory or field spectrometer, because the s and b terms are still in Equation 5.2 that links $L(i)'$ and $\rho(i)$, yet are removed from Equation 5.5. Therefore, it is easier to predict a spectral ratio than to predict a single-band radiance from laboratory or field reflectance spectra. Thus, spectral ratio images are easier than single-band images to interpret for compositional information about materials on Earth's surface. Spectral ratioing also makes possible the employment of laboratory or field reflectance spectra of rocks and soils as training sets for supervised classification of multispectral remote sensing data. This topic will be discussed in a later section of this chapter.

In Appendix C is a list of brightness codes and ratio codes of a library of laboratory reflectance spectra of minerals for the LANDSAT MSS and LANDSAT TM sensors. These brightness codes (for single-band reflectance) and ratio codes (for spectral ratios), which compress the spectral information into deciles of each single-band or spectral ratio, aid considerably in the interpretation of color composite images of single bands or color ratio images. They also can be used to select the most effective single-band color composites or color ratio images for enhancing a particular mineral with one of the above three sensor packages. However, the ratio codes are much more directly related to spectral ratio images than the brightness codes are related to single-band color composites, for reasons given in the previous two paragraphs.

A second desirable trait of spectral ratios is that they are more robust than single-channel radiances because all illumination, atmospheric, and electronic gain effects are convolved into one parameter, $C(i, j)$, for the entire image. In a spectral ratio image, a particular rock or soil will appear the same whether it is on a hillside facing the sun or facing away from the sun, even though the brightness differences between the two hillsides may be remarkably different. It is this trait that makes spectral ratioing a desirable preprocessing step prior to automatic multispectral classification (either supervised or unsupervised), as long as compositional information is the primary goal. Because spectral ratios suppress brightness effects, only one training set per target class is needed for the entire image, whereas the use of single-band radiance inputs to multispectral classification requires training sets on both sun-facing and sun-opposed topographic slopes for each target class.

Those terrain elements that differ from others chiefly in brightness, however, will be less effectively classified by multispectral classification programs with spectral ratio inputs than with single-band radiance inputs. One example is water. Another example is plant species differentiation, where brightness differences are important. Spectral ratioing is compelling to the user for mapping color differences in the presence of strong illumination differences, such as correctly recognizing the color of a red rock, whether in bright sunlight or in the shadow of a tree.

A third desirable trait of spectral ratios, especially for geological remote sensing, is that spectral ratios tend to separate the effects of grain size and chemical composition in rocks, soils, and minerals by suppressing brightness differences. Chapter 2 demonstrated that the primary effect of decreased grain size in the visible and reflective infrared wavelength regions is to increase reflectance at all wavelengths in transparent materials, such as quartz and calcite. Spectral ratios are best adapted toward the detection of absorption bands because they compare reflectance at one wavelength with reflectance at another wavelength. In fact, the greatest utility of spectral ratio imaging can be attained for mapping a particular target by selecting spectral ratios of spectral bands that are located inside and outside of one or more absorption bands of the target material. This choice enhances the effect of chemical composition on the final spectral ratio image, while suppressing effects of grain size, topographic slopes, sun position, and atmospheric state.

An example (Salmon and Vincent 1974) of spectral ratio imaging of the Wind River Basin, Wyoming is shown in Figure 5.4a as $R(5,4)'$, produced from LANDSAT MSS spectral bands 4 (0.5–0.6 μm, the visible green) and 5 (0.6–0.7 μm, the visible red), which are shown in Figure 5.4b and 5.4c, respectively, and a color ratio image of $R(7,6)'$, $R(6,5)'$, and $R(5,4)'$ as blue, green, red in Color Plate CP4. As seen in Chapter 3, ferric oxides (such as hematite) have a greater rise in reflectance from the visible green to the visible red wavelength regions than most other natural materials. This would imply that its $R(5,4)'$ spectral ratio would be relatively high, or that ferric oxides would be bright in an $R(5,4)'$ ratio image. That is just the case, as shown in Figure 5.4a, where the brightest areas surrounding the Wind River Basin are outcrops of Triassic redbeds of the Chugwater Formation, containing gypsiferrous units with as much as 5% ferric oxide (Picard 1965). These redbeds are orange in the color ratio image because $R(5,4)'$ is displayed in red. In fact, careful field work over the years has failed to turn up a single bright area in Figure 5.4a that was not well covered by ferric oxides. This figure illustrates well the point made earlier about the suppression of topographic slope/illumination effects by spectral ratio images. Wind River Basin is surrounded by mountains; the location creates Triassic redbed exposures in slopes that trend practically every direction of the compass, yet all of the Triassic redbed exposures are approximately equally bright in the $R'(5,4)$ ratio image, no matter how they are oriented with respect to the position of the sun. (During this LANDSAT I satellite overpass on August 5, 1972, about 9:30 A.M., local time, that orientation was approximately in the Southeast, or the lower-right corner of the image.)

Temporal Ratioing and Vegetation Indices

Ordinarily, recent experiments and examples are the most desirable citations. However, an old experiment that employed temporal ratios of spectral ratios will be cited here because it offers a quantitative methodology that can be employed even in regions where no ground truth information is available. The experiment (Vincent et al. 1975) was performed to evaluate the effectiveness of spectral ratioing of LANDSAT

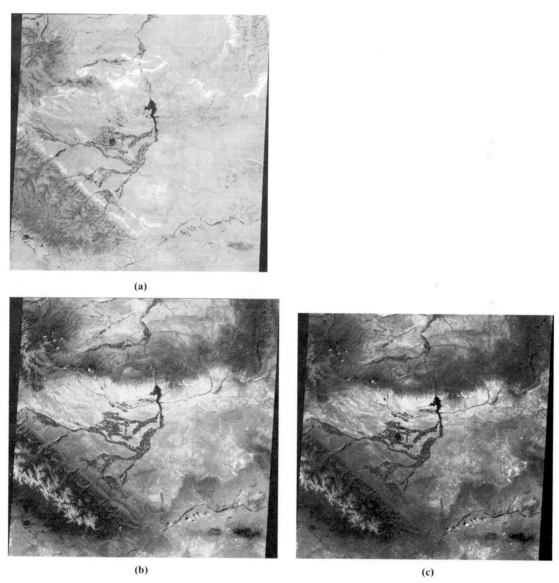

Figure 5.4 LANDSAT MSS images (covering 185 km × 185 km) of the Wind River Basin and Range, Wyoming. A dark objected-corrected spectral ratio image (a) of MSS band 5 divided by band 4, showing ferric oxides as bright. Single band images are shown in (b) and (c) for MSS bands 4 and 5, respectively. *(Courtesy of GeoSpectra Corp.)*

MSS data for suppressing atmospheric and solar illumination variations, using LANDSAT MSS frames for the Wind River Range, Wyoming for two different LANDSAT I overpasses that were 72 days apart. The first frame, mentioned above, was from an overpass on August 5, 1972 (Frame E-1013-17294), and the second frame was from an overpass on October 16, 1972 (Frame E-1085-17300). A small, subframe area (25,125 pixels) near the Atlantic City, Wyoming iron mine at the southern tip of the Wind River Mountains was chosen as the test site. The August 5 frame had 0% cloud cover and was collected with sun elevation of 54.9° and sun azimuth of 130.2°. The October 16 frame had 20% cloud cover, including cloud cover over approximately 25% of the test area, and was collected with a sun elevation of 34.0° and azimuth of 153.4°. The two frames were co-registered correctly within two pixels of one another.

The method selected for testing invariance to atmospheric and solar illumination differences of a given spectral parameter was temporal ratioing; the method involved dividing the DN of the tested parameter for time 1 by that of time 2 on a pixel-by-pixel basis. In other words, the images for times 1 and 2 of the tested parameter were ratioed. The number of pixels changing by less than ±5 percent of the resulting temporal ratio for the tested parameter were then counted for a restricted group of pixels called ground invariant (GI) pixels. GI pixels were chosen on the basis of having a dark object-corrected $DN(7,5)'$ (see Equation 5.6) smaller than 1.20, thereby eliminating pixels with 50% or more vegetation cover. Cloud cover was eliminated by taking a temporal ratio of MSS band 7 on the two dates and eliminating areas that more than doubled in brightness (clouds) on the second date. The spectral parameter with the most pixels that changed by less than ±5 percent between the two dates of overpass was the most invariant spectral parameter to solar illumination and atmospheric changes.

The results of this temporal ratio test for different spectral parameters are summarized in Table 5.1.

Since the band 7 path radiance was the same for both overpasses, the fact that band 5 changed less than band 7 (as indicated by the greater number of pixels changed for band 5 in column 2 of the table) may be an indication that the solar illumination changes were greater than the atmospheric changes between the two dates. The table clearly shows, however, that both the noncorrected spectral ratio and the dark object-corrected spectral ratio of bands 7 and 5 were more invariant to solar illumination and atmospheric effects than either single band, taken individually. Ten pixels in the inactive part of the Atlantic City iron mine open pit that had no vegetation were used for ratio normalization in the temporal ratio of the corrected $R(7,5)'$ spectral ratio, to normalize the $q(i,j)C(i,j)$ multiplicative factors in Equation 5.6 between the two dates.

The last sentence can be explained more fully by considering the temporal ratio $TR_{t1, t2}$ for times t_1 and t_2 of the dark object-corrected spectral ratio $DN(i, j)'$ from Equation 5.6 as follows:

$$TR_{t_1,t_2} = \left\{ \frac{[q(i,j)C(i,j)]_{t_1}}{[q(i,j)C(i,j)]_{t_2}} \right\} \left\{ \frac{\left[\frac{\rho(i)}{\rho(j)}\right]_{t_1}}{\left[\frac{\rho(i)}{\rho(j)}\right]_{t_2}} \right\} \qquad \text{(Eqn. 5.7)}$$

where t_1 and t_2 denote times of the first and second overpasses, respectively. If a normalization area is chosen that is likely to have its spectral ratio unchanged on the ground between the two overpasses—a condition that demands an unvegetated area

TABLE 5.1 COMPARISON OF INVARIANCE TO SOLAR
ILLUMINATION AND ATMOSPHERIC CHANGES FOR MSS BANDS 7
AND 5 AND SPECTRAL RATIOS $R(7,5)$ (NOT CORRECTED) AND $R(7,5)'$
(DARK OBJECT-CORRECTED AND RATIO NORMALIZED) FOR THE
ATLANTIC CITY, WYOMING TEST AREA, OUT OF A TOTAL OF 11,032
GROUND INVARIANT (GI) PIXELS.

Spectral Parameter	No. of GI Pixels Changed by Less Than ± 5 Percent
Band 7	361
Band 5	1122
$R(7,5)$	2320
$R(7,5)'$	4115

that has not been disturbed between overpasses—the last bracketed term in Equation 5.7 is equal to 1.0, and the measured temporal ratio over the normalization area, $TR^N_{t_1,t_2}$, becomes equal to the first bracketed term. After substitution, Equation 5.7 becomes

$$TR_{t_1,t_2} = TR^N_{t_1,t_2} \left\{ \frac{\left[\dfrac{\rho(i)}{\rho(j)} \right]_{t_1}}{\left[\dfrac{\rho(i)}{\rho(j)} \right]_{t_2}} \right\} \qquad \text{(Eqn. 5.8)}$$

The normalized temporal ratio of the $R(7,5)'$ spectral ratio, $TR^N_{t_1,t_2}$, as defined by the following equation

$$TR'_{t_1,t_2} = \frac{TR_{t_1,t_2}}{TR^N_{t_1,t_2}} = \left\{ \frac{\left[\dfrac{\rho(i)}{\rho(j)} \right]_{t_1}}{\left[\dfrac{\rho(i)}{\rho(j)} \right]_{t_2}} \right\} \qquad \text{(Eqn. 5.9)}$$

is now proportional only to the on-the-ground changes in the $R(7,5)'$ spectral ratio, and it is now invariant with regard to changes that have occurred in sun position, atmospheric conditions, and electronic gain or offset of the sensor between the two overpasses. It should be noted that Equation 5.9 does not require a knowledge of the $R(7,5)'$ spectral ratio for the normalization area; it only requires that the normalization area does not change in its $R(7,5)'$ spectral ratio between overpasses. This is usually true of unvegetated rock or soil exposures that have not been covered by wind-blown material between passes. There should also be no ground freezing or frost differences in the normalization area between the two overpass times.

The 4115 pixels that changed by less than ± 5 percent for $R(7,5)'$ between the two dates represented 37.3% of the GI pixels. It was found that 70.2% of the GI pixels changed by less than ± 10 percent, and 96.83% of the GI pixels changed by less than ± 15 percent for $R(7,5)'$ between the two dates. This shows that spectral ratios corrected for dark object subtraction for each date and ratio normalized between dates is a robust method for minimizing solar illumination and atmospheric changes between overpasses. The above numbers are conservative, considering that some of the GI pixels could have actually changed on the ground between the two dates (especially pixels with small amounts of vegetation) and that the spatial registration of the images was only within two pixels, a margin which is substandard by today's correction methods.

Equation 5.9 and the ensuing discussion introduce the possibility that $R(7,5)'$ for LANDSAT MSS data and $R(4,3)'$ for LANDSAT TM data would be robust vegetation indices, the purpose of which are to measure the amount of vegetation biomass on Earth's surface. A glance at Figure 3.3 will refresh your memory about the distinctive rise in the reflectance of green vegetation, as wavelength increases from visible red to reflective infrared. This rise is caused by the selective absorption of red light by chlorophyll, on which vegetation indices are based. Vegetation indices can be helpful to geologists who employ geobotanical methods for mineral and hydrocarbon exploration or to those who employ GIS models, such as surface water run-off models that call for knowledge of the amount of vegetation cover present. Both of these applications will be discussed in a later chapter.

Equations 4.7 and 5.1 also raise a point about the robustness of those vegetation indices in current use that involve the difference divided by the sum of two spectral bands, one in the visible red and one in the reflective infrared wavelength region. The transformed vegetation index (*TVI*) for LANDSAT TM (using TM bands 3 and 4), is given by Lillesand and Kiefer (1994) as follows:

$$TVI = (100)\sqrt{\left[\frac{DN(4) - DN(3)}{DN(4) + DN(3)}\right] + 0.5} \qquad \text{(Eqn. 5.10)}$$

where
  $DN(4)$ and $DN(3)$ are digital numbers of TM bands 4 and 3.

The normalized vegetation index (*NDVI*) is defined in terms of the visible red (band 1) and reflective infrared (band 2) spectral bands of the weather satellite AVHRR, as follows (Lillesand and Kiefer 1994):

$$NVDI = \frac{DN(2) - DN(1)}{DN(2) + DN(1)} \qquad \text{(Eqn. 5.11)}$$

where $DN(2)$ and $DN(1)$ are digital numbers of AVHRR bands 2 and 1.

If there is no correction for atmospheric path radiance, the spectral band differences divided by the spectral band sums in both Equations 5.10 and 5.11 will include some additive terms that will not divide out. After substitution of Equation 5.2 into Equations 5.10 and 5.11, it can be seen that if dark object subtraction is carried out first, the shadow factor(*s*) will divide out of both of these vegetation indexes, just as it will in a spectral ratio as defined by Equation 5.5. However, the multiplicative factors in $DN(4)$ and $DN(3)$ of Equation 5.10 and $DN(2)$ and $DN(1)$ of Equation 5.11 will no longer factor out separately from the target reflectances of each spectral band. Thus we need to normalize the vegetation index separately for each different target on the ground when we are comparing vegetation targets between two data collection dates. If either the $R(4,3)'$ spectral ratio of LANDSAT TM data or the $R(2,1)'$ spectral ratio of AVHRR data, as defined by Equation 5.5, were used as a vegetative index, temporal ratio normalization like that described by Equation 5.9 would be possible as long as any part of the ground scene (ground-invariant pixels) could reliably be known to be unchanged between the two data collection dates, without the need for determining the spectra of the ground-invariant pixels. Research is needed to investigate whether either of these two spectral ratios will provide the same amount of biomass information as the transformed (*TVI*) or normalized (*NDVI*) vegetation indexes. If so, the $R(4,3)'$ spectral ratio would be superior as a vegetation index because of its greater robustness.

Thermal Infrared Spectral Ratioing

The benefits of spectral ratioing are not so immediately apparent in the thermal wavelength region, as is shown by the following spectral ratio equation for wavelengths longer than 4.0 μm:

$$R(i, j)' = C(i, j)\left[\frac{1 - b'\rho^c(i)}{1 - b'\rho^c(j)}\right]\left[\frac{L_{bb}(i, T)}{L_{bb}(j, T)}\right] \qquad \text{(Eqn. 5.12)}$$

where

$$C(i, j) = \frac{q(i)g(i)\tau(i)f(i)}{q(j)g(j)\tau(j)f(j)}$$

Once again, $C(i, j)$ is a constant over the whole image and the compositional information is wholly contained in the spectral reflectances of the *i*th and *j*th bands in the first bracketed term, which can change with every pixel. Unlike the reflective wavelength region case in Equation 5.5, the last term in the brackets in Equation 5.12 still contains temperature information that can change with every pixel. However, although the last term in brackets can change with every pixel, this ratio of black-body emission for two overlapping spectral bands (for instance, 8.2–10.9 μm and 9.4–12.1 μm) tends to change less than the second bracketed term, which contains the information we seek. It would take a temperature difference of about 5.2 K to mask a spectral reflectance difference of about 1% between the *i*th and *j*th bands, if one band was spectrally located in a rock or mineral reststrahlen band and the other was not (Vincent 1973).

Figure 5.5 (Vincent et al. 1972) shows single-band thermal infrared images for an 8.2–10.9 μm band and a 9.4–12.1 μm band produced by an airborne multispectral scanner over a sand quarry near Mill Creek, Oklahoma. Figure 5.6 (Vincent et al. 1972) shows the first spectral ratio image ever made (August, 1970) and an aerial photo of the same area. The spectral ratio image, which employs the same two spectral bands as Figure 5.5, dramatically discriminates quartz sandstone and quartz sand (both are dark in the spectral ratio image) from surrounding grass, water, topsoil, limestone, and shale outcroppings (all are bright in the ratio image). The two single-band images display practically no compositional information and show few or no differences between images. Therefore, though the case for spectral ratioing to enhance compositional information is not as mathematically clear-cut in the thermal wavelength region as in the reflective wavelength region, practical applications show

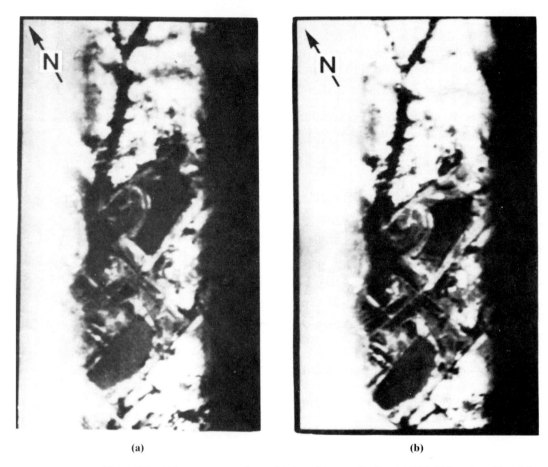

(a)   (b)

Figure 5.5   Airborne scanner thermal infrared images for band 1 (8.2–10.9 μm) and band 2
(9.4–12.1 μm), shown respectively in (a) and (b), for a sand quarry near Mill Creek, Oklahoma.
Cold objects are dark and warm objects are light. Hot and cold reference plates are at the left
and right margins of each image, respectively. Image dimensions are roughly 1 km × 2 km.
(Vincent et al. 1972. Reprinted with permission of the American Geophysical Union.)

that spectral ratioing of multispectral thermal infrared spectral bands reveals unique
compositional information about silicate rocks and soils, even in daytime, when temperature differences across the scene are relatively large.

   In earlier chapters, a shift to shorter wavelengths of the strongest reflectance
maximum in the 8.0–12.0 μm wavelength region spectra of more felsic silicate rocks
was called the felsic shift. This phenomenon begs the question of what compositional
parameters of silicate rocks best correlate with the spectral position and shape of silicate reflectance maxima in the 8.0–12.0 μm wavelength region.

   When a rock is analyzed by a wet-chemistry process commonly referred to by
geologists as *rapid rock analysis,* the chemical composition of the rock is given in

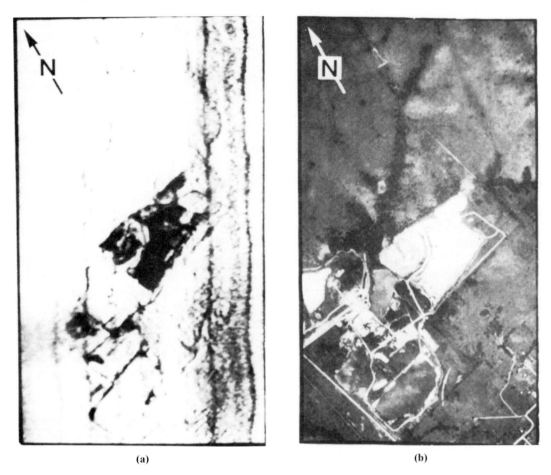

(a)   (b)

Figure 5.6   An $R(1,2)$ spectral ratio image of band 1 (8.2–10.9 μm) and band 2 (9.4–12.1 μm) in Figure 5.5 and an aerial photo, shown respectively in (a) and (b), for a sand quarry near Mill Creek, Oklahoma. Quartz-rich sandstone is dark in the spectral ratio image, which is believed to be the first spectral ratio image ever made (August, 1970) for any wavelength region. The infrared scanner data and the aerial photo were collected at the same time. Image dimensions are roughly 1 km × 2 km. *(Vincent et al. 1972. Reprinted with permission of the American Geophysical Union.)*

weight percentages of oxides, such as %SiO_2, %Fe_2O_3, %FeO, %CaO, and so forth (Vincent 1973). These oxides may not exist as separate entities in the rock, but rather reflect the relative amounts of each oxide cation in the rocks, e.g., %SiO_2 is a measure of the amount of silicon. Oxide weight percentages alone do not indicate the mineral composition of a rock.

The most common method for determining mineral composition, called modal analysis, involves the direct measurement of mineral content by microscopic examination of polished thin sections of the rock. Weight percentages of minerals measured this way are called *modes*. In a less common method, the approximate mineral weight percentages are determined theoretically from a normative calculation. This calculation employs a series of equations that convert the oxide weight percentages from rapid rock analysis to approximate volume percentages (or *norms*) of normative minerals (Johannsen 1931; Rodgers et al. 1970). Though similar, the two methods are not equivalent. On an a priori physical basis, the mineral parameters (modes or norms) would seem superior to chemical parameters (oxide weight percentages) for correlation with the reflectance maxima location and shape because the particular ionic bonds involved should depend more on lattice structure than simply on quantity of cations present in the rock. This prediction is borne out in the next two figures, Figures 5.7 and 5.8.

When chemical and mineralogical parameters of 26 igneous silicate rocks were sought that would correlate well with the same thermal infrared spectral ratio that lead to Figure 5.6, i.e., a spectral ratio of an 8.2–10.9 μm wavelength band 1 with a 9.4–12.1 μm wavelength band 2 (Vincent 1973), Figures 5.7 and 5.8 resulted (Vincent 1975a). Figure 5.7 shows the correlation of this two-band spectral ratio, here called $R(1,2)$, with one of the best chemical parameters found, which is the percentage of SiO_2 minus the percentage of Al_2O_3, as determined from rapid rock analysis, where set A is all of the igneous rocks except sample 26, which is a peridotite, and set B is all of the igneous rocks except two granites and a rhyolite (samples 1, 3, and 5), as well as the peridotite (sample 26). More recently, Walter and Salisbury (1989) devised an index that is more closely associated with the polymerization of silicate minerals than previous chemical indexes. The Walter and Salisbury index is equal to the ratio of % SiO_2 (or silica) to the sum of silica, plus the oxides of calcium, iron, and magnesium. Silica content is almost synonymous with quartz content.

Figure 5.8 shows the best linear combination (called M_{16}) of volume percentages of normative minerals that best correlated by multiple regression with the $R(1,2)$ spectral ratio described above. The normative minerals were calculated from the oxides that had been determined from rapid rock analysis for set C, which is all the 26 igneous rocks except for one basalt (sample 22). The standard error in M_{16} was 0.009 (Vincent 1973, 1975a). As these two figures show, the felsic shift in igneous silicate rocks can indeed be measured with an $R(1,2)$ thermal infrared spectral ratio image.

Figure 5.9 (Vincent 1975a) shows single-band images from the same two airborne multispectral scanner spectral bands as Figure 5.5, and an $R(1,2)$ spectral ratio image produced from these bands (same as $R(1, 2)$ for Figures 5.7 and 5.8) for a region containing a variety of igneous silicate rocks near Pisgah Crater, California. The $R(1,2)$ spectral ratio image shows basaltic lava and a basaltic cinder cone that are bright because they have low quartz content, as both Figures 5.7 and 5.8 predict. The surrounding dacitic fanglomerate and wind-blown, quartz-rich sand across part of the lava flow are dark because the quartz reststrahlen band causes a reflectance maximum (emittance minimum) in the 8.2–10.9 μm band, thereby reducing the $R(1,2)$ spectral ratio. As in the previous example, the single-band thermal infrared images are controlled primarily by temperature variations across the scene, whereas the spectral ratio of the two bands is controlled by compositional differences across the scene. The white line connecting a few white dots and running slightly left of top-to-bottom across the scene is a power line connecting metal-framework pylons. It

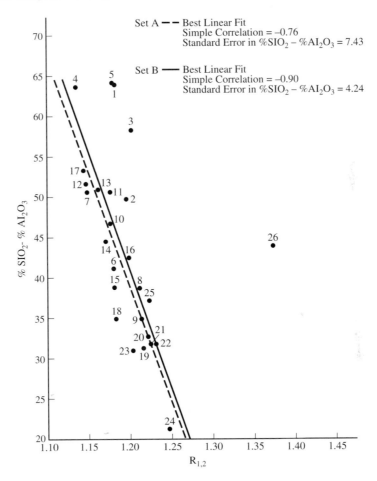

Figure 5.7 Plot of $R(1,2)$ spectral ratio of lab spectra of 26 igneous silicate rocks for aerial thermal infrared scanner bands 1 (8.2–10.9 μm) and 2 (9.4–12.1 μm) versus the ratio of the percentage of SiO_2 to the percentage of Al_2O_3. Correlations are given for Set A, consisting of all of the 26 rocks except peridotite (#26), and Set B, consisting of all rocks except peridotite (#26), rhyolite (#5), and two granites (#1 and #3). *(Vincent 1973. Reprinted with permission.)*

shows up in Figure 5.9c as bright because it has no quartz, not because it is hotter or cooler than the desert background.

More recently, papers by Salisbury and D'Aria (1992), Gillespie (1992), Watson (1992), and others were published in a special technical journal issue about the extraction of information from multispectral thermal infrared data, including the use of spectral ratio imaging. It is fair to say that spectral ratio images are noisier than many other types of multispectral images, sometimes requiring spatial filtering, but spectral ratios have been directly related to chemical and mineral composition of rocks and soils, as the above examples demonstrate. This relationship is responsible for much of the enthusiasm expressed about spectral ratios in this chapter.

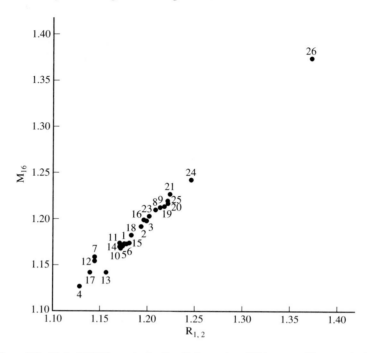

Figure 5.8 Plot of $R(1,2)$ spectral ratio of lab spectra of 26 igneous silicate rocks for aerial thermal infrared scanner bands 1 (8.2–10.9 μm) and 2 (9.4–12.1 μm) versus an M_{16} parameter, which represents the best linear combination of normative minerals that best correlated with $R(1,2)$ for all of the igneous silicate rocks in Figure 5.7 except for sample #22 (a basalt). *(Vincent, 1973. Reprinted with permission.)*

Decorrelation contrast stretch images (Gillespie 1992), which will be discussed again in the subsection on principal component transformations, show more details about the scene than spectral ratio images show from multispectral thermal infrared scanner data, but these images cannot be as easily interpreted for composition by the use of lab or field spectra. A decorrelation stretch image mixes the effects of temperature with compositional variations across the scene to a greater extent than spectral ratio images mix the two. This difference is the principal reason why more details are apparent in the former type of image. In other words, spectral ratio images separate the effects of temperature and compositional variations across the imaged scene better than decorrelation contrast stretch images separate them.

Another advantage of spectral ratio imaging is that it facilitates the use of multispectral data from the thermal wavelength region with data from the reflective wavelength region. Single-band images from the thermal region and reflective region are dominated primarily by temperature and brightness, respectively. These two characteristics make it difficult to combine single bands from the two drastically different wavelength regions. However, spectral ratio images produced from spectral band pairs that are subsets contained within each of these wavelength regions, respectively, can be readily combined because they both map the same quantity: chemical composition (Vincent et al. 1984).

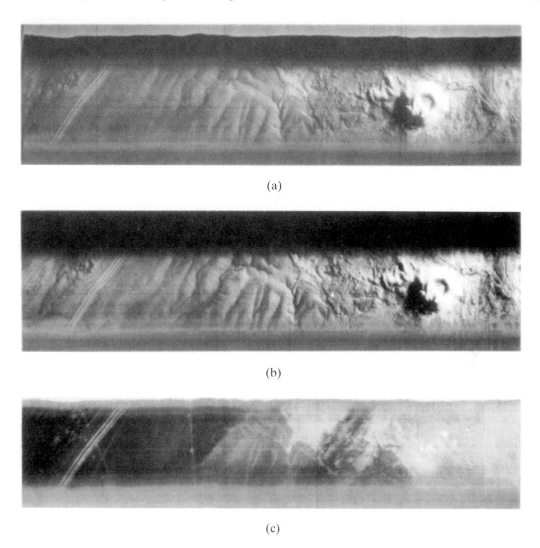

(a)

(b)

(c)

Figure 5.9 Airborne scanner thermal infrared images for band 1 (8.2–10.9 μm) and band 2 (9.4–12.1 μm), shown respectively in (a) and (b), and an $R(1,2)$ spectral ratio image (c) for a region near Pisgah Crater, California. Quartz-rich rocks and soils are displayed as dark in the spectral ratio image. Time of day was just after sunrise. *(Vincent 1973. Reprinted with permission.)*

Imaging of Gases

A final advantage of thermal infrared spectral ratioing is that it can be used for imaging gases that have characteristic absorption bands inside wavelength regions called atmospheric windows, where the atmosphere absorbs least. Consider the case where a t-meter-thick plume of a certain gas is mixed with air at 1 atmosphere of pressure near Earth's surface. Assume that the plume gas absorber concentration is w, which is a product of the partial pressure of the absorbing plume gas times the plume thickness t over which this partial pressure exists, with dimensions of torr-cm. Further assume that absorption coefficient of the plume gas is given by $\alpha(\lambda)$, a wavelength-dependent number with dimensions of $(\text{torr-cm})^{-1}$. The total infrared radiation power per unit area that is radiated into the hemisphere above the plume over the ith spectral interval (defined by the upper and lower wavelength limits of λ_u and λ_l), is a quantity called the *total exitance* $M^T(i)$, which has dimensions of watts/cm^2. However, $M^T(i)$ actually consists of three terms, each representing a different contribution, as given in the following equation:

$$M^T(i) = M^{be}(i) + M^e(i) + M^{br}(i) \qquad \text{(Eqn. 5.13)}$$

where

$M^T(i)$ = total exitance in the ith spectral band

$M^{be}(i)$ = exitance emitted by the background of spectral emittance $\epsilon(i)$ (taken from Equation 2.7 and assumed to be constant over the wavelength interval λ_u to λ_l) and temperature T_b that has been transmitted once through the plume

$M^e(i)$ = exitance emitted by the plume of temperature T_p (assuming that plume emission absorbed by the ground is re-emitted and absorbed by the plume)

$M^{br}(i)$ = exitance from reflection of sunlight and atmospheric emission off the background with attenuation of both incoming and outgoing radiation by the plume.

Figure 5.10 depicts the three types of radiation described by Equation 5.13. The $M^{br}(i)$ term in Equation 5.13 is the solar exitance and atmospheric emission exitance impinging on the ground in the ith spectral band, which can be calculated from the LOWTRAN atmospheric model (Wolfe and Zissis 1978). Appendix A

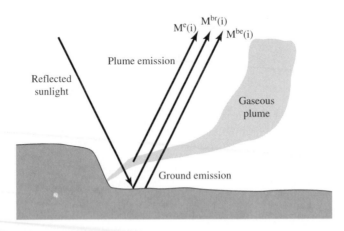

Figure 5.10 Schematic of three types of infrared radiation coming from a gaseous plume of methane exiting a solid waste landfill: ground emission, $M^{be}(i)$; reflected sunlight and atmospheric emission, $M^{br}(i)$; and plume emission, $M^e(i)$. *(Courtesy of R. K. Vincent.)*

gives the integral version of Equation 5.13, in which the spectral exitance from a "black body" is implied in the integrand as π times the Planck radiation function $L_\lambda^{bb}(T)$, taken from Equation 4.5, as follows:

$$M_\lambda^{bb}(T) = \pi L_\lambda^{bb}(T) \qquad \text{(Eqn. 5.14)}$$

where M_λ^{bb} is radiation power per unit area radiated into a hemisphere per unit wavelength interval, with dimensions of watts/(cm^2 $-$ μm).

Examination of the three terms in Equation 5.13 reveals that both the ground emission term, $M^{be}(i)$, and the ground reflection term, $M^{br}(i)$, decrease with increasing plume gas concentration (w) in a spectral band that encompasses an absorption band of the plume gas (where $\alpha(\lambda)$ is of appreciable magnitude), whereas the cloud emission term, $M^e(i)$, increases with increasing plume gas concentration in the same spectral band. This relationship makes common sense, since the plume is acting as a transmission filter for the ground emission and ground reflection terms, and decreasing amounts of each of the two terms are transmitted through the plume as concentration of the plume gas increases. The plume emission term, however, increases with increasing plume gas concentration. The relative size of these terms will, of course, depend on the conditions at the time of the observation. For instance, the ground reflection term, $M^{br}(i)$, will be relatively small at night for wavelengths short of 4 μm, when the sun is down and the atmosphere is relatively cool. Atmospheric emission that reflects off the ground can become appreciable at these shorter wavelengths if the air is humid, however, and is usually larger than reflected sunlight at wavelengths in the 8–14 μm atmospheric window, even in the daytime. Whether the ground emission term or the plume emission term dominates will depend primarily on the relative temperatures of the background (T_b) and the plume (T_p). For T_b appreciably greater than T_p, background emission will dominate plume emission. When the reverse is true, the plume emission term will dominate. When T_b and T_p are about equal, the ground emission term, $M^{be}(i)$, and the plume emission term, $M^e(i)$, tend to neutralize one another, and the absorption band of the plume cannot be detected from these two terms.

Consider the situation where three equal-bandwidth, thermal infrared spectral bands (bands 1, 2, and 3, from lowest to highest wavelength region, respectively) are available, and consider that band 2 is centered on an absorption band of the plume gas, whereas bands 1 and 3 are located at nearby wavelengths that are outside that absorption band of the plume gas. Now consider a three-band spectral ratio defined as follows:

$$R(1 + 3,2) = \frac{L^T(1) + L^T(3)}{2L^T(2)} = \frac{M^T(1) + M^T(3)}{2M^T(2)} \qquad \text{(Eqn. 5.15)}$$

where the total exitances come from Equation 5.13 for i = 1, 2, and 3, and band 2 encompasses an absorption band of the plume gas. If a spectral ratio image is made from this three-band ratio, the plume gas can be imaged; this image not only detects the plume, but also helps the observer to estimate its areal extent and to locate its source. Since places in the spectral ratio image that are on and off the plume can be viewed as increased and decreased plume gas concentrations, respectively, and since the band that contains the plume gas absorption band is in the denominator of Equation 5.15, a plume whose exitance is dominated by the ground emission term, $M^{be}(i)$, and/or the reflected sunlight term, $M^{br}(i)$, will appear in the spectral ratio image as a bright feature, compared to the background with no plume. Conversely, a plume whose exitance is dominated by the cloud emission term, $M^{e}(i)$, will appear in the spectral ratio image as a dark feature, compared to the background with no plume.

From the deductive reasoning in the previous paragraphs and the fact that band 2, which contains the plume gas absorption band, is in the denominator of Equation 5.15, it is possible to make some general conclusions about what viewing conditions will permit the best imaging of the plume gas. These conclusions are generalized in Table 5.2.

Methane gas has a strong absorption band near a wavelength of 3.314 μm, which is spectrally located within a transparent atmospheric window, a necessary requirement for gas imaging. The imaging of methane gas, which is the principal constituent of natural gas and of gaseous emissions from solid-waste landfills (Vincent 1994), is particularly compelling in light of its applications to both exploration and environmental geology. Methane concentrations required for imaging methane plumes under various environmental conditions are discussed in Chapter 9. Imaging of other industrial gases, such as carbon monoxide, nitrous oxide, and benzene, is possible in the same manner as methane, because these gases also display characteristic absorption bands in the atmospheric windows from 3.0–4.1 μm and 4.3–5.0 μm (Erley and Blake 1964).

TABLE 5.2 GENERALIZED FAVORABLE VIEWING CONDITIONS FOR IMAGING GAS/AIR PLUMES BY NADIR VIEWING IN AN $R(1 + 3,2)$ SPECTRAL RATIO IMAGE OF THERMAL INFRARED SPECTRAL BANDS 1, 2, AND 3, WHERE BAND 2 IS SPECTRALLY LOCATED TO INCLUDE AN ABSORPTION BAND OF THE PLUME GAS, AND BANDS 1 AND 3 ARE LOCATED AT NEARBY WAVELENGTHS THAT ARE OUTSIDE THE ABSORPTION BAND OF THE PLUME GAS.

I. Moderate-to-warm ground, with a cool plume under clear, daytime skies. This condition likely will prevail in early-to-mid-mornings or in late afternoons of autumn. These conditions minimize emission from the plume and emphasize ground emission and reflected sunlight. Thus, the plume will be observed as a bright feature in the $R(1 + 3,2)$ spectral ratio image.

II. Cold ground and warm plume, under clear, nighttime skies. These conditions are most likely to occur in winter or early spring, when the ground surface is frozen. They minimize ground emission and reflected sunlight and emphasize plume emission. Thus, the plume will be observed as a dark feature in the $R(1 + 3,2)$ spectral ratio image.

III. Cold ground and cold plume, under clear, daytime skies. These conditions are most likely to occur in the coldest part of winter in midlatitudes and in Arctic regions during late autumn and early spring. They minimize ground emission and plume emission, but emphasize reflected sunlight. Thus, the plume will be observed as a bright feature in the $R(1 + 3,2)$ spectral ratio image.

IV. Cold water and warm plume, under clear skies, just after dark. These conditions are most likely to occur in offshore regions during most of the year. They minimize background emission and sunlight reflection, but emphasize plume emission. Thus, the plume will be observed as a dark feature in the $R(1 + 3,2)$ spectral ratio image.

PRINCIPAL COMPONENT TRANSFORMATION AND MULTISPECTRAL CLASSIFICATION

The display of information in image processing becomes a more difficult task when the number (n) of spectral bands exceeds three because there are only three primary colors in a color image display. How can information from more than three bands be shown simultaneously in only three primary colors? Besides the combination of three spectral ratios into a color ratio image, which is a simple extension of what was discussed in the previous section, there are two different types of answers to that question. One involves image enhancement with n-dimensional spectral parameters called principal components and the second involves image classification by use of all n spectral bands to divide the image into several discrete categories of scene elements. Image enhancement results in a nearly continuous-toned image, i.e., each of the three primary display colors will have values between 0 and 255. Image classification results in a classified image that assigns a discrete color to each category that was classified, and there will be only as many colors as there are categories.

Principal Component Transformation

Principal components (also called by the names Hotelling, discrete Karhunen-Loeve, or eigenvector) transformation is an image enhancement technique for displaying the maximum spectral contrast from n spectral bands with just three primary display colors, and it involves a two-step process that is briefly described here. (A more complete description of principal component transformations is given in Hall 1979.) The first step inputs n histograms of the scene to be imaged, one histogram per spectral band, to the principal component algorithm. This algorithm calculates n principal components, which are actually orthogonal vectors in n-dimensional space that are oriented along directions of maximum remaining variance. Outputs of this first step are the projection coefficients of the n spectral bands onto the final n principal components, yielding the series of equations given below:

$$P_k = \sum_{i=1}^{n} a_{i,k} DN(i) \qquad \text{(Eqn. 5.16)}$$

where P_k is the kth principal component value for a given pixel, $DN(i)$ is the digital number of the ith spectral band (out of a total of n spectral bands) for a given pixel, and $a_{i,k}$ are the projection coefficients calculated by the principal component algorithm for projecting digital numbers from the n spectral bands onto the n principal component vectors.

The first principal component is a vector that is in the direction of the maximum variance of pixels in the scene. It accounts for more of the spectral variance in the data than any other principal component. The second principal component, orthogonal to the first component in n-dimensional space, is a vector that is aligned along the maximum remaining variance in the data (within the constraint that it is also orthogonal to the first principal component), thereby removing the maximum amount of variance in the data that remains after the first component is applied.

Each subsequent principal component, orthogonal to all the other principal components, removes the maximum amount of remaining variance, which becomes smaller as the order of the principal component increases. The nth component, contains all of the remaining variance and separates the most spectrally unique pixels from the rest of the pixels in the scene. All n principal components account for 100% of the variance in the data; however, the first three principal components usually account for a majority (50%–95%) of the variance, when $n > 3$.

The second step is the transformation of the imaged scene to principal component space. This is done by using Equation 5.16 to calculate all of the n principal component values for each pixel in the scene. An image of each of the n principal components can then be produced, usually after contrast stretching each of the principal components, resulting in a different image of the same scene for each of the n principal components. Each of these principal component images, however, contains information from each of the n spectral bands, because of Equation 5.16. The greatest variance in the scene is explained by the first principal component, which chiefly displays brightness information at wavelengths shorter than 4 μm and temperature information at longer wavelengths (thermal infrared). This component tends to highlight bright (or warm), sunlit slopes rather than shadowed (or cool) areas, thus yielding the most topographic information of any of the principal components, regardless of the number of spectral bands involved. The nth principal component image often produces a fairly homogeneous image, interrupted by a few bright and dark pixels that are the most spectrally unique pixels in the scene. If the data contain sporadic electronic noise, the noise will usually show up in the nth principal component image. However, because LANDSAT TM and SPOT multispectral scanner data have already had most of the systematic electronic noise removed from the data before a user receives it, and because some important remote sensing tasks, such as mineral exploration or environmental contaminant mapping, require searches for relatively rare occurrences of spectrally unique targets, the remote sensing user is wise to carefully examine the higher-order principal component images, including the nth one, before discarding any of them. Ignoring the higher-order principal components for such tasks is frequently the equivalent of throwing the baby out with the bath water.

Any three principal components can be used to create a color composite image that contains information from all n spectral channels. When $n > 3$ and the first three principal components are imaged together, most of the variance from all n spectral bands can be displayed in one color composite image; this variance is greater than can be displayed in a color composite made from any three of the n spectral bands. This advantage can sometimes help the geologist or environmentalist observe boundaries between terrain units that might not have been readily observed in color composites of any three single-band images. Likewise, when three higher-order principal components are displayed as a color image, it is possible for the user to detect small areas on the ground that are spectrally quite different from most of the scene. This image, therefore, can be especially helpful in areas about which little information is available.

Color Plate CP9 (a–d) shows a series of four LANDSAT TM images of the Minas Plomosas area in the northern Mexican state of Chihuahua (Torres et al. 1989).

A natural color image (TM bands 1, 2, and 3 displayed as blue, green, and red, respectively) is shown in part (a), an iron oxide ratio image (TM ratios of $R(5,7)$ shown in blue, $R(4,3)$ in green, and $R(3,2)$ in red) is shown in (b), a thermal image (TM band 6) is shown in (c), and a color principal components image (with principal components P_1, P_2, and P_3 displayed in blue, green, and red, respectively) is shown in (d). The narrow, blue feature in the iron oxide ratio image of Color Plate CP9b, extending from approximately D1 to G3 in the image coordinates, is the Cuchillo Formation. This color in the iron oxide image usually designates calcium sulfate (gypsum or anhydrite) or calcium carbonate (limestone), though the composition of the Cuchillo Formation at this location has not been confirmed in the field. There are known limestone outcrops near image coordinates C6 (just right of a pink region, judged to be ferric-oxide-rich, in the iron oxide ratio image) that have a darker, purplish-blue color, lending credence to a gypsum or anhydrite composition for the light-blue color of the Cuchillo Formation in the iron oxide image. Note, however, that the principal components image in Color Plate CP9d makes the Cuchillo Formation distinctively black, along with a V-shaped feature (with the V pointing to the left) at image coordinates G4. This V-shaped feature occurs at the left edge of a local topographic high (an anticline?) with a left–right orientation, that has Las Vigas Formation rocks exposed on top of it. This same V-shaped feature is seen as blue in the iron oxide image, but is not as distinct from the surrounding terrain as it is in the principal components image. The thermal infrared image of TM band 6 (Color Plate CP9c) shows this V-shaped feature as white, which means that it is warmer than almost all other parts of the image. The V-shaped feature has been field-confirmed as a region with thick caliche-type deposits (Torres et al. 1989) of unspecified composition, which (with the higher temperature and calcium sulfate or calcium carbonate multispectral signature) led those investigators to interpret this as a region of shallow-depth hydrothermal activity.

As the above example shows, principal components imaging can produce improved spectral contrast between spectrally unique terrain elements and the surrounding terrain. However, there are two disadvantages of principal component images that limit their utility for many remote sensing tasks. One is that principal components, unlike spectral ratio images, are a scene-dependent imaging technique. The colors in a principal component image will not mean the same thing from image to image, unless the areas covered by the two images have similar geological outcrops and terrain cover (vegetation and water). More fundamentally, it is difficult to interpret the colors of a principal component image by using reflectance spectra of rocks, soils, and minerals as guides. When principal components imaging is employed for the detection of rocks or soils with unusual spectral properties, it is usually necessary to employ spectral ratio images for identifying what the composition is, at least to the extent afforded by the spectral ratios that can be produced from the available spectral bands. Spectral ratio images are more environmentally invariant (robust) and more relatable to laboratory spectra of minerals, rocks, and soils than are principal component images. Nonetheless, the two can be used together, as described above, where detection and identification of unusual rock or soil exposures are performed as a two-step process.

An inverse form of principal component transformation that was mentioned in the previous section is more robust (keeps the colors more nearly the same for a given target in images collected on different dates) than principal component trans-formations described above. Decorrelation contrast-stretch images (Gillespie 1992) are produced by applying a principal component transformation to a multispectral image data set, contrast stretching three principal component images (also called decorrelation images because the principal components are uncorrelated with one another), and transforming these stretched principal component images back to the original *n*-dimensional spectral space with the inverse of the principal components transformation. The resulting images are more like spectral ratio images than prin-cipal component transformations in that they downplay overall brightness (or tem-perature) effects and are more robust than principal component transformations. However, they mix brightness (or temperature) with compositional information to a greater extent than spectral ratio images and are not as easily relatable to laboratory or field spectra. They do, however, make a more contrast-rich image than spectral ratio images.

Color Plate CP11a shows a color composite of three thermal infrared spectral bands, Color Plate CP11b shows a color spectral ratio image of three thermal in-frared spectral ratios, and Color Plate CP12 shows a decorrelation stretch image of the Death Valley, California region (Gillespie 1992). The data for these images were collected by a TIMS (Thermal Infrared Multispectral Scanner) airborne scanner, which has six spectral bands in the 8–14 μm atmospheric window. The decorrelation stretch image shows more contrast than the spectral ratio image because it mixes some of the effects of temperature and emittance (Hook et al. 1992). This mixing is evidenced by the fact that the decorrelation stretch image in CP12 looks like a com-bination of the color composite CP11a of three thermal infrared spectral bands (controlled mostly by temperature) and the color ratio CP11b image (controlled by emittance, or composition). The numbers in these images refer to the following ter-rain features (Gillespie 1992):

Numbers	Ground Description
1 and 2	Piedmonts adjacent to Death Valley
3	Part of Panamint Range mountains
4	Valley floor with shallow standing water over calcite, halite, and silty sediments
5	Dolomite outcrop
6	Quartzite outcrop
7	Shale outcrop
8	Volcanic rocks

Vegetation is generally sparse, with pinon-juniper woodlands in the mountains, above creosote-bush plains, and an unvegetated valley floor (Gillespie 1992).

Hook et al. (1992) employed two newer techniques, thermal log residuals and alpha residuals, to extract compositional information from thermal infrared scan-ner data. Their techniques, however, also mixed effects of temperature with effects of emittance (composition), although to a lesser degree than decorrelation stretch-ing, owing to their use of Wein's approximation in Planck's function. Wein's approximation is incorrect to the same extent that the effects of temperature and spectral emittance are mixed. Thermal spectral ratios appear to mix the two effects less than any of these methods, though spectral ratios result in lower-contrast images because of their suppression of temperature effects.

Multispectral Classification

Multispectral classification algorithms attempt to categorize each pixel as belonging to one of a number of unique spectral classes. These classes are each characterized by a common range of values for each of the spectral parameters that are used as input to the classification algorithm. The common ranges of values for all of the spectral parameters for a given spectral (or target) class is called the *spectral signature* of that class. With this definition, the spectral signature of a class applies to one data collection event and has no implied robustness with regard to temporal changes in environmental conditions.

As diagrammed in Figure 5.11 there are basically two types of classification algorithms: supervised and unsupervised. The first step in Figure 5.11 is often a preprocessing step, sometimes involving different forms of spectral ratioing, such as division of each spectral band by the sum of all spectral bands or dark object-subtracted spectral ratioing of pairs of spectral bands, as will be discussed later as a preprocessing step for multispectral classification. Supervised classification requires

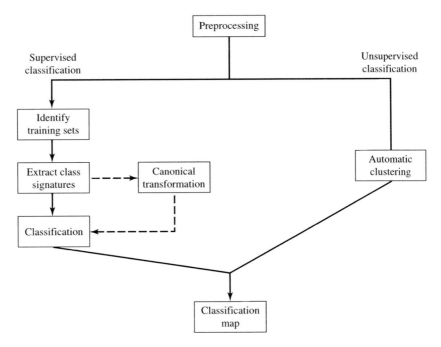

Figure 5.11 Schematic diagram of supervised and unsupervised classification procedures. *(Courtesy of R. K. Vincent.)*

the user to identify a group of pixels (called a training set) in an image that all belong to one target class. For instance, pixels located on a known outcrop of Triassic redbeds of the Chugwater Formation could be selected by the user as a training set for a target class called Triassic redbeds. This requirement is repeated for all target classes that the user wishes to classify in the image, such as Cody shale, Dakota sandstone, and so on. The automatic classification algorithm then extracts a spectral signature from each of the training sets and proceeds to automatically compare the spectral properties of every pixel in the image with the spectral signature of each target class. Every pixel in the image is then classified as belonging to one of the supervised target classes or as a class entitled "Other." This label simply means "none of the target classes." The Other class is inhomogeneous and can designate other rock or soil types, vegetation (possibly divided into deciduous and coniferous target classes), standing water, snow, and others.

Each pixel of the image can be represented by a point in n-dimensional multispectral space, defined by its $DN(i)$ value for each of the n spectral bands. For ease of graphic demonstration in the following explanations of different types of supervised classification algorithms, assume that a multispectral sensor has three spectral bands ($n = 3$) and that there are three training areas on the ground, each representative of a different class of target for which you are searching. Each pixel in the training sets is represented in Figure 5.12 as a number between 1 and 3 (the target class number) in an x, y, z coordinate system comprised of the $DN(i)$ for each of spectral bands 1, 2, and 3, respectively. The pixel marked as U in Figure 5.12 is an unknown pixel that we would like to classify as belonging to one of the three target classes or to a fourth class entitled Other. Eventually, every pixel in the image that is not inside one of the training set areas will take a turn being U, until the entire scene in the image is classified as one of seven classes.

Surface materials in a well-selected training set that are relatively homogeneous will have very similar reflectances, meaning that the $DN(i)$ for all the pixels in that supervised training set will bunch together, or cluster, in n-dimensional space. If surface materials at different training sites are spectrally distinct from each other, the clusters of their $DN(i)$ values will not overlap. Such is the case in Figure 5.12 for three spectral bands ($n = 3$), where three supervised training classes exhibit spectral signatures that form tight (well-defined), nonoverlapping clusters. It is the challenge of the classification algorithm to define this degree of "separateness" of the spectral signatures of training sets and to choose which cluster all other pixels fall into.

There are several different supervised classification algorithms, the simplest of which is a minimum-distance-to-means classifier. Figure 5.13 shows the same graphical plot as the previous figure, with x's denoting the mean of each of the three supervised target classes and vectors between each of these means and the unknown pixel U, each of which represents the distance of U from the mean of each target class. For this algorithm, each of these distances to the mean of each class is calculated, and U is labeled with the same class number as the class that it is nearest (for which this distance is minimum), unless this minimum distance to a target class is greater than some maximum value that is selected by the user beforehand, in which case U is labeled the class number 4, for Other. In other words, each pixel is classified according to the target class that it is nearest in n-dimensional multispectral co-

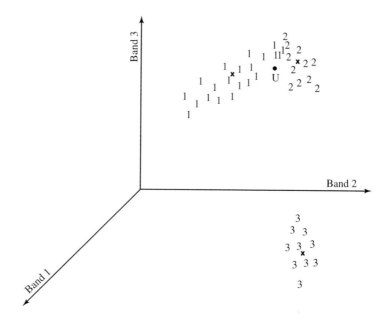

Figure 5.12   Plot of an unknown pixel (U) in spectral band 1, 2, and 3 space, with three target training sets. Pixels denoted by 1, 2, and 3 are in the training sets for classes 1, 2, and 3, respectively. *(Courtesy of R. K. Vincent.)*

ordinate space. If it is not acceptably close enough to any of the supervised classes, it is classified as Other. In this way, every pixel of the image is classified as one of the target classes or as Other, and a classification map with each class represented by a different color is produced. In any supervised or unsupervised classification program, the resulting classification map consists of discrete colors, one for each of the discrete target classes for which a training set was identified by the user.

The minimum-distance-to-means classifier has the advantage of simplicity, but it has the disadvantage of ignoring the variance within each supervised class, i.e., the shape of the "cloud" of pixels in each class is ignored. Only the mean of the class, which is a between-class measure, is employed in this method. It is possible for a pixel to be a member of one class but be located nearer to the mean of another class than to the mean of its own class. This condition is especially true of "elongated cloud" classes such as class 1 in Figure 5.13. If U happens to fall in the lower-left edge of class 1, it would be incorrectly labeled as class 2 because it is closer to the class 2 mean than to the class 1 mean.

The simplest supervised classification scheme that takes target class variance into account is the parallelepiped method, also called AND-gating. In this method, the spectral signature of a target class consists of the maximum and minimum *DN* numbers found in the training set for each of the spectral bands. For instance, if a multispectral data set with only three spectral bands were available, the *DN* range of band 1 might be 10–18, of band 2 might be 21–32, and of band 3 might be 35–42 for the Triassic Redbed training set (that group of pixels that the user has identified as being located on a known Triassic Redbed outcrop). For every pixel in the image, the

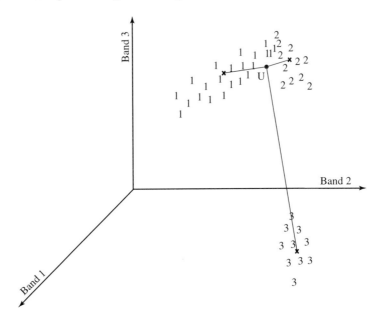

Figure 5.13 Minimum-distance classification. Same plot as Figure 5.12, with target class means denoted by black dots. A minimum-distance classifier would make the unknown pixel (U) a member of class 2, because it is located in *n*-dimensional space nearer the mean of class 2 than to the means of classes 1 and 3. *(Courtesy of R. K. Vincent.)*

parallelepiped method would then compare the $DN(1)$ number of a given pixel to see whether it is in the 10–18 range of $DN(1)$; if it is, the $DN(2)$ is compared to see whether it is in the 21–32 range of $DN(2)$; if it is, the $DN(3)$ of that same pixel is compared to see whether it is in the 35–42 range of $DN(3)$. If all three "gates" are affirmatively satisfied, that pixel is classified as a member of the Triassic Redbed target class and is given that target class number (assume Triassic Redbed is target class number 1). If any one of the AND-gates fails, that pixel is not a member of class 1, and it is similarly tested for membership in the other target classes, each with their own set of DN_i ranges for each spectral band. If that pixel fails the AND-gate test for all of the target classes, it is labeled as a member of the Other class. The reason AND-gating logic is called the parallelepiped method is that the various $DN(i)$ ranges for each of the *n* spectral bands forms a parallelepiped volume in *n*-dimensional spectral space. The two-dimensional equivalent of a parallelepiped is a parallelogram.

Figure 5.14a shows the same graphical plot as Figure 5.12, with the $DN(i)$ maxima and minima forming a parallelepiped around each of the target classes. The above procedure determines whether U falls into one of these parallelepipeds. If so, it is given the classification of that target class; if not, it is given the classification of 7, or Other. Figure 5.14b shows a slightly different data set than Figure 5.14a for three different spectral classes. Notice that some of the classes overlap one another in Figure 5.14b, that is, their parallelepipeds intersect one another. What to do?

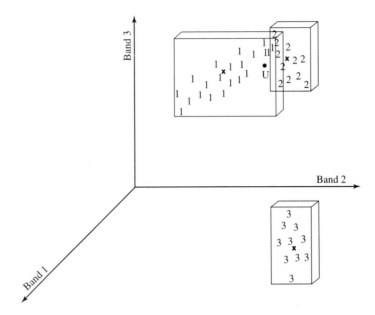

Figure 5.14 Parallelepiped classification. (a) Same plot as Figure 5.12, with *n*-dimensional space divided into parallelepipeds that have edges parallel to band 1, 2, and 3 axes. Here, the parallelepiped classification method has classified the unknown pixel (U) as a member of class 1. (b) Same plot, except that class 2 pixels overlap class 3 pixels. The overlap is usually resolved by placing the boundary between two classes halfway between the overlap region in each spectral band. *(Courtesy of R. K. Vincent.)*

There are two simple solutions to this dilemma. One is to classify U as the first target class tested into whose parallelepiped it falls. If the target classes are tested in decreasing order of likelihood of occurrence in the scene, this procedure makes some sense because the most likely occurring class will win all of the unknown pixels in the overlap region with a sparser class. The second simple solution is Solomonic: divide the overlapping region into equal parts, with each class obtaining its closest part. Thus, if class 2 has a range in $DN(2)$ of 15–20 and class 3 has a range in $DN(2)$ of 11–16, their class $DN(2)$ ranges are changed to 16–20 for class 2 and to 11–15 for class 3, and so on for the other $DN(i)$.

As shown in the previous section of this chapter, spectral ratios are more independent of variations in solar illumination and atmospheric transmission, thus more robust, than single-band intensity (image brightness) values. This robustness makes it desirable to use spectral ratios, rather than single-band intensities, as inputs to multispectral classification algorithms for those target classes for which color is a more unique signature than brightness. Ferric oxides, for example, are a target class that has more nearly unique color, than brightness. When spectral ratioing is performed as a preprocessing step, followed by parallelepiped multispectral classification, the resulting multispectral classification method is called ratio gating logic, which is just the AND-gating of spectral ratios. An example of parallelepiped classification with spectral ratio inputs is given for LANDSAT MSS data of the Wind River Basin and Range, Wyoming, in Color Plate CP6 (Vincent 1975b); CP4 is a color ratio image of the same area. Note that the Triassic redbeds, which are dark orange in the ratio image and red in the automatic classification image, can be distinguished from other ferric-oxide-rich rocks and soils in the image, which appear lighter orange in the ratio image and white (none of the above classes) in the classification image. This classification image was useful for intraformation mapping, where discriminations could be made among some formation members and facies that were not delineated on the best-published geologic map of the Wind River Basin at that time.

Ratio gating logic is especially important if the $q(i)C(i,j)$ term in Equation 5.6 has been determined, and the user wishes to employ $\rho(i)/\rho(j)$ from laboratory or field spectra as the training set. In this case, the maximum and minimum values of each spectral ratio (which defines the AND-gate for that spectral ratio) can be taken as 5% less and 5% more than the average spectral ratio value for the target class, as determined from laboratory or field reflectance spectra. Broadening the variance from 5% to 10% (or higher values) usually results in more of the target class being recognized, but more false alarms (false positives) may also result from broadening the spectral signature. An example of the employment of reflectance ratios from laboratory spectra for parallelepiped classification with ratio inputs is given by R. D. Dillman and R. K. Vincent (1974) for airborne data collected near Halloran Springs, California by an analog multispectral scanner. In that paper, the classification scheme is called ratio gating logic (RAGAL). This scheme will likely become more important as data from hyperspectral sensors with many, narrow spectral bands become more available. This availability is likely to increase because selection of spectral bands that coincide with and fall adjacent to absorption bands of the targeted mineral, rock, soil, or gas will be possible, resulting in a more unique "spectral signature" of the target. In addition, this form of parallelepiped classification is computationally much faster than other classification schemes described in this section.

There is a similar type of multispectral classification scheme called multiple discriminant analysis that involves a canonical component transformation of the training sets for all the target classes prior to application of the parallelepiped method. The canonical components are similar to the principal component axes described in the previous subsection, in that there are n orthogonal canonical components (where n is the number of spectral bands) that are all orthogonal to one another, and the mathematical calculation of both are done with the same algorithm. However, the inputs employed for calculation of the principal components are all of the pixels in the scene to be imaged (or a randomly subsampled set of pixels), whereas inputs for the canonical components are just the pixels in the supervised target classes. Instead of imaging each principal (or canonical) component, as is done in principal component imaging, ranges of each canonical component are AND-gated for each of the target classes. Figure 5.15 shows the parallelepipeds that are formed for each of the same target classes that were included in Figure 5.12. The unknown pixel, U, would be classified as a member of target class number 1. The advantage of the multiple discriminant analysis over the simple parallelepiped method is that the former results in greater classification accuracy than the latter because the canonical components separate the target classes better. There is the disadvantage,

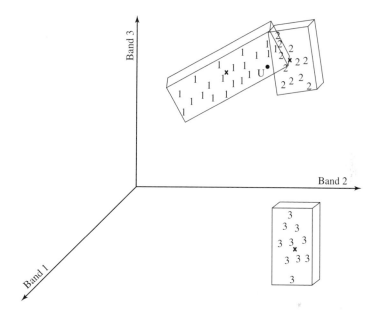

Figure 5.15   Canonical component transformation. Plot similar to Figure 5.12, with *n*-dimensional space divided into parallelepipeds that have been transformed to canonical coordinates before parallelepiped classification. The parallelepiped edges are parallel to the principal component axes of the training sets, in the directions of greatest variances for the training sets. *(Courtesy of R. K. Vincent.)*

however, that the canonical components are dependent on the target classes in an interactive way, such that if even one target class were to be added or subtracted, a different set of canonical components would most likely result. By contrast, there would be no effect on the simple parallelepiped method of changing the number of target classes, since the target classes are not interactive (each stands alone) in that method. Multiple discriminant analysis takes about the same amount of computer time as the simple parallelepiped method.

Maximum likelihood (Hall 1979) is a more complex multispectral classification algorithm that creates *n*-dimensional ellipsoids, rather than parallelepipeds, for each target class, as are shown in Figure 5.16 for the same three-dimensional ($n = 3$) case as in Figure 5.12. The *n*-dimensional ellipsoid for each target class is calculated to enclose 95% (or some other chosen value) of the pixels of a training set, such that the probability is 95% that any new pixel falling within that ellipsoid is a member of that particular target class. Ellipsoids do a better job of describing the envelope around a target class in multispectral space than the previously described parallelepipeds, primarily because the corners of the parallelepipeds are rounded off, thereby reducing the overall volume of the target class's multispectral signature in *n*-dimensional space. Maximum likelihood is especially helpful when the ellipsoids of two or more different target classes overlap in *n*-dimensional multispectral space, because it provides a statistical test for automatically deciding whether a pixel in the overlap region belongs to one class or another. The added step of weighting each target class

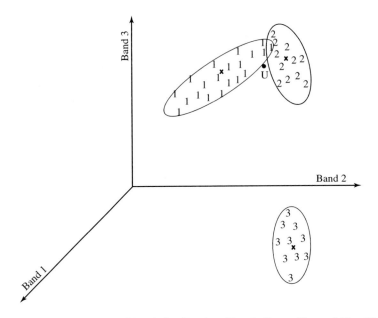

Figure 5.16 Maximum-likelihood classification. Plot similar to Figure 5.12, with ellipsoids defining the target classification space when maximum-likelihood classification is employed. Here the unknown pixel is classified as Other, and is not a member of any of the three target classes. *(Courtesy of R. K. Vincent.)*

according to expectations of how often it will occur in the scene and/or how costly it would be to misclassify that class converts a maximum-likelihood classifier into a Bayesian classifier. The additional class weightings improve upon the accurate classification of pixels in overlapping target ellipsoids.

Both maximum-likelihood and Bayesian classification methods have practical drawbacks. One disadvantage is that the maximum-likelihood and Bayesian algorithms require enough pixels in a training set with which to fit a Gaussian population curve, whereas some targets are so small that they tend to cover only a few pixels in areal extent and are sometimes entirely limited to only one pixel. Examples of this situation are unique rock outcrops associated with mineral deposits or limited areas affected by a contaminant spill. To qualify for maximum-likelihood or Bayesian classification, each target class must have at least $n + 1$ pixels, where n is the number of spectral bands. Such sparse training sets will not work with either of the two more complex methods. Another disadvantage, which happens to be shared by multiple discriminant analysis, is that the maximum-likelihood and Bayesian methods treat target classes in an interactive manner, such that changing one training set requires a recalculation of all of the n-dimensional ellipsoids for all target classes. A final disadvantage is that both maximum-likelihood and Bayesian classification take significantly more computer time (up to a factor of 10) to perform than parallelepiped or multiple discriminant analysis classification. This difference in computer run-time becomes ever greater as the number of spectral bands is increased.

Geologists will find that the parallelepiped method, especially with spectral ratio inputs, and the multiple discriminant analysis method are more practical supervised classification methods for geological mapping than the maximum-likelihood or Bayesian methods with data from multispectral sensors that have six or more spectral bands, which includes LANDSAT TM. For tasks like mineral exploration, where training sites are likely to be only a few pixels in areal extent, neither maximum-likelihood nor Bayesian classification can usually be applied.

All of the foregoing supervised classification schemes depend on the user's identification of class identities via the selection of a group of pixels (a training set) that represents each target class. Multispectral signatures of those identified classes are extracted, and the remainder of the scene is classified as one of the pre-identified classes or as Other. There is another approach, called unsupervised classification, or clustering, whereby the $DN(i)$ values of randomly sampled pixels from the scene are plotted in n-dimensional space (for n spectral bands) and clusters of spectrally similar pixels are identified by a mathematical procedure. The multispectral signatures of these clusters are extracted, and all the pixels in the scene are then classified as one of the clusters or as Other, if the number of clusters is truncated at some reasonable limit.

The most common clustering method, which requires the user to specify the number of clusters to be classified (sometimes confusingly called a supervised mode), is called K-means approximation. The first step in the K-means method is to locate a specified number of cluster centers. The second step is to assign each pixel in the image to the cluster whose mean vector is closest to the n-dimensional vector of that pixel, a process similar to the minimum-distance-to-mean method. Revised mean vectors are computed for each cluster, such that the sum of the squared distances from all pixels in a cluster to the new cluster center is minimized, and the image is then reclassified. This procedure is iterated until the class mean vectors stop changing significantly between iterations. Although the K-means algorithm does not actually converge, it is possible for the user to choose a practical upper limit or maximum for the number of iterations.

It is also possible to perform automatic clustering without a requirement for user selection of the number of clusters. If initial clustering according to the K-means method is performed using a guess for the number of clusters, the user can employ an iso2 clustering algorithm (Sauer and Koechner 1992) on the initial K-means cluster centers to automatically bring the iterated K-means/iso2 procedure to convergence. The iso2 algorithm discards clusters whose memberships (number of member pixels) are below a user-specified threshold, splits clusters that exceed a certain variance, and merges clusters together that are closer than a certain minimum distance. The K-means algorithm and the iso2 algorithm are reiterated until no clusters are discarded, split, or merged, or until the maximum number of iterations has been met. The splitting and merging requirements are made relative to the data set, and the iso2 algorithm is therefore more generally useful and more automatic than the ISODATA clustering algorithm from which it came. As a last step, the image is automatically classified into those final cluster classes.

This method is referred to as unsupervised because the user is not required to select either training sets or the number of clusters. However, after unsupervised classification is complete, the user must still identify what each spectral cluster represents by comparing pixels of a certain cluster class with "ground truth" from field trips or geological maps. Sometimes cluster class identifications can be performed by examination of the average DN value of each spectral band for each cluster class and comparison of those values with laboratory or field reflectance spectra. This comparison is easier to do if spectral ratios, rather than single-band intensities, are used as inputs to the K-means and iso2 algorithms.

The advantages to clustering are its automatic nature and the fact that the clusters are separated according to similar spectral properties in all n spectral bands, whereas supervised classification uses only identifications that were made on the basis of human eyes, which are limited to the visible wavelength region. The disadvantages include the requirement for identifying the clusters after the classification is completed and the long computer run-times required. (Several hours are typical for whole LANDSAT frames, with a work station computer.) The K-means/iso2 method takes much more computer time than the parallelepiped method and can often take considerably longer than even the maximum-likelihood method.

Additionally, there is one general disadvantage to multispectral classification as a whole, by any of the above methods, for geological remote sensing. Geological targets, usually rock outcrops and exposed soils, normally display considerable variability in vegetation cover and chemical composition within each target class and often can blend gradually into another adjacent rock or soil type. This characteristic makes continuous-toned color composite images and color ratio images suitable for interpretation by the user for drawing boundaries between lithologic units. Multispectral classification maps (sometimes called automatic recognition maps) divide the image into a limited number of discrete target classes. Usually, the division between adjacent lithologic members is incomplete in that pixels of the Other class (where mixtures of classes will often end up) normally exist between adjacent units on the ground. If the spectral signatures are broadened to make adjacent units contiguous in the multispectral classification map, many false alarms can and will occur in the rest of the image.

Nonetheless, two of the greatest utilities of multispectral classification methods for environmental scientists and geologists include the classification of vegetation land cover for surface water run-off models (see Chapter 9) and the search for outcrops of a few spectrally unique target classes, such as hydrothermal alteration products (e.g., alunite) for precious metal exploration or marker beds (e.g., redbeds) for stratigraphic mapping.

REVIEW

Equations from the previous chapter for spectral radiance coming from Earth's surface contained additive terms for atmospheric path radiance and electronic additive offset of the sensor that contained no information about Earth's surface. In the reflective wavelength region, short of 4.0 μm, the additive terms for the ith spectral band can be empirically determined and practically eliminated by the histogramming of an imaged scene, determination of the digital number $DN(i)$ of the darkest object in the scene, and subtraction of this digital number from every pixel in the scene. In the thermal infrared wavelength region beyond 4.0 μm, the atmospheric path radiance is assumed negligibly small, and the additive electronic offset of the sensor is measured by the comparison of radiances from hot and cold calibration plates of known temperatures in the sensor package.

There are two broad categories of multispectral image processing methods, image enhancement and image classification. The former yields an almost continuous-toned image, whereas the latter yields an image with only a relatively few discrete categories. Both methods, however, may display information from more than three spectral bands, even though there are only three primary colors that can be used to display the resulting images, such that the human eye can perceive them. Spectral ratio imaging and principal components transformation are two important types of image enhancement.

Once additive terms have been removed, multispectral data contains a great deal of information about the chemical composition of Earth's surface, especially when the additive-corrected radiance of one spectral band is divided by that of another spectral band for all the pixels in the imaged scene to form what is called a spectral ratio. Spectral ratio imaging can be used to reduce the variability of environmental parameters, such as solar illumination and atmospheric transmission, and to reveal the pixel-to-pixel variability of chemical composition across the scene being imaged. The production of images from spectral ratios that are selected for their compositional information (in or around spectral reflectance minima or maxima for a rock, mineral, soil) can be particularly useful for mapping certain geological targets of the user's choice. Spectral ratio imaging can also be useful for imaging gases, such as methane.

Principal components transformation is another method for image enhancement. It is useful for target discrimination, but it is less useful than spectral ratioing for target identification because it produces colors for a particular target that usually vary from scene to scene. These images are unlike spectral ratio images, where the same materials appear a more uniform color, regardless of the scene from which the images are produced.

Multispectral classification methods, both supervised and unsupervised, produce classification maps that display discrete colors for each target class. Supervised classification methods most frequently employed are the minimum-distance-to-means classifier; the parallelepiped classifier, which can be applied either to n-dimensional multispectral space (n is the number of spectral bands) or to canonical components; and the maximum-likelihood classifier, which can be intensified by the addition of Bayesian weighting rules. Unsupervised classification is most often performed by the K-means clustering method, which can be made automatic by the addition of the iso2 algorithm. The minimum-distance-to-means classifier does not address variance in the data, though it has relatively fast computer run-times. The parallelepiped multispectral classification method, which does account for variance in the data, is much faster, though somewhat cruder, than maximum-likelihood and K-means classification methods. Parallelepiped classification is also easier to implement with laboratory or field spectra as inputs, especially if spectral ratios instead of individual spectral bands are employed. With multispectral data sets that include six or more spectral bands, the speed of computation advantage of parallelepiped classification over maximum-likelihood and *K*-means classification becomes increasingly pronounced.

Generally speaking, image enhancement methods are more useful than multispectral classification methods for most geological remote sensing tasks. Nevertheless, multispectral classification methods are useful for geological remote sensing when a few spectrally unique targets are being sought, as in some cases of mineral exploration, or when vegetation cover maps are required for surface water run-off models.

Exercises

(*Note:* Appendix C contains LANDSAT TM and MSS brightness codes and ratio codes for laboratory reflectance spectra of many mineral samples and a few vegetation and snow samples. Explanations of those codes are also given in Appendix C. A high brightness code for a given target in a given spectral band means that the target will be bright in the image of that band. Likewise, a high ratio code for a given target means that the target will be bright in the corresponding spectral ratio image. This information is needed for some of the exercises below.)

1. **(a)** If you wanted vigorous grass to be red in a color composite of three LANDSAT TM spectral bands, what bands would you assign to the blue, green, and red display colors? (*Hint:* The red color should be assigned to a highest brightness-code spectral band for vigorous grass, and the blue and green colors should be assigned to spectral bands with lowest brightness codes for vigorous grass.)

   (b) What else would appear red in this image? (*Hint:* The bands displayed as blue and green must have very low codes, 0 or 1, and the band displayed as red must have a code of 3 or greater for a target to appear red in the display image.)

   (c) What color would hematite be in this image?

   (d) What TM spectral ratio would best separate hematite and vigorous grass?

2. What three LANDSAT TM spectral ratios would yield a color ratio image that made alunite red, yet that best separated alunite from hematite, kaolinite, and vigorous grass? Give the TM ratio codes of each of these four targets for the three TM spectral ratios in this color ratio image.

3. **(a)** What single band or single spectral ratio of either LANDSAT TM or MSS best separates the three grain sizes of snow: coarse, fine, and frost, i.e., very fine?

   (b) Answer the same question for separating conifers from vigorous and senescent (dry) grass.

Cited References

AVERY, T. E., and G. L. BERLIN. 1992. *Fundamentals of Remote Sensing and Airphoto Interpretation*. 5th ed. New York: Macmillan Publishing Co.

CRANE, R. B. 1971. Preprocessing Techniques to Reduce Atmospheric and Sensor Variability in Multispectral Scanner Data. In *Proceedings of the Seventh International Symposium on Remote Sensing of Environment*, 1345–1354. Ann Arbor: Environmental Research Institute of Michigan.

Note: Captions for Color Plates located on p. 367.

Color Plate 1

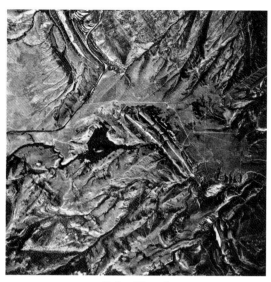

Color Plate 2

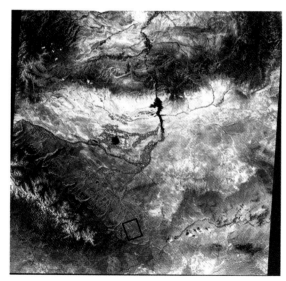

Color Plate 3

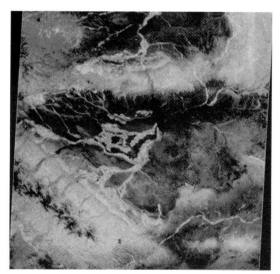

Color Plate 4

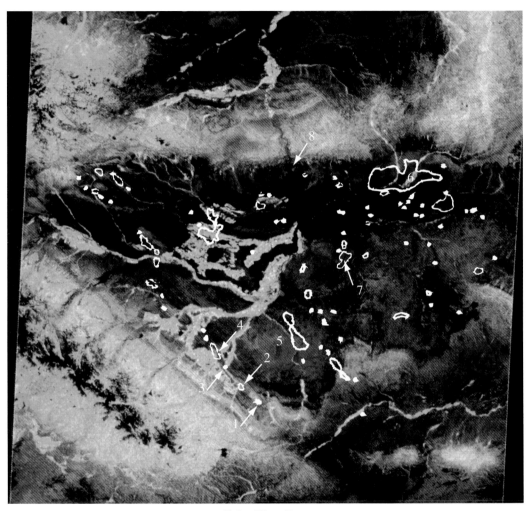

Color Plate 5

Color Plate 6

Color Plate 7

Color Plate 8

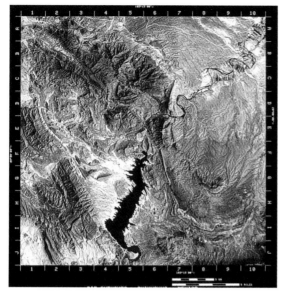

Color Plate 9a

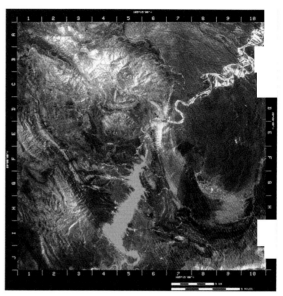

Color Plate 9b

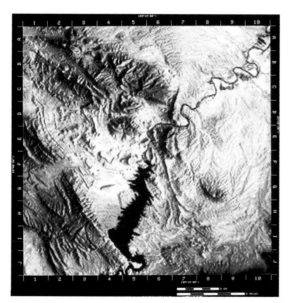

Color Plate 9c

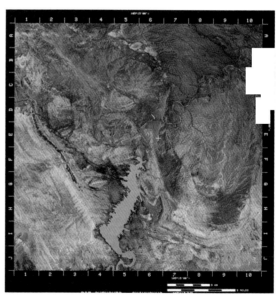

Color Plate 9d

Color Plate 10

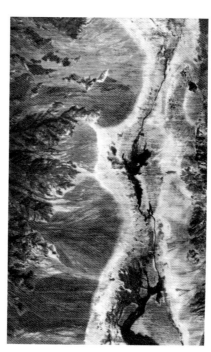

Color Plate 12

Color Plate 11a

Color Plate 11b

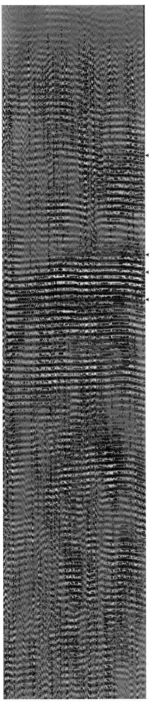

← Dundee Fm. (Devonian)

← A-2 Carbonate (Silurian)
← A-1 Carbonate (Silurian)

← Trenton Fm. (Ordovician)

Color Plate 13

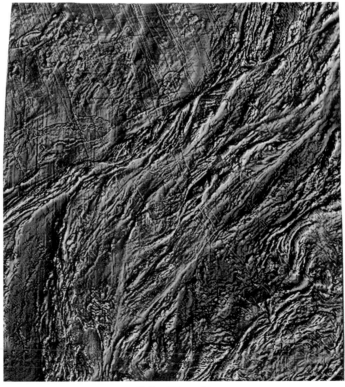

Color Plate 14

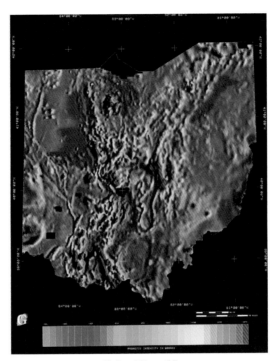

Color Plate 15

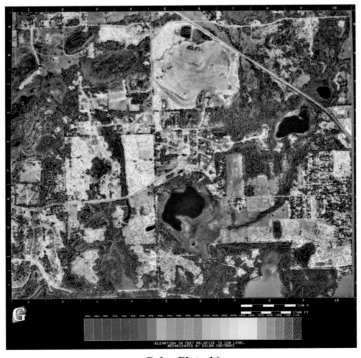

Color Plate 16

ELEVATION IN FEET REPRESENTED BY COLOR CONTOURS

Color Plate 17

ELEVATION IN METERS RELATIVE TO SEA-LEVEL.

Color Plate 18

CRIPPEN, R. E. 1986. The Regression Intersection Method of Adjusting Image Data for Band Ratioing. In *Proceedings of the Fifth Thematic Conference on Remote Sensing for Exploration Geology*, vol. 1, 407–416. Ann Arbor: Environmental Research Institute of Michigan.

DILLMAN, R. D., and R. K. VINCENT. 1974. Unsupervised Mapping of Geologic Features and Soils in California. In *Proceedings of the Ninth International Symposium on Remote Sensing of Environment*, 2013–2025. Ann Arbor: Environmental Research Institute of Michigan.

EARTH RESOURCE MAPPING PROPRIETARY LTD. 1995. *ER Mapper 5.0 Reference Manual.* West Perth, Australia: Earth Resource Mapping.

ERLEY, D. S., and B. H. BLAKE. 1964. *Infrared Spectra of Gases and Vapors, Spectra No. 4, 7, and 104.* Midland, Mich.: Chemical Physics Research Lab, Dow Chemical Company.

GILLESPIE, A. R. 1992. Enhancement of Multispectral Thermal Infrared Images: Decorrelation Contrast Stretching. *Remote Sensing of Environment* 42, no. 2:147–155.

HALL, E. L. 1979. *Computer Image Processing and Recognition.* New York: Academic Press.

HOOK, S. J., A. R. GABELL, A. A. GREEN, and P. S. KEALY. 1992. A Comparison of Techniques for Extracting Emissivity Information from Thermal Infrared Data for Geologic Studies. *Remote Sensing of Environment* 42, no. 2:123–135.

JOHANNSEN, A. 1931. *A Descriptive Petrography of the Igneous Rocks,* vol. 1. Chicago: University of Chicago Press.

LILLESAND, T. M., and R. W. KIEFER. 1994. *Remote Sensing and Image Interpretation.* 3rd ed. New York: John Wiley & Sons.

PICARD, M. D. 1965. Iron Oxides and Fine-Grained Rocks of the Red Peak and Crow Mountain Sandstone Members, Chugwater (Triassic) Formation, Wyoming. *Journal of Sedimentary Petrology* 35, no. 2:464–479.

RODGERS, K. A., R. H. A. COCHRANE, and P. C. LECOUTEUR. 1970. FORTRAN II and FORTRAN IV Programs for Petrochemical Calculations. *Mineralogical Magazine* 37:952–953.

SALISBURY, J. W., and D. M. D'ARIA. 1992. Emissivity of Terrestrial Materials in the 8–14 μm Atmospheric Window. *Remote Sensing of Environment* 42, no. 2:83–106.

SALMON, B., and R. K. VINCENT. 1974. Surface Compositional Mapping in the Wind River Range and Basin, Wyoming by Multispectral Techniques Applied to ERTS-1 Data. *Proceedings of the Ninth Symposium on Remote Sensing of Environment,* 2005–2011. Ann Arbor: Environmental Research Institute of Michigan.

SAUER, T., and D. KOECHNER. 1992. KHOROS Reference Manual, vol. 3, release 1.0, 2-305–2-308. University of New Mexico.

TORRES, V., R. K. VINCENT, and P. J. ETZLER. 1989. Integrated Mineral Exploration in Northern Mexico Using LANDSAT, Aerial Photography, and Ground Work. In *Seventh Thematic Conference on Remote Sensing for Exploration Geology,* 1227–1237. Ann Arbor: Environmental Research Institute of Michigan.

VINCENT, R. K. 1972. An ERTS Multispectral Scanner Experiment for Mapping Iron Compounds. In *Proceedings of the Eighth International Symposium on Remote Sensing of Environment,* vol. 2, 1239–1247. Ann Arbor: Environmental Research Institute of Michigan.

———. 1973. *A Thermal Infrared Ratio Imaging Method for Mapping Compositional Variations Among Silicate Rock Types.* Ph.D. University of Michigan, Ann Arbor.

———. 1975a. The Potential Role of Thermal Infrared Multispectral Scanners in Geological Remote Sensing. (Invited paper.) *Proceedings of the IEEE* 63:137–147.

———. 1975b. Commercial Applications of Geological Remote Sensing. *Proceedings of the IEEE Conference on Decision and Control,* 258–263. Houston, Tex.

————. 1977. Geochemical Mapping by Spectral Ratioing Methods. In *Remote Sensing Applications for Mineral Exploration,* ed. William L. Smith, 251–278. Stroudsburg, Pa.: Dowden, Hutchison, & Ross.

————. 1994. Remote Sensing for Solid Waste Landfills and Hazardous Waste Sites. *Photogrammetric Engineering and Remote Sensing* 60, no. 8:979–982.

VINCENT, R. K., and E. B. SINGLETON. 1994. *Methane Gas Concentrations Required for Infrared Imaging.* Final Report for Grant Sponsored by Hughes Santa Barbara Research Center, Department of Geology. Bowling Green State University, Bowling Green, Ohio.

VINCENT, R. K. and F. THOMSON. 1971. Discrimination of Basic Silicate Rocks by Recognition Maps Processed from Aerial Infrared Data. In *Proceedings of the Seventh International Symposium on Remote Sensing of Environment,* 247–252. Ann Arbor: Environmental Research Institute of Michigan.

VINCENT, R. K., P. K. PLEITNER, and M. L. WILSON. 1984. Integration of Airborne Thematic Mapper and Thermal Infrared Multispectral Scanner Data for Lithologic and Hydrothermal Alteration Mapping. In *Proceedings of the International Symposium on Remote Sensing of Environment,* vol. 1, 219–226. Third Thematic Conference, Remote Sensing for Exploration Geology. Ann Arbor: Environmental Research Institute of Michigan.

VINCENT, R. K., F. THOMSON, and K. WATSON. 1972. Recognition of Exposed Quartz Sand and Sandstone by Two-Channel Infrared Imagery. *Journal of Geophysical Research* 77: 2473–2477.

VINCENT, R. K., B. C. SALMON, W. W. PILLARS, and J. E. HARRIS. 1975. Surface Compositional Mapping by Spectral Ratioing of ERTS-1 MSS Data in the Wind River Basin and Range, Wyoming. In NASA Report CR-ERIM 193300-32-F, prepared by the Environmental Research Institute of Michigan, Ann Arbor.

WALTER, L. S., and J. W. SALISBURY. 1989. Spectral Characterization of Igneous Rocks in the 8–12 μm Region. *Journal of Geophysical Research* 94, no. B7:9203–9213.

WATSON, K. 1992. Spectral Ratio Method for Measuring Emissivity. *Remote Sensing of Environment* 42, no. 2:113–116.

WOLFE, W. L., and G. J. ZISSIS, eds. 1978. *The Infrared Handbook.* Prepared for the Office of Naval Research, Department of the Navy, by the Infrared Information and Analysis (IRIA) Center of the Environmental Research Institute of Michigan (ERIM), Library of Congress Catalog Card No. 77-90786, U.S. Government Printing Office.

Additional References (Uncited)

DILLMAN, R., and R. K. VINCENT. 1974. Unsupervised Mapping of Geologic Features and Soils in California. *Proceedings of the Ninth Symposium on Remote Sensing of Environment,* 875–895. Ann Arbor: Environmental Research Institute of Michigan.

VINCENT, R. K., 1973. Ratio Maps of Iron Ore Deposits, Atlantic City District, Wyoming. In *Symposium on Significant Results Obtained from the Earth Resources Technology Satellite-1,* 1:379–386.

————. 1973. Spectral Ratio Imaging Methods for Geological Remote Sensing from Aircraft and Satellites. In *Proceedings of American Society of Photogrammetry Management Utilization of Remote Sensing Data Conference,* 377–397. Sioux Falls, S. Dak.

VINCENT, R. K., and W. W. PILLARS. 1974. Skylab S–192 Ratio Codes of Soil, Mineral, and Rock Spectra for Ratio Image Selection and Interpretation. In *Proceedings of the Ninth*

International Symposium on Remote Sensing of Environment, 875–896. Ann Arbor: Environmental Research Institute of Michigan.

VINCENT, R. K., and F. THOMSON. 1972. Rock Type Discrimination from Ratioed Infrared Scanner Images of Pisgah Crater, California. *Science* 175:986–988.

———. 1972. Spectral Compositional Imaging of Silicate Rocks. *Journal of Geophysical Research* 77:2465–2471.

6

Spatial Image
Processing Methods

▼ ▼ ▼ ▼ ▼ ▼ ▼ ▼ ▼ ▼ ▼ ▼ ▼ ▼ ▼

INTRODUCTION

The previous chapters have been devoted to the gleaning of chemical composition information from multispectral data, information that requires the use of several to many spectral bands of data that cover the same area of interest on the ground. Spatial information that can be obtained from one spectral band alone gives other important information to the geologist and to the environmental scientist, particularly for mapping features that are surface expressions of underground structures. Structural mapping has practical applications in petroleum exploration, ground-water exploration, mineral exploration, engineering geology, and environmental geology. Spatial information does not have to come from cameras, linear arrays, or scanners, however. Automatic spatial image processing algorithms can extract useful information about geological structure from gridded geophysical data (topographic, magnetic, seismic, and gravity data) that twenty years ago would not have been considered candidates for image processing.

The most universal uses of spatial image processing involve geometric corrections, spatial filtering, and texture classification, which are described in another text (Jensen 1996) but will not be covered here. There are, however, three special types of spatial image processing applications to geological and environmental remote sensing that will be discussed in this chapter: automatic linear feature mapping, geophysical imaging, and automatic elevation mapping from digitized stereo image pairs, which is part of a relatively new field of study called digital photogrammetry.

The mapping of linear features suspected of occurring along fracture traces in Earth's crust was one of the earliest geological applications of image data collected from aircraft and satellite platforms; this mapping was performed primarily by manual photo interpretation. There are two types of fractures: faults and joints. Faults are cracks in rocks along which some motion has occurred. Joints are cracks along which no motion has occurred. Automated linear mapping does not replace the need for manual linear mapping, but it provides a faster, more objective way of producing statistical information about fracture density (such as the cumulative length of fractures per unit area on the ground) and regional trends of geological structures.

Contour maps produced from topographic, magnetic, seismic, and gravity data have for a long time been standard tools for mapping underground geological structures, with surface topography employed mainly for terrain corrections of gravity and seismic data. In the 1970s, shaded relief mapping, later called directional gradient-enhancement imaging, of gridded topographic data became possible, followed a few years later by similar treatment of well-gridded magnetic, seismic, and gravity data. Since then, geophysical imaging of all four types of data has become an important exploration and research tool.

Manual stereo viewing of stereo image pairs was also an early method for mapping geological structures, especially those that exhibit a pronounced three-dimensional surface expression. Semiautomated stereo viewing has also been important for the creation of maps of all scales. Digital photogrammetry, including the automatic creation of digital orthophotos and digital elevation models (DEM), is creating new technologies that will likely spill out of both the mapping and remote sensing worlds from which it came. Automation of the process for extracting elevation data from every pixel of a digitized stereo image pair is an important advancement for photogrammetric applications, for the mapping of geological structures that have subtle surface expressions, and for applications in several other technologies, including flight simulation, construction, and environmental monitoring.

All three of those applications can now be enriched by spatial image processing, the subject of the first three subsections of this chapter. The last subsection discusses the integration of remote sensing data sets with geographic information systems (GIS). This is an important consequence of digital photogrammetry, acting as a bridge between images of the remote sensing domain and digitized maps of the GIS domain.

AUTOMATIC LINEAR FEATURE MAPPING

Photo interpretation of linear features will probably never be totally replaced by automatic methods because of the daunting amount of artificial intelligence required to automatically separate linear features associated with important fractures from less geologically significant fractures or other linear features. However, there is a class of fracture mapping that is more suitable for automatic mapping than manual photo interpretation, namely, mapping as a search for statistically important fracture trends. When fractures are treated statistically for regional structural trend information, the significance of a particular fracture is less important than the number of fractures that trend in a particular direction. In the search for trends, although operator bias is reduced in statistical applications, photo interpretation is at some disadvantage because automated methods can be more objective and reproducible.

The first commercially available software package for the automatic mapping of linear features in a digital image was LIRA (*Li*near *R*ecognition and *A*nalysis). The following discussion comes primarily from a 1979 paper about LIRA (Coupland and Vincent 1981).

It is important to keep in mind the difference between a linear feature and a fracture. A linear feature is a straight line in an image. Some of those linear features are true surface expressions of fractures in Earth's crust, and some are not. Roads, power-line cuts, linear edges of forests, and other manmade linear features often do not have any relationship to faults and must be eliminated from analysis of structural trends. Many such manmade linear features, however, are much narrower than linear features associated with fractures, or they tend to fall into North–South and East–West directions that can be discriminated against automatically by automatic spatial processing methods. Especially in urbanized areas, discriminating other features requires limited editing by a photo interpreter before the automatic linear feature mapping results can be used reliably for statistical purposes related to structural trends.

LIRA operates on one spectral band of geometrically corrected digital image data and operates on differences in brightness, not color. LIRA divides the digital image, which is a rectangular raster matrix of brightness data, into an array of square subframes, the size of which is selected by the user, usually ranging from 50 pixels to 300 pixels square. Each subframe is searched for linear features separately. If a linear feature is shorter than a subframe, it will be extended to cross the subframe, and if a linear feature is longer than the selected subframe size, it will appear in several adjacent subframes and be expressed as a group of shorter linear segments. In practice, the subframe size is selected to match the approximate length of the fractures to be mapped.

For the purpose of automatic linear recognition, a linear feature is defined as a two-dimensional step function in brightness. A linear feature location is examined by correlating the subframe of image data with a linear model (in matrix form). That model is essentially a two-dimensional step function of the same azimuth and orientation as the linear features for which a search is conducted. To make the computations tractable, the azimuth range is searched in $2°$ steps. If the magnitude of the correlation exceeds a certain threshold specified by the user, a linear feature is recognized. By varying the correlation threshold required to recognize a linear feature, the user can control the strength of linear features to be mapped. Linear features with higher correlation coefficients (positive or negative) are said to have a greater linear strength.

The "detectability" of a linear feature depends on the ratio of the linear strength to the standard deviation of the background noise in the correlation numbers calculated for that subframe. This ratio is proportional to the square root of the subframe size in pixels. Thus, "weak" linear features (less brightness contrast, leading to lower linear strength) are more easily detected in larger subframes, if the linear feature crosses the entire subframe in the actual data. Conversely, short linear features become less detectable in larger subframes. Hence, there is a trade-off between spatial resolution and the ability to find weak or diffuse linear features, which can help in the discrimination of most manmade linear features (usually narrow, with high contrast) from diffuse linear features that are more likely to be associated with fractures.

The resulting linear features detected by LIRA can be translated to linear density information by an automatic calculation of the total line-length of linear features detected per subframe, divided by the area of that subframe. The result is often expressed as a linear density contour map, which is produced from the data grid consisting of the calculated linear density per subframe. The interpreter can then examine the linear density contour map to find the broad structural or lithologic trends in the area being investigated.

Figures 6.1 and 6.2, respectively, show a band 5 (covering the visible red wavelength region) LANDSAT MSS image of the La Ronge frame, in Canada, and a generalized geological map of the area, taken from existing geological maps. The most important faults, particularly the Needle Falls shear zone, are not easily identifiable in the LANDSAT image. Figure 6.3 shows a linear density contour map of the La Ronge frame, produced from linear features recognized by the LIRA computer program. The numbers on the contour map are pixels of linear feature length per subframe, which can be converted to meters of linear features per square kilometer by multiplying them by 0.3659. Figure 6.4 is a manual interpretation of the linear density contour map, with darker areas having higher linear density. Comparison of Figures 6.4 and 6.2 shows that the most important fault zones, including the Needle Falls shear zone, are associated with linear density highs (Coupland and Vincent 1981).

The linear features detected by LIRA can also be segregated according to direction, yielding directional linear densities, or can be used to create Rose diagrams for different subframes or groups of subframes in the image. Both of these types of

Figure 6.1 A LANDSAT MSS band 5 image of the La Ronge frame in Northwest Territories, Canada. This covers an area of approximately 185 km × 185 km. *(After Coupland and Vincent 1981. Reprinted with permission of the Environmental Research Institute of Michigan.)*

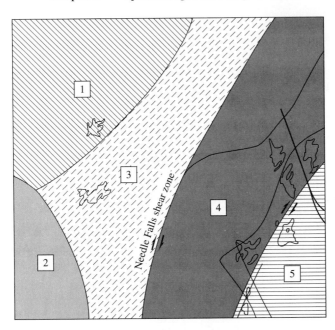

Figure 6.2 Generalized geology of the La Ronge, Canada frame. Geologic regions 1 to 5 are described in the text (same region as shown in Figure 6.1). Heavy lines indicate previously known faults. *(After Coupland and Vincent 1981. Reprinted with permission of the Environmental Research Institute of Michigan.)*

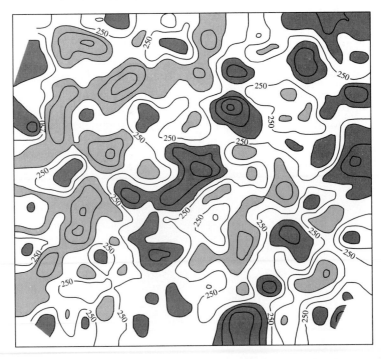

Figure 6.3 Linear density contour map of the La Ronge, Canada frame (same region as shown in Figure 6.1), from linear features automatically mapped by the LIRA software package. *(After Coupland and Vincent 1981. Reprinted with permission of the Environmental Research Institute of Michigan.)*

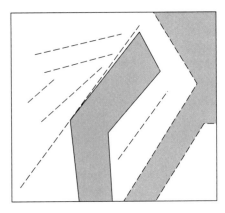

Figure 6.4 Manual interpretation of the linear density contour map in Figure 6.3, produced from LIRA linear features (same region as shown in Figure 6.1). Dashed lines are prominent trends in linear density. Shading indicates areas of high linear density, which areas have lower linear density. *(After Coupland and Vincent 1981. Reprinted with permission of the Environmental Research Institute of Michigan.)*

information can be useful in finding boundaries where structural trends change. Sometimes these changing trends indicate underlying lithologic boundaries or they may provide information related to lateral differences in structural stress.

Automatic linear feature mapping can be applied to any type of raster image, including images from topographic, magnetic, and gravity data of the type to be discussed in the next section. The statistical analysis of linear features discussed above has also been applied with good results to photo-interpreted linear features that have been digitized. For example, Marrs et al. (1984) and Martinsen and Marrs (1985) used various directional trends of photo-interpreted linear features to determine regional structure and facies control in the Powder River Basin of Wyoming and Montana.

It is a difficult research task to prove the validity of an automatic linear mapping program because many more linear features are typically mapped by programs like LIRA than are known to be actual mapped faults or joints. However, LIRA linear density highs in terrain shallowly underlain by the Devonian-aged Antrim shale (about 600 m deep) in Otsego County, Michigan have been found by experience to contain more natural gas than other areas in the same county, with increased production of 100%–400% higher than surrounding production from the same shale. Because the Antrim shale is an oil shale that is impermeable and nonporous except for where it is fractured (breccia reservoirs), this is an indication that many of the linear features mapped by LIRA are true fractures. Research is underway to find independent geophysical confirmation of linear features mapped by LIRA in the Bowling Green, Ohio area, which includes known traces of the Bowling Green Fault Zone.

GEOPHYSICAL IMAGING

Magnetic and gravity data collected by aerial magnetometer surveys and ground or airborne gravimeter surveys are measurements collected along parallel lines or at randomly spaced points. Traditionally, such data are used to create a contour plot from the collected data points, with interpolation methods used to determine magnetic or gravity field strength for points in between data collection points. Topographic data (terrain elevations) traditionally have been collected directly in the form of contour plots from stereo photos. The resulting contour plots of gravity, magnetic, or topographic data are difficult to interpret visually, usually requiring the assistance of an experienced geophysicist or photogrammetrist if gradient or slope information is desired.

The first attempts at imaging geophysical data simply involved the creation of an evenly spaced grid by the calculation of a value for each grid element through interpolation of the nearest data collection points to each grid element. An image was then produced of the resulting raster grid (with evenly spaced grid elements in two dimensions), with high-to-low values of the geophysical datum displayed as bright-to-dark for a black-and-white image, or as red-to-purple for a color image. In this type of image, level-slicing the geophysical datum value replaces contouring and only marginally simplifies interpretation. The determination of gradients in the geophysical datum, which is the most difficult part of contour map interpretation, is not improved by this type of imaging.

The first imaging method designed for improving the interpretation of gradients was applied to digital elevation data. In 1975, R. M. Batson, K. Edwards, and E. M. Eliason of the U.S. Geological Survey published a method for producing shaded relief images of government-provided digital elevation models, or DEM, that turned a raster grid (data values aligned in a two-dimensional grid) of elevation postings into what looked like a low-sun-angle photograph or a radar image. The user was allowed to select the azimuth and elevation angle above the horizon of the "source of illumination," and the shaded relief algorithm produced an image that displayed topographic slopes facing toward the "source of illumination" as bright features, while displaying slopes facing away from that direction as dark. The resulting image could also be called a directional gradient-enhancement image, with the user selecting the direction of gradient to be enhanced. The highest positive gradients were displayed as bright portions of the image, and the highest negative gradients were displayed as dark, with flat areas (gradient near zero) displayed as a medium tone of gray.

Examples of directional gradient-enhancement images of topographic data are shown in the next four figures. Figure 6.5 is a directional gradient-enhancement image produced from a 100-meter-posted DEM of the Charleston $1° \times 2°$ quadrangle map (designated NJ 17-5 by the U.S.G.S.) covering 38°–39° north latitude and 80°–82° west longitude. More than 90% of the area covered in Figure 6.5 is in West Virginia, with Charleston located in the lower-central part of the westernmost half. The "sun" position was specified to be 30° above the horizon in the Northwest (upper left), and NW-facing slopes are bright. The DEM from which Figure 6.5 was produced is said to be 100-meter-posted because it is a raster grid of posted elevation values, each separated from its nearest neighbors in both x and y directions by 100 meters. However, most of the posted elevation values have been interpolated between elevation contour lines and are not measured values. A 100-meter-posted, digital elevation data file is called a Level 1 DEM in the civilian sector, or Level 1 DTED (for digital terrain elevation data) in the military services. These files can now be downloaded free via the Internet in digital form from the U.S.G.S. for the entire United States.

Figure 6.5 Directional gradient-enhancement image (with "illumination" from the lower-right corner, or SE) of the Charleston, West Virginia, 1° × 2° quadrangle topographic map (lat. 38–39° N and long. 80–82° W), with image dimensions of 110 km × 175 km and north toward the top. This image was made from a digital elevation model (DEM) with 100-meter posting. *(Courtesy of GeoSpectra Corp.)*

More detailed DEMs with 30-meter-postings are also available in much of the United States for 7.5-minute quadrangle maps, each comprising exactly 1/128th of the area covered by a 1° × 2° quad map. Figure 6.6 shows a directional gradient-enhancement image of the Fayetteville 7.5-minute quad map, which covers the region between 38°0'–38°7.5' north latitude and 80°52.5'–81°0' west longitude, with the "sun" direction selected at 30° above the horizon in the NE direction (upper-right corner), or 90° from the sun direction in the previous figure. The New River diagonally bisects this image, with Fayetteville, West Virginia located near the center of the lower-right (SE) quarter of the image. Note the much finer detail that can be seen in Figure 6.6, compared to the same area in the lower-right 1/64th of the westernmost half of Figure 6.5. Figure 6.7 shows a directional gradient-enhancement image of the same Fayetteville 7.5-minute quad area, but with the selected "sun" direction in the SE (lower right). A comparison between Figures 6.6 and 6.7 demonstrates that a directional gradient-enhancement image primarily enhances linear features that trend perpendicular to the selected "sun" direction (the direction in which the gradient is calculated). All linear trends that are within 45° of being perpendicular to the "sun" azimuth are enhanced well enough to be seen clearly, provided that the data has sufficient spatial resolution

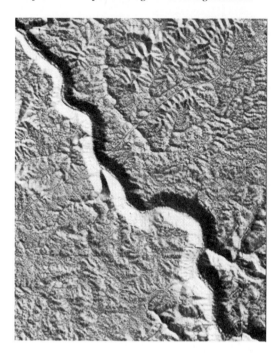

Figure 6.6 Directional gradient-enhancement image (with "illumination" from the upper-right corner, or NE) of the Fayetteville 7.5-minute quad topographic map, which covers 1/128th of the map in Figure 6.5, just left of the center of the bottom of Figure 6.5. Coverage is lat. 38°0′–38°7.5′ N, long. 80°52.5′–81°0′ W.

Figure 6.7 Directional gradient-enhancement image (with "illumination" from the lower-right corner, or SE) of the Fayetteville 7.5-minute quad topographic map, which covers 1/128th of the map in Figure 6.5, just left of the center of the bottom of Figure 6.5. Coverage is lat. 38°0′–38°7.5′ N, long. 80°52.5′–81°0′ W.

to make them visible by this type of imaging. Therefore, the employment of two directional gradient-enhancement images, with the two "sun" directions 90° apart, is a wise choice for each map area of interest because all linear trends will be enhanced in one image or the other.

In the previous section of this chapter, it was remarked that the automatic linear feature mapping program called LIRA could be applied to any raster grid of digital data. As proof, Figure 6.8 is the same image as shown in Figure 6.6, except that it is overlain by linear features that were automatically recognized and extracted by LIRA from the Fayetteville 7.5-minute quad DEM. Since LIRA already searches for linear features that trend in every 2°-increment between 0° and 180° (since 180°–360° is redundant with 0°–180°) and it produces its own local contrast-stretching, it works better on DEM directly than on directional gradient-enhancement images; these would limit the linear features that LIRA recognizes. It is always possible to segregate linear features found by LIRA into different classes of azimuthal range after LIRA has recognized linear features with little or no bias.

In 1980, GeoSpectra Corporation produced the first directional gradient-enhancement image of gravity data for a region in Pennsylvania and one of magnetic data for a region in the Wollaston Basin, Canada. The procedure involves gridding the gravity or magnetic data into a relatively fine (small-pixel) raster grid and applying the directional gradient-enhancement algorithm as if the magnetic or gravity data were elevation data. Figure 6.9 shows two black and white directional gradient-enhancement images of the state of Pennsylvania (the first gravity image ever produced) for two orthogonal gradient directions. Likewise, Figure 6.10 shows two black and white directional gradient-enhancement images of aeromagnetic data for two orthogonal gradient directions in the MacDonald Fault area of Northwest Territories, Canada. The narrow, "wormy" lineations are actually diabasic dikes intruded into slightly less magnetic country rock, none of which are observable at the surface because of glacial soil cover. The MacDonald Fault itself can be seen to be a right-lateral fault trending SW–NE near the middle of the image. The dikes are clearly younger than the fault because they are not displaced on opposite sides of it. It is clear from Figures 6.9 and 6.10 that a geologist can interpret a geophysical image of this type for geological structural information without requiring the assistance of an experienced geophysical data interpreter. A color version of this image, shown in Color Plate CP14, was later produced with a hue-saturation-intensity algorithm, where color (red, grading through yellow and green to purple) was controlled by the total magnetic field strength and image brightness was controlled by the black-and-white directional gradient-enhancement image.

Another example of a color aeromagnetic directional gradient-enhancement image is shown in Color Plate CP15 for the entire state of Ohio. The North-trending Bowling Green Fault can be seen as a nearly vertical linear feature in the upper-left quadrant of this image. The Bowling Green Fault has been proposed to be controlled by and coincident with the western boundary of the Grenville Front, a Precambrian-aged basement feature (Onash and Kahle 1991).

Geophysical images of the types demonstrated above are extremely useful for mapping deep-seated structures (Herman et al. 1991) or for mapping lithologic

Figure 6.8   The same image as in Figure 6.6 for Fayetteville 7.5-minute quad with LIRA linear features overlain on the image. The LIRA computer program was applied to the Fayetteville DEM directly, not to the directional gradient-enhancement image. Coverage is lat. 38°0′–38°7.5′ N, long. 80°52.5′–81°0′ W. *(Courtesy of GeoSpectra Corp.)*

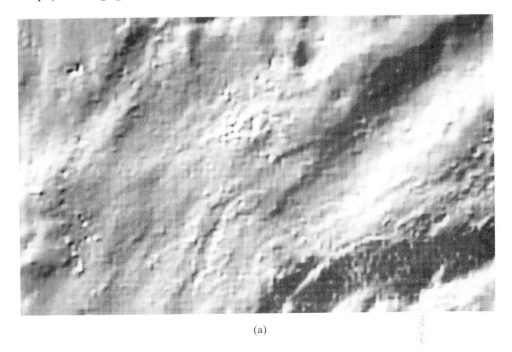

(a)

(b)

Figure 6.9 Gravity directional gradient-enhancement image of the state of Pennsylvania with (a) "illumination" from the SE (lower-right corner) and (b) from the NE (upper-right corner). *(Courtesy of GeoSpectra Corp.)*

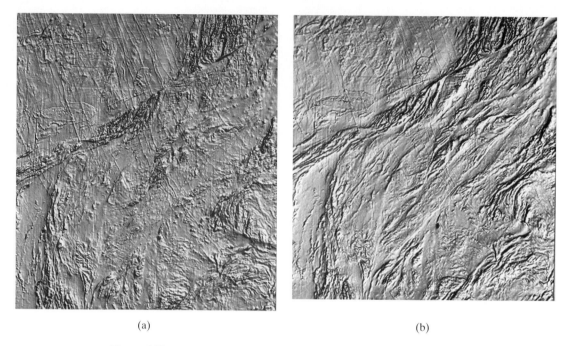

(a)   (b)

Figure 6.10 Magnetic directional gradient-enhancement image of the McDonald Fault, Northwest Territories, Canada, with (a) "illumination" from the NE (upper-right corner) and (b) from the SE (lower-right corner). The area covered is lat. 60°–64° N and long. 104°–112° W. North is toward the top. The McDonald Fault trends NE-SW in the top half of the image. The "wormy" features almost perpendicular to the fault are diabasic dikes that have more iron content than the older rocks into which they intruded. This area is covered by glacial till, which hides these features from visual observation.

contacts in basement rocks, either onshore or offshore. In fact, aeromagnetic gradient-enhancement imaging is the most cost-effective prebidding tool for assessing the occurrence of underground structures favorable for oil and gas accumulation on offshore tracts. If geophysical imaging is applied prior to seismic data collection, which is at least an order of magnitude more costly and far more time-consuming, seismic lines can be selected that will optimally cover the most important structures in the tract, thereby avoiding areas where there is little evidence of subsea structure. More about the use of geophysical images in exploration will be discussed in later chapters.

Even the current methods applied to three-dimensional seismic surveys owe their roots to the application of image processing techniques to seismic data. Perhaps the earliest published example of gridding seismic returns from the top of a targeted formation and producing a three-dimensional representation of the surface of that formation is shown in Figure 6.11 (Vincent and Coupland 1980). These surface portrayals of the Devonian-aged Dundee Formation, taken from gridded seismic return times of the top of the Dundee from a few seismic lines, were used to confirm the pres-

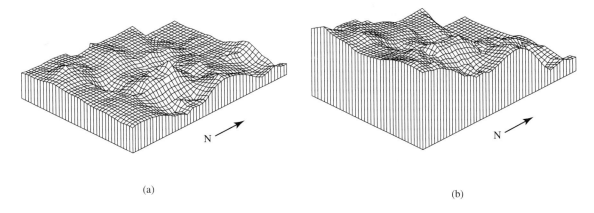

(a)

(b)

Figure 6.11 Three-dimensional plots of the topography of the Devonian Dundee Formation (a) and a Cambrian sedimentary unit (b) as calculated from seismic data for the Bay County, Michigan study area. These Devonian and Cambrian units are estimated to be approximately 1 km and 3.5 km below the surface, respectively. The area covered by these plots is approximately 9 km × 10 km. *(After Vincent and Coupland 1980. Reprinted with permission.)*

ence of a fault mapped from photo interpretation of a LANDSAT image in Bay County, Michigan. A multifrequency version (called Spectraseis by GeoSpectra) of a 4.5-mile-long, North-trending seismic line in the northern part of Bay County, Michigan (Fraser Township) is shown in Color Plate CP12, where frequency bands of 30–34 Hz, 34–38 Hz, and 46–50 Hz from Vibraseis data are displayed in colors of green, red, and blue, respectively (Vincent 1979). The region between the A-2 carbonate of Silurian age and the Trenton Formation of Ordovician age along the southern half of the seismic line shows lack of blue, the highest frequency band, and this finding may indicate gas at that depth. Although Devonian rocks were the deepest ones tested along this seismic line, there have been subsequent deep gas discoveries very near there (within a mile or two). Intriguingly, in CP12 there appears to be a "plume" of similar regions of high-frequency loss extending downward from the larger, shallower high-frequency loss region described above, probably extending to Cambrian-aged rocks, and perhaps even below them. More is said about CP12 in Chapter 8 under the subsection entitled "Structural Mapping," in conjunction with Figure 8.5.

DIGITAL PHOTOGRAMMETRY

Photogrammetry, which involves the extraction of geographic information from photographs, has long been an important source of the information required for map making. Elevation data is one fundamental type of geographic information that is extracted from photographs by photogrammetric methods. The United States military

(Defense Mapping Agency, or DMA) and civilian (U.S. Geological Survey, or U.S.G.S.) mapping agencies have provided digital topographic maps and digital raster files of elevation data for public use at low cost in the continental United States since the 1970s. The raster files of elevation data are called digital elevation models (DEM) by the U.S.G.S. and digital terrain elevation data sets (DTED) by the DMA. Those elevation data sets were extracted with a high degree of manual labor from high-resolution images. The elevation data and source images were separated, after the spatial resolution of the elevation data was considerably lowered from that of the source images.

Digital photogrammetry is the digital extraction of geographic information from photographs. The automatic extraction of elevation data from digitized stereo images is a method that is less than two decades old. Its development began on special-purpose computers; its development was greatly accelerated when general-purpose minicomputers (small mainframes) became available and work was begun for a software solution to the problem. Digital elevation extraction software is now available for off-the-shelf workstation computers and personal computers. This availability greatly facilitates the use of the software for remote sensing and geographic information system (GIS) database generation.

Such software is especially important for adding remote sensing images to GIS databases, a topic of the next section of this chapter. The software adds important information because no aerial photo or multispectral scanner image can be made to overlay directly on digitized map data (a GIS database) with accuracies anywhere close to one pixel of the digitized photo or scanner image (usually less than 3 ft and often less than 0.5 ft) without rigorous correction for parallax , which can vary from pixel to pixel. (Parallax is discussed in the following paragraphs.) Even high-order polynomial transformations, such as rubber-sheet stretching, cannot make a remote sensing image (photo or scanner image) overlay a digitized map base with an error as low as one pixel of the image, unless the pixel size in the digital image is so large that small, high-relief features, such as individual buildings, cannot be well resolved. Before further GIS and geological significance of digital photogrammetry can be discussed, however, it is necessary to take a brief detour into the technical world of photogrammetrists.

Two photographs or images that overlap each other, called a stereo pair, view the overlapped region from two different portions of the same camera lens, giving rise to parallax in two different directions. Parallax is the apparent displacement of an object in the x-dimension and y-dimension of the image coordinates caused by the object's elevation, or z value, above a selected datum plane, such as the average elevation of the overlapped region. How those directions of parallax differ from one image to the other image in a stereo pair depends on the geometry of the lens and imaging apparatus. In aerial photography (film) or with two-dimensional solid-state arrays, central perspective geometry prevails, and parallax is always in a direction away from the center of the lens in each of the two stereo pair images, as sketched in Figure 6.12. When a human operator views a stereo pair in a device called a stereo-comparator, the eye–brain combination correlates the two images and translates parallax into a three-dimensional image. This process is similar to the way that humans achieve depth perception by correlating images produced from the left eye

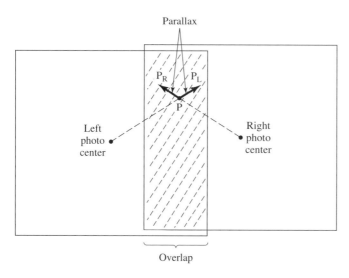

Figure 6.12 Left and right photos of a stereo pair, with stereo overlap region denoted as diagonally lined. Point P, if it has an elevation above the median plane of the ground area in the overlap region, will be displaced to P_L in the left photo and P_R in the right photo. Displacement (called parallax) is away from the center of the respective photos. *(Courtesy of R. K. Vincent.)*

and right eye, translating parallax almost instantly to relative distances away from the center between the eyes to all points in the scene. The smaller the distance an object is from the center between a human's eyes, the greater the parallax.

If the two images of the stereo pair are digitized (by a laboratory scanner, when the original images are photos instead of digital images), each image is broken up into picture elements called pixels. For instance, if a typical 9-in.-square aerial photo is scanned with a 25 μm spot size (about 1000 dots per inch—dpi), the photo is broken up into a raster array of about 9000 by 9000 pixels (81 million in all), each pixel representing the image brightness in a 25 μm $\times$ 25 μm area on the photo. The scale of a photo is given by the ratio of the actual distance on the photo between two points and the distance on the ground between those same two points. For instance, if 1 in. on the photo represents 10,000 in. on the ground, the scale (S) is given by the ratio 1/10,000, which is often written as 1:10,000. Therefore, the linear pixel dimension on the ground, X, is determined by multiplying the linear dimension of the pixel on the photo, x, which is also the scan spot size, by the reciprocal of the scale, or $X = x/S$. For example, if x is 1/1000 in. (for 1000 dpi), the linear dimension in inches of a pixel on the ground is $X = (1/S)(1/1000)$. Thus, for a 1:10,000 scale photo scanned at 1000 dpi, a pixel on the ground is a square 10 in. by 10 in. in size.

It is quite helpful, both conceptually and computationally, to first transform the coordinate system of each of the digitized stereo pair images into a coordinate system called epipolar space, which is a coordinate system where all parallax is in the x-dimension only, with no y-parallax. Not all digital photogrammetry software takes this step first, but the best ones do. Of course, a different mathematical model for epipolar space is required for every type of lens-sensor geometry. A discussion of the various models is beyond the scope of this text. For the remaining explanation of digital photogrammetry included here, it will be assumed that both images of the stereo pair have been co-registered (mapped onto one another) and transformed into epipolar space. The left and right images will be referred to as "epipolarized" when this condition is true. The following discussion is intended for a user of digital photogrammetry, not for the developer of such a system.

Automatic elevation extraction software in a computer automatically correlates the left image with the right image, after an operator assists the computer by picking a few (usually from six to ten) pixels representing the same photographic features in the right image and the left image. These pixels are called match points. An operator usually also finds control points in the two images for which the ground coordinates (including the elevation) are known. Finding these points is a requirement if the resulting elevation database is to be tied to a given map coordinate system. The control points also are commonly used as match points. The left and right images are then mapped onto one another with a resampling algorithm that would overlay the images perfectly in epipolar space if all points in the scene were to be on the same flat datum plane and had no elevation differences.

An automatic correlation algorithm, and there are many from which to choose, correlates the left image to the right image by taking a small window of pixels (called a correlation window) in the right image as the master and finding the same-sized matching window in the left image (the slave) that best correlates with the master. The difference between where the matched pixel (the center of the correlation window) is actually located and where it would have been located if the pixel was located on the datum plane is the parallax, which is directly proportional to the elevation of that pixel above or below the datum plane.

The next window selected in the master image is separated from the first master window by an interval of pixels called the posting interval. Parallax measurement will occur only once for each posting, which is separated by the posting interval from the next posting. One pixel of parallax, P, is the unit of elevation quantization corresponding to one pixel of displacement in the x-dimension of the left epipolarized image of the stereo pair, compared to the same photographic feature on the same line of pixels (the y-dimension) in the right epipolarized image. The relationship between one pixel of parallax and the X-pixel dimension is given by the following equation:

$$\frac{P}{X} = \frac{H}{B} \qquad\qquad \text{(Eqn. 6.1)}$$

where
  P = the elevation difference represented by one pixel of parallax
  X = the linear pixel dimension on the ground
  H = the height, or altitude of the camera above the ground
  B = the base, or distance between the camera centers when the two stereo images were collected

It is clear that as long as H and B are in the same units of length, P will have the same units of length as X.

Most digital elevation extraction software packages use a posting interval of many pixels, usually ranging from 10 to 100 pixels, because their correlation algorithms are computationally slow, usually on the order of 100–200 pixels of elevation extraction per second on a Sun Sparc II workstation. This can be referred to mathematically as a sparse-net representation of the topography of a surface. One software package, called ATOM (a registered trademark of GeoSpectra that stands for Automatic Topographic Mapper), can extract about 5000 elevations per second on a Sun Sparc II workstation. This makes it possible to attempt an elevation extraction for every pixel (a posting interval of 1 pixel) in the overlap region of a digitized stereo pair that has been scanned with a density of 1000 dots per inch in about 9000 seconds, which is less than 3 hours. ATOM yields a dense-net representation of the topography of a surface.

The root-mean-square error in elevations measured by automatic correlation of the left and right images is directly proportional to the size of the pixels of the digitized images. Empirically, the following relationship has been found to be approximately correct for dense-net data sets, when the control points are known to an accuracy within one pixel:

$$z(rms) = \left(\frac{3}{4}\right)P = \left(\frac{3}{4}\right)\left(\frac{XH}{B}\right) \qquad \text{(Eqn. 6.2)}$$

where

$z(rms)$ = the root-mean-square error in elevations measured by automatic correlation

Of course, this error in elevation measurement applies only to the measured elevations at each posting. If interpolation algorithms are used to estimate the elevations between postings, as sparse-net representations usually require, there will be an additional error of interpolation that depends on how much the surface undulates between posting intervals. If, however, the digital elevation model (DEM) is produced with a 1-pixel posting interval (a dense-net representation of the surface topography), there will be little or no error of interpolation. Of course, some pixels on the ground will be featureless and will not be correlated by any posting interval, thereby requiring some limited interpolation in the best of cases. Equation 6.2 was empirically derived from elevation data sets produced for several scales of stereo images by ATOM, which employs a 1-pixel posting interval.

In many situations, the user will be presented with a requirement for a certain elevation error, or $z(rms)$, and will be given a choice in the scale of stereo photos to be collected. The purpose of the following discussion is to provide the user with guidelines for selecting the scale of photography (or, alternatively, the height H of the platform that contains a metric camera with lens focal length f) such that a certain $z(rms)$ can be achieved by automatic correlation, if a standard base-to-height (B/H) ratio (usually about 0.6 for optimum results) is used in data collection. The relationship between the scale and the f/H ratio is as follows:

$$\frac{x}{X} = \frac{f}{H} \qquad \text{(Eqn. 6.3)}$$

In Equation 6.3, x, X, f, and H are all in units of length.

If Equation 6.2 is solved for H and Equation 6.3 is used to substitute for X in Equation 6.2, the result can be written as

$$H = \left(\frac{4}{3}\right)\left(\frac{f}{x}\right)\left(\frac{B}{H}\right)z(rms) \qquad \text{(Eqn. 6.4)}$$

In Equation 6.4, H is allowed to be written on both sides of the equation to permit easy substitution for the base-to-height ratio (B/H). In common practice this ratio is usually $3/5 = 0.6$ for stereo photos collected by an aerial survey firm.

Here is practical example of how Equation 6.4 can be used. Suppose the operator, who can order photos collected by a metric camera with a 6-in. focal length ($f = [1/2]$ ft) and plans to have the photos scanned with a 1000 dpi scanner ($x = [1/1000]$ in. $= [1/12,000]$ ft), is told by a customer that a DEM with accuracy of 1 ft, or $z(rms) = 1$ ft, is required. What height above ground should the camera be flown to achieve that accuracy, assuming that the standard B/H ratio is to be employed ($B/H = 3/5$)? Substituting these values into Equation 6.4 and solving for H yields a height above ground of 4800 ft. The values for $f = [1/2]$ ft and $B/H = 3/5 = 0.6$ are common for metric cameras that are typically used for aerial photography. However, even for a metric camera on a tethered balloon, Equation 6.4 can be used to determine the height of the balloon required for a specified $z(rms)$, though the focal length of the camera is usually less than for an aerial camera and the photo scanner spot size can be much less than 1000 dpi (such as 300 dpi $= [1/300]$ in.). A B/H ratio of $3/5$ is optimal for the automatic correlation of stereo photos, regardless of photo scale. It should be remembered, however, that f and x should be in the same units of length. H will have the same units of length as was specified for $z(rms)$.

There are two important digital photogrammetry products that result from the processing of a digitized stereo pair by ATOM software. One is an every-pixel-posted DEM and the other is a digital orthophoto, both perfectly co-registered and in orthonormalized map space (if a few known ground-control points are available in the desired map coordinates). Orthonormalization, which removes parallax, and map projection resampling causes both the DEM and the digital orthophoto to be overlaid within about one pixel of accuracy on the digitized map data, assuming the latter is correct. Thus, every-pixel-posted DEM has become the bridge between remote sensing image data, in this case aerial photography, and GIS databases.

Two examples of digital orthophoto images with overlaid color contour maps will now be given that were generated from every-pixel-posted DEM and that ATOM extracted from two digitized aerial stereo pairs of different spatial resolutions. The first, in Color Plate CP16, shows a digital orthophoto overlaid with a 10-ft-interval contour map of a manmade ski slope called Mt. Brighton, near the town of Brighton, Michigan, about 18 miles north of Ann Arbor (James 1990). Original scale of the stereo photos was about 1:18,000, and $z(rms)$ was approximately 2 ft.

The second example, shown in Color Plate CP17, was produced for Gary Stonerock, owner of Air-Land Survey in Clarkston, Michigan from two photos that were at 1:4800 scale for an area in a northern suburb of Detroit. The $z(rms)$ was estimated to be about 6 in., even though the contour intervals shown in the figure are 4 ft. The highest areas (bright red) are the tops of trees that had not lost their foliage; these images could have been edited and interpolated in another hour or two of work in front of a computer screen. As in the previous example, an image of the DEM would show considerably more detail than the contour map, which is a form of data compression. In fact, when Gary Stonerock viewed the elevation (DEM) image of the same area on a computer screen, he was able to discern a dumbbell-shaped region trending left to right in the upper-central part of this figure that was slightly flatter (at least within the 6-in. accuracy of the DEM) than the rest of the area. He interpreted this area as someone's earlier aborted attempt to develop this acreage as a subdivision. The dumbbell would have become two cul-de-sacs, connected by a short street, had it been completed.

INTEGRATION OF REMOTE SENSING DATA AND GEOGRAPHIC INFORMATION SYSTEMS

The principal difference in the past between remote sensing technology and geographic information systems technology was the difference between images and digitized maps. Images are based on measured electromagnetic wave reflections (or emissions) from real-world surfaces and are most often stored in raster format (with pixels). Maps are artistic representations of what is located on, above, or below the surface and are most often stored in vector format because the boundaries between categories of classifications are polygons that can be stored as a group of head-to-tail vectors. Maps can contain categories that are invisible at any wavelength, like political boundaries. Though less than half the age of remote sensing technology, GIS technology has become a more widely used management tool than remote sensing, chiefly because of its ability to manipulate digitized information from known maps for the purpose of creating higher forms of information. However, maps have to be updated as Earth's surface changes from both natural and manmade causes, changes which remote sensing images can efficiently record. Similarly, to become a widely used management tool, remote sensing information must be combined and manipulated with known information. The future of both technologies lies in their integration. Digital orthophotos, which are images that can accurately overlay maps, are a bridge between the two technologies.

The integration of remote sensing data with GIS data has many practical advantages, especially if a dense-net digital orthophoto and DEM are included. First, images contain a great deal more information than map data, and the ability to accurately overlay digitized map data onto an image can reduce the need for digitizing some elements of the scene. For instance, suppose a digitized map of underground wire cables and gas lines exists and the user wishes to see where those underground features are located with respect to buildings and streets. In a purely GIS world, it would be necessary to digitize the corners of all the buildings and the curbs of all the streets in order to display this relationship. The locations of building corners and street curbs in the past came from detailed surveys made by civil engineers. Consider, however, what happens if the digitized map of underground utility lines is displayed on an orthophoto image that has been accurately co-registered with the digitized map. The locations of buildings and streets become obvious from the orthophoto, and they need not be digitized to display their spatial relationships to the underground utility lines. This overlay greatly reduces the time and cost of GIS database generation.

A second advantage of integrating GIS and remote sensing data arises when some of the features at Earth's surface need to be classified into categories that are needed by a user who is trying to calculate from some model that uses a GIS database as input. Some of these classifications can be made far more efficiently with multispectral image data than can be made by a field worker with personal observations on the ground. For example, if a surface water run-off model uses information from a digitized soil map as part of its input, but also needs amounts and types of vegetation cover as input for its calculations, it would be possible to obtain the soil classification from the digitized soil map and the vegetation cover classification from LANDSAT multispectral information.

Third, if the digital orthophoto is produced from a dense-net DEM that has every-pixel posting, it is possible to determine the x, y, and z (elevation) of any pixel just by moving a cursor to the place on the computer display that corresponds to a point of interest. This technique can take the place of surveying when the desired accuracy of the location is within the accuracy of the digital orthophoto. For instance, if the aerial photography from which the digital orthophoto was produced is a 1:4000 scale, a pixel size of 4 in. by 4 in. in x, y and about 6 in. in z would result if dense-net elevation extraction were employed. A manhole cover or a large transformer box could be located in the image and its position determined well enough to avoid the expenditure of a field survey for that purpose.

Fourth, volume changes associated with piles, excavations, or subsidence can be determined by subtraction of the DEM produced by dense-net methods from stereo images taken before and after the change in volume occurred in a far more accurate manner than any other method now in practice. Cut and fill estimates, a common task for civil engineering firms, could become much more accurate and faster to perform.

The dense-net DEM, therefore, has great potential for geological applications (Vincent et al. 1987a, 1988) from two viewpoints, that of operational support and exploration. From the aspect of the most rudimentary operational support, it is possible to produce an acceptably accurate DEM (and contour maps) from satellite stereo images in foreign areas of the world where topographic maps are either nonexistent or too poor to be trustworthy. This greatly improves the logistical tasks of deciding where mine sites, processing sites, seismic lines, roads, pipeline routes, and even refinery sites are to be located.

For example, Color Plate CP18 shows a color elevation contour map of Mt. Ertsberg, Irian Jaya, Indonesia, that was created for a mining company as an aid to logistical planning. The contour map, which has a contour interval of 84.685 m, was produced by GeoSpectra's ATOM software in 1989 from two SPOT satellite images, one of 10-meter resolution and the other of 20-meter resolution. Persistent cloud cover prevented data collection by the 10-meter band for two overpasses over a reasonable time period, necessitating the use of a 20-meter image as the first overpass. Although the contour interval was selected primarily for the best color balance in the image, it also represents between 2.5 and 3.0 times the estimated $z(rms)$. Ground control points consisted of fewer than a dozen places along logging trails to which the mining company had relatively easy access. The official topographic map of the region that this contour map replaced listed an estimated maximum height for Mt. Ertsberg, but had no contours drawn over the mountain.

If two 10-meter SPOT stereo images had been used, $z(rms)$ would have been much lower. Vincent et al. (1987b) reported achievement of a measured $z(rms)$ elevation error equal to 18.4 meters for two 10-meter SPOT images with an H/B ratio of 1.55 for a test area comprised of the Big Bear City 7.5-minute quadrangle map region in California, where the test points were withheld from the control points and were selected randomly from the ATOM-derived DEM. After the software was improved to aid the operator in automatic location of match points by image correlation, such that all the match points were assured to be within one pixel between the two stereo images, the same procedure on the same data set (with the same control points and withheld test points as before) resulted in $z(rms) = 12.5$ meters. Therefore, if ground control points can be located in the images to within about 1 pixel, and if the absolute x, y, and z coordinates of those ground control points are known to within one pixel or better (10 meters in x and y and slightly more than that in z), the $z(rms)$ error can be expected to be about 12.5 meters for SPOT stereo images that have an H/B ratio of 1.55. A number of different H/B ratios are possible with SPOT, which achieves stereo coverage by looking to opposite sides on two different overpasses, yielding convergent stereo images.

The dense-net DEM is also helpful from the aspect of exploration aids for both petroleum and minerals. Directional gradient-enhancement images of the elevation data can enhance topographic effects of underground geological structures and faults that have subtle surface topographic expressions. An example (Vincent et al. 1988) will be given for a frontier petroleum exploration area in the Philip Smith Mountains, which are northern foothills of the Brooks Range, Alaska and part of the Alaska North Slope Borough and the Arctic National Wildlife Refuge. Two SPOT panchromatic scenes (each with 10-meter resolution) were collected with look angles of $-3.0°$ (looking right, or West) and $19.8°$ (looking left, or East), respectively, with a B/H ratio of 0.6323. A decimated look-right image of the overlapping region is shown in Figure 6.13, which covers an area of 33 km $\times$ 37 km, with North toward the top. The ATOM software package was used to extract a high-resolution, every-pixel-posted

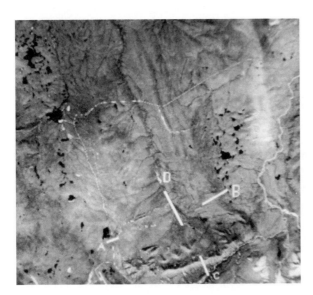

Figure 6.13   SPOT 10-meter panchromatic image (looking right or West by $-3°$ off nadir) of Phillip Smith Mountains, Alaska. Transects A, B, C, and D denoted in the image are shown in Figure 6.16. The image covers an area of approximately 33 km $\times$ 37 km, with North toward the top. *(After Vincent et al. 1988. Reprinted with permission of the Environmental Research Institute of Michigan.)*

DEM, using points selected from a 15-minute topographic map (1:62,500 scale) for ground control points. Elevation profiles of the undecimated data set were produced from the DEM for the transects denoted in this figure as A, B, C, and D. Figures 6.14 and 6.15 show directional gradient-enhancement images produced from the high-resolution DEM, with the "sun" located in the NE and SE corners, respectively. An "elephant's foot" structural feature can be seen in Figure 6.14, yet it is poorly seen, if at all, in the SPOT image of Figure 6.13. This is a graphic demonstration of the value of high-resolution DEM data for enhancing structural information that may

Figure 6.14   Shaded relief image of the high-resolution DEM produced by the ATOM software applied to SPOT 10-meter stereo images for the same area as covered by Figure 6.13, with "illumination" from the NE corner (upper-right) of the image. The "elephant's foot" structural feature is likely an uplifted fault block. The image covers an area of approximately 33 km × 37 km, with North toward the top. *(After Vincent et al. 1988. Reprinted with permission of the Environmental Research Institute of Michigan.)*

Figure 6.15   Shaded relief image of the same area as Figure 6.13, with "illumination" from the SE (lower-right) corner. The image covers an area of approximately 33 km × 37 km, with North toward the top. *(After Vincent et al. 1988. Reprinted with permission of the Environmental Research Institute of Michigan.)*

prove essential for successful preliminary screening of potential petroleum exploration targets. In addition, the elevation profiles from Figure 6.13 transects A, B, C, and D are shown in Figure 6.16, from which the structural dips were determined to be 43.25° down toward the SW for transect A; 35.7° down toward the NE for transects B and C; and 28.7° down toward the NW for transect D. Therefore, the "elephant's foot" is likely an uplifted fault block that plunges to the NW with an apparent dip of 28.7°, along a stream gradient.

Because the high-resolution DEM is co-registered with the SPOT image of Figure 6.13, it is possible to create a simulated perspective view image with the observer in any position above ground that the user selects, looking in any direction. Figure 6.17 shows such a perspective view for an observer located 750 meters above ground, a few kilometers East of the Galbraith Lake Landing Strip, looking Northwest. The two white lines are the Dalton Highway (the leftmost) and the Trans-Alaska Pipeline, which is less straight than the highway. The image appears in low relief because the top of the hill is only 500 meters above the valley floor, and this vertical relief has not been exaggerated. Another image, shown in Figure 6.18, was produced with a factor of 2 vertical relief exaggeration that emphasizes the topographic features but distorts the pipeline. However, it clearly shows exposed, uplifted sedimentary strata that contributed to the structural interpretation of the "elephant's foot" as an uplifted faulted block that plunges toward the NW.

Geology is an observational science, and the geologist can produce a better geological model of a geological structure by making observations from more viewpoints. Better yet, this technology can also assist greatly in the communication of interpretive results to managers. For instance, simulated perspective views like Figures 6.17 and 6.18 can help convey to decision-making managers what the geological structure and the environmental conditions are like in a prospective exploration area, even prior to the commencement of field work activity.

Environmental monitoring is likewise an observational science. Perspective views of the pipeline are just as useful for determining where a pipeline is leaking oil as they are for determining structural dip. However, the pipeline monitoring job requires repeated observation over the lifetime of the pipeline, whereas the structural dip measurement requires only one good observation.

When dense-net DEM and digital orthophoto production are combined with simulated perspective viewing, as in the case of the last example above, the result has been called digital holography (Vincent 1989), which is defined as the capability of creating a nearly infinite set of perspective views from one dense-net DEM and digital orthophoto database produced from the same stereo source images, with the same spatial resolution as the original digitized stereo images. Because the DEM is calculated from the stereo images directly, no information is required for digital holography other than the source digitized stereo images, unless it is desirable for the simulated perspective views to be tied to a map, whereupon a few (6 to 10) ground control points (with known x, y, and z coordinates) are required within the stereo overlapped region.

Figure 6.19 shows two stereo aerial photos of Mount Everest, collected by a Lear jet flying approximately 10,000 ft over the mountain's summit. Figure 6.20 is a simulated perspective view made from these two digitized photos (scanned with a 25-μm spot

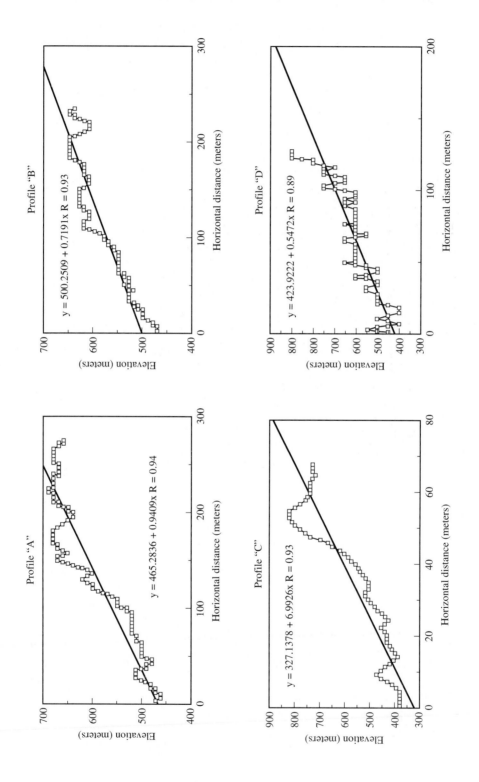

174

Figure 6.16 *(opposite)*   Four elevation profiles of transects A, B, C, and D, respectively, as shown in Figure 6.13. Profile A has a slope of 0.9409, which translates to a slope angle of 43.25°, sloping down toward the SW. The slope of profile B (sloping down toward the NE) is 35.7°, and the slope of profile C (down toward the SE) is 81°. The slope of profile D (down toward the NW) was calculated to be 28.7°, and may be considered to be the apparent structural dip of the "elephant's foot" horst. Profile D was selected along a stream gradient. *(After Vincent et al. 1988. Reprinted with permission of the Environmental Research Institute of Michigan.)*

Figure 6.17   Simulated perspective view of Phillip Smith Mountains, Alaska looking NW for an observer 750 m above the ground, just SW of the "elephant's foot" and a few km East of the Galbraith Lake landing strip. No vertical exaggeration has been used. The leftmost white line is the Dalton Highway and the white line parallel to it (less straight) is the Trans-Alaska Pipeline. The hills to the upper right of this image are about 500 m above the valley floor. This image was made from an SPOT stereo image pair, with no other information necessary. *(After Vincent et al. 1988. Reprinted with permission of the Environmental Research Institute of Michigan.)*

size), showing a panoramic side-view of the world's highest mountain that would be impossible to photograph with a camera, because other mountain peaks would block the view of all but the highest parts of Mount Everest. This simulated perspective view (displayed in the November, 1988 issue of *National Geographic*), which utilizes the image of the mountainside and valley as seen from directly above the mountain, was used by the Coulours mountain-climbing team in their assault of Mount Everest, because it showed a better panorama of the southwest side of Everest than any photo ever taken. A computed "fly-around" of Mount Everest made from this same data set was produced and

Figure 6.18 Simulated perspective view of Phillip Smith Mountains, Alaska looking North, with a factor of 2 vertical relief exaggeration, which emphasizes topographic relief, but distorts the highway and pipeline. Notice the exposed and uplifted sedimentary strata, common in uplifted fault blocks. *(After Vincent et al. 1988. Courtesy of GeoSpectra Corp.)*

displayed as part of a PBS-TV "Nova" science program in November, 1990. Figures 6.17 and 6.20 also demonstrate the potential of digital holography for environmental monitoring of wilderness areas from high-resolution stereo images of the kind that will soon be available from commercial satellites. Software packages are available for personal computers, including laptops, that can produce simulated perspective views and "fly-arounds" of high-resolution databases like this one within seconds to minutes of processing time.

More recently a French company named ISTAR developed a DEM generator for SPOT satellite stereo images (Renouard 1989, 1991), which was used by Pisot et al. (1993) to study sedimentary basins with orthophoto and perspective view images. They also made a GIS database from digitized maps that were draped over the SPOT DEM and found that interpretation of their integrated GIS/RS (Geographic Information Systems and Remote Sensing) database in three-dimensional space added accuracy, consistency, and quantification to their geological remote sensing efforts.

Besides the geological, GIS, and environmental applications discussed above, digital holography has wide-ranging applications in flight simulators and mission

Figure 6.19   Stereo photos of Mt. Everest taken by a Lear jet approximately 3300 m above the summit. The area covered is approximately 8 km × 8 km. *(Courtesy of GeoSpectra Corp.)*

Figure 6.20   Simulated wide-angle perspective view of Mt. Everest produced solely from the two stereo photos of Figure 6.19 and no other information. It would be impossible to take this photo from any side direction because the view would be blocked by intervening mountains. This perspective view was generated from down-looking stereo photos, which can observe the valleys as well as the mountain summit. *(Courtesy of GeoSpectra Corp.)*

planning systems for the military services, as well as in medical imaging, education (especially in electronic publishing), and entertainment, perhaps including the future of television as a form of "Perspectivision," where the viewer might decide whether to watch a football game from the 50-yard-line, the end zone, or up above the stadium seats.

REVIEW

Spatial image processing is as important as multispectral image processing to geological and environmental remote sensing. Automatic linear feature mapping is a powerful, objective method for mapping regional structural trends and fracture density of terrain. Although it does not replace manual interpretation of linear features that are related to fractures, it is superior to manual methods for statistical treatments of groups of linear features.

Geophysical imaging is an important tool for enhancing deep-seated structural features and lithologic contacts. Both magnetic and gravity images present the geologist with a data representation that is far easier to interpret than contour maps. Imaging of airborne magnetic data is perhaps the most cost-effective tool for exploring offshore sea-floor structures and, if used as a preleasing reconnaissance tool, can provide quick, useful information for bidding on offshore tracts, as well as for data collection planning and interpretation of offshore seismic data. Image processing for seismic data has already found its place in three-dimensional seismic surveying; this field might be further enhanced by multifrequency imaging of seismic data.

The integration of remote sensing images with GIS digitized map data improves the performance of both types of data as a management tool and is made possible by digital photogrammetry. Digital photogrammetry involves the digital extraction of geographic information from photographs, including dense-net DEM and digital orthophotos; undoubtedly it will play an increasingly important role in geological and environmental remote sensing in the future. Digital orthophotos can be overlain with digitized map data very accurately, thereby decreasing the complexity and cost of GIS database production. Highly accurate topographic information in the form of dense-net DEM data improves the logistical aspects of the energy and mining industries and adds new structural mapping tools for petroleum and mineral exploration. Simulated perspective viewing of data sets produced by digital photogrammetry has given birth to the field of digital holography. This new field is making contributions in structural interpretation and environmental monitoring. It also has promising potential in such disparate fields as military mission planning and flight simulation, medical imaging, education, and advanced television.

Exercises

1. **(a)** Describe and contrast the differences that would result between the application of an automatic linear feature mapping program to a 100-meter-posted digital elevation model (DEM) and to a LANDSAT MSS band 5 image of the same geographical area.

(b) How would the automatic linear feature mapping results differ if applied to a directional gradient-enhancement image with the "sun" in the Southeast, as opposed to being applied to the original DEM?

(c) An automatic linear mapping program is applied to two directional gradient-enhancement images with the same "sun" angle, one of aerial magnetometer data and one of DEM data. Contrast and compare the two results.

2. (a) If a stereo pair are collected at an altitude of 4000 ft with a 6-in. focal lens on film that is 9 in. × 9 in. in size, what is the approximate area covered on the ground by each photo (assuming no photo enlargement)?

(b) If the photos described in part (a) were scanned with a 500 dpi scanner, what would be the size of one pixel on the ground and what would be the expected rms error in z (elevation), if the distance between photo centers was 2400 ft on the ground, in a DEM produced by ATOM software?

Cited References

BATSON, R. M., K. EDWARDS, and E. M. ELIASON. 1975. Computer-Generated Shaded Relief Images. *Journal of Research, U.S. Geological Survey* 3, no. 4:401–408.

COUPLAND, D. H., and R. K. VINCENT. 1981. Automatic Linear Recognition and Analysis Using Computer Program LIRA. In *Proceedings of the Fifteenth International Symposium on Remote Sensing of Environment,* 499–508. Ann Arbor: Environmental Research Institute of Michigan.

HERMAN, J. D., R. K. VINCENT, and B. DRAKE. 1991. Geological and Geophysical Evaluation of the Region Around Saginaw Bay, Michigan (Central Michigan Basin) with Image Processing Techniques. In *Early Sedimentary Evolution of the Michigan Basin,* 221–240. Geological Society of America, Special Paper #256.

JAMES, D. 1990. Master's Thesis, Eastern Michigan University, Ypsilanti.

JENSEN, J. R. 1996. *Introductory Digital Image Processing.* Upper Saddle River, N. J.: Prentice-Hall.

MARRS, R. W., R. S. MARTINSEN, and G. L. RAINES. 1984. Regional Structure and Facies Control in the Powder River Basin Wyoming and Montana, Interpreted from Satellite Imagery. In *Proceedings of the International Symposium on Remote Sensing of the Environment, Third Thematic Conference, Remote Sensing for Exploration Geology,* vol. 1, 25–34. Ann Arbor: Environmental Research Institute of Michigan.

MARTINSEN, R. S., and R. W. MARRS. 1985. Comparison of Major Lineament Trends to Sedimentary Rock Thicknesses and Facies Distribution, Powder River Basin, Wyoming. In *Proceedings of the International Symposium on Remote Sensing of the Environment, Fourth Thematic Conference, Remote Sensing for Exploration Geology,* vol. 1, 9–19. Ann Arbor: Environmental Research Institute of Michigan.

ONASCH, C. M., and C. F. KAHLE. 1991. Recurrent Tectonics in a Cratonic Setting: An Example from Northwestern Ohio. *Geological Society of America Bulletin* 103:1259–1269.

PISOT, N., J. P. XAVIER, V. MIEGEBIELLE, D. COQUELET, and P. LEYMARIE. 1993. Geological Study of Sedimentary Basins Using SPOT Data. *ISPRS Journal of Photogrammetry and Remote Sensing* 48, no. 6:2–15.

RENOUARD, L. 1989. Creation Automatique de MNT a Partir de Couples D'Images SPOT. *SPOT1 Utilisation des Images, Bilan, Resultats,* 1347–1356. Cepadues Editions, Toulouse, France.

————. 1991. Restitution Automatique du Relief a Partir de Couples Stereoscopiques des Images du Satellite SPOT. Ph.D. Thesis, Ecole Polytechnique, Paris.

VINCENT, R. K. 1979. Integration of Data from Space, Aircraft, and Ground Platforms for Resources Planning. Presented orally at the American Society of Photogrammetry–ACSM Fall Meeting in Sioux Falls, South Dakota, Sept. 18.

————. 1989. Digital Holography Provides the Ultimate in Three-Dimensional Image Processing. *The Image Society Newsletter* fall issue: pages 13 and 16.

VINCENT, R. K., and D. H. COUPLAND. 1980. Petroleum Exploration with LANDSAT in Bay County, Michigan—An Interim Case Study. In *Proceedings of the Fourteenth International Symposium on Remote Sensing of the Environment,* vol. 1, 379–387. Ann Arbor: Environmental Research Institute of Michigan.

VINCENT, R. K., M. A. TRUE, and D. V. ROBERTS. 1987a. Automatic Extraction of High-Resolution Elevation Data Sets from Digitized Aerial Photos and Their Importance for Energy Mapping. *NCGA's Mapping and Geographic Information Systems 1987 Proceedings,* 203–210. San Diego, Calif. Meeting.

VINCENT, R. K., M. A. TRUE, and P. K. PLEITNER. 1987b. Automatic Extraction of High-Resolution Elevation Data from SPOT Stereo Images. In *Proceedings of the SPOT 1 Image Utilization, Assessment, and Results International Conference,* Centre National d'Etudes Spatiales (CNES), Paris.

VINCENT, R. K., W. T. LEHMAN, R. L. HENRY, J. D. HERMAN, M. E. STIVERS, M. L. WILSON, and P. J. ETZLER. 1988. The Application of High-Resolution Digital Elevation Models to Petroleum and Mineral Exploration and Production. In *Proceedings of the International Symposium on Remote Sensing of the Environment, Sixth Thematic Conference, Remote Sensing for Exploration Geology,* 293–301. Ann Arbor: Environmental Research Institute of Michigan.

7

Remote Sensing Applications to Mineral Exploration

▼ ▼ ▼ ▼ ▼ ▼ ▼ ▼ ▼ ▼ ▼ ▼ ▼ ▼ ▼

INTRODUCTION

Unusually colorful rocks are a distinguishing feature of many metallic ore deposits. The reason is that base metals and most precious metals are concentrated naturally by oxidation-reduction geochemical phenomena, and colorful ferric and ferrous compounds are common by-products of those same phenomena. Because the mapping of surface soil and rock colors, including "colors" that are invisible to the naked eye, is the principal capability of multispectral remote sensing, one of the earliest expectations of this new technology was that it would offer important surface geochemical clues related to the presence of near-surface metallic ore bodies. The first section of this chapter gives examples of how this expectation has been confirmed for a number of metals, including gold, which is especially susceptible to remote sensing exploration. This section will be considerably longer than the remaining two sections of this chapter because remote sensing has been applied more often to exploration for metal deposits than to all other types of mineral exploration combined.

The second section deals with geological remote sensing applications for diamonds and rare earth mineral exploration. The latter is included because of the increased importance of rare earth minerals in the manufacture of high-technology tools, such as lasers and superconductors. The third section describes a much more mundane problem, the use of geological remote sensing in the exploration of industrial minerals such as sulfur, sand, and gravel. However mundane, the economic impact of finding construction materials nearest the sites of current construction is likely to exceed the economic importance of all other types of mineral exploration combined in developed countries like the United States.

THE USE OF REMOTE SENSING IN EXPLORATION FOR METALS

The following section is organized in the approximate chronological order in which metals were explored with digital satellite data, and the final subsection is reserved for geobotanical remote sensing for metal exploration. Economic conditions are ever-changing, and the earliest applications of satellite remote sensing to exploration for metallic ores were for metals that are now not as economically desirable as they once were. However, there are valuable lessons to be gained from recording those chronological explorations. The demand for nonrenewable resources has always been cyclical, and that implies that metals of less interest now will once again be important exploration targets in the future. Perhaps the most important technical lesson from this historical perspective is that remote sensing usually is applied to rocks, minerals, and structures associated with a particular ore, and not the ore itself. This approach is used for several reasons. The ore is not always exposed at the surface, and it often is not as spectrally unique nor as widely disseminated as the minerals and rocks that are associated with the ore body. Thus, the indirect approach is favored for geological remote sensing that is applied to exploration for most metals.

Uranium Exploration

The earliest use of satellite remote sensing for metals was for uranium for two reasons, one economic and the other technical. First, the market for uranium was strong in the early 1970s when LANDSAT I was launched, and the major oil companies became active in its exploration about that time. Second, ferric oxides, which are associated with the geochemical cells that are important for uranium concentration in sandstone, could be mapped with data from the first MSS scanner of the LANDSAT series.

Uranium occurs in natural association with feldspar-rich granites and the clastic sediments derived from them. Arkosic sands and sandstone, especially those of Eocene age, are enriched in uranium, albeit in noneconomic quantities. The uranyl (a +6 valence) ion is soluble in oxygenated ground water, which carries it down the hydrostatic gradient until a reductant, such as pyrite, is encountered. Uranium compounds are then precipitated out of solution in the form of uranium oxides and salts that contain uranium in the reduced state (with a valence of +4). Natural gas escaping from underground hydrocarbon reservoirs is a possible causal agent for near-surface pyrite, as will be discussed in the next chapter. Thus, there is a close correlation between gas fields and uranium deposits in regions where underground natural gas reservoirs are overlain by uranium-rich arkosic sandstone. Pyrite is also commonly associated with near-surface occurrences of oil shales, coal, and organic carbon from decayed vegetation in old stream beds and swamps. When a geochemical cell occurs in an arkosic sandstone, it consists of zones, including an oxidized zone that is enriched in ferric oxides, such as limonite and hematite. Uranium ores are found in the reduced zone, which is usually surrounded by an oxidized zone.

When colorful ferric oxides are localized in an otherwise bland-colored arkosic sandstone, the presence of a geochemical cell that overlies uranium ore is possible. A number of early LANDSAT investigators (Salmon and Pillars 1975; Spikaris and Condit 1975; Offield 1976; Vincent 1977; and Raines et al. 1978) successfully employed spectral ratio images of LANDSAT MSS data. The $R_{5,4}$ ratio, most prominently mentioned, was used in mapping ferric oxides in exposed sandstones for uranium exploration. A 1974 LANDSAT exploration project for Utah International Mining Company provides a case study that demonstrates how remote sensing offers a welcome addition to traditional exploration methods for uranium.

Utah International had, to its satisfaction, already completely explored the Wind River Basin in Wyoming for uranium. The company owned and operated the Lucky Mac mine, located in the SE corner of the Wind River Basin. At that time, Lucky Mac was the largest open pit mine in the state. An airborne scintillometer survey of the entire basin had been completed within the previous year that had found no new areas of anomalous radiation. The LANDSAT study included the production and interpretation of a black-and-white $R_{5,4}$ spectral ratio image, shown in Chapter 5 as Figure 5.4a; it displayed ferric oxides as bright, vegetation and water as dark, and everything else as a medium shade of gray. A study of that ratio image resulted in the delineation of 19 areas of locally enriched ferric oxides in Eocene-aged arkosic sands of the Wind River Basin that were suspected of harboring uranium deposits. When the 19 suspect areas from the LANDSAT study were compared with the airborne scintillometer survey results, none of them showed up as anomalously radioactive. However, company geologists decided to field-check the satellite study results anyway.

It turned out to be a wise decision. All of the 19 areas had ferric oxides at the surface, some of which were obvious neither from the ground nor from visual observations from a low-flying aircraft, which was subsequently chartered. Eleven of the 19 areas had anomalous ground-measured radioactivity, from 3 to 10 times above background levels. These values where taken to be the average of more than 20 random measurements of the suspected sites. One site was a new prospect being cored by another company, previously unknown to Utah International. Another of those 11 sites was actively cored shortly thereafter by Utah International and was found to be uranium-enriched, but not economic at then-current uranium prices. One of the suspect areas was on an Indian reservation just a few miles from the company's Riverton, Wyoming office. The Riverton office geologists were eager to find out why they had not observed these sites with the airborne scintillometer survey. After more study, they concluded that the threshold of radioactivity, taken as a certain percentage of the scintillometer readings over the Lucky Mac mine, had been set too high for exploration purposes. The scintillometer data was of invisible radiation and in the form of a contour map, neither of which were conducive to visual verification. In contrast, the LANDSAT spectral ratio data were in image form, which made it easier for geologists to separate features of possible geological importance from those that were not. Fortunately, it was also possible for them to visually verify the ferric oxides in the field. The collapse in uranium commodity prices within the next few years kept all of those prospect areas from becoming economic, but a fast, inexpensive, new exploration tool for uranium was born.

Copper Exploration

The story of copper exploration with satellite remote sensing is somewhat similar to that of uranium. Porphyry copper deposits generally form in active continental margins or island-arc regions. Thus, they are associated with magmas that are formed from both upper mantle and crustal rocks as ocean plates are subducted. Lowell and Guilbert (1970) describe the typical porphyry copper deposit as being associated with an elongated stock of about 1.5 km in linear dimension and containing a distinctive alteration pattern that resembles a series of concentric shells that are coaxial with the ore body, as diagrammed in Figure 7.1. The outermost shell, called the propylitic zone, contains epidote, calcite, and chlorite. The second shell from the outside is a narrow zone called the argillic zone, containing quartz, kaolinite, and montmorillonite, which is near the contact of the propylitic and the phyllic zone. The phyllic zone contains quartz, sericite, and pyrite. The innermost alteration zone, called the potassic core, consists of potassium feldspar, biotite, and quartz. The highest-grade ore body usually occurs near the contact between the potassic core and the phyllic zone. Ferric oxides often occur around the outside of the propylitic zone and can occur in parts of the other zones, also. Somewhat like the geochemical cells associated with uranium deposits, the alteration zones of a copper porphyry deposit

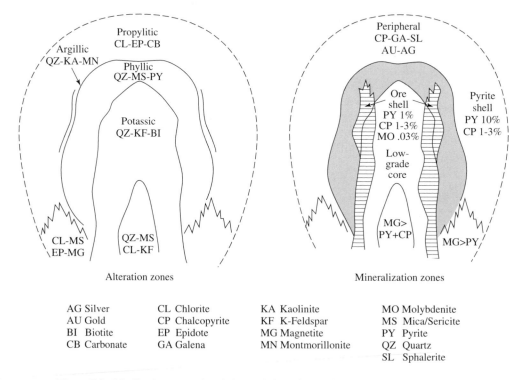

AG Silver	CL Chlorite	KA Kaolinite	MO Molybdenite
AU Gold	CP Chalcopyrite	KF K-Feldspar	MS Mica/Sericite
BI Biotite	EP Epidote	MG Magnetite	PY Pyrite
CB Carbonate	GA Galena	MN Montmorillonite	QZ Quartz
			SL Sphalerite

Figure 7.1 Idealized cross-sectional views of alteration zoning and mineralization zoning for the southwestern U.S. porphyry copper deposits *(M. J. Abrams and D. Brown, 1985. Reprinted by permission of the American Association of Petroleum Geologists, from Lowell and Guilbert, 1970.)*

can display very distinct multispectral patterns associated with the different mineral assemblages of the various zones, except in greater variety.

Abrams and Siegal (1976) employed a 24-channel aircraft multispectral scanner (designated by NASA as MSDS) to map a copper porphyry deposit at Red Mountain, Arizona. The 24 spectral bands of the Bendix-built MSDS multispectral scanner extended from 0.34–13.0 μm, but the thermal infrared bands worked only occasionally, and this was not one of those occasions. Table 7.1 shows just the 11 MSDS spectral bands that are within the 0.4–2.5 μm wavelength range and that are numbered in a convenient manner for purposes of this book.

At Red Mountain, they found that a spectral ratio image of spectral bands 10 and 11 ($R_{10,11}$) clearly delineated some of the alteration associated with the deposit. The reason for this limited success was that the $R_{10,11}$ spectral ratio image mapped the presence of clays, some micas, and alunite as a group of minerals with reflectance spectra that exhibit reflectance minima caused by bending vibrations of the Al—O—H bond (Hunt et al. 1971, 1973).

Abrams et al. (1977) used data from the same MSDS aircraft scanner to map alteration patterns in the Cuprite mining district in Nevada. This area currently a more interesting prospect for precious metals than for copper. Color Plate CP19 is a simplified geologic map of the Cuprite mining district, according to Ashley (Abrams et al. 1977). Small amounts of argillized alteration, displayed as green in CP19, are found mostly in contact with the less altered Thirsty Canyon Tuff, a 7-million-year-old rhyolitic ash flow (Ashley and Silberman 1976), displayed as orange. Areas of opalized and silicified alteration are displayed as light blue and dark blue, respectively, in CP19. Argillized rocks, containing kaolinite and montmorillonite clays and glass altered to opal, are the least intensely altered in this suite of altered rocks. The opalized rocks contain abundant opal and up to 40% alunite and kaolinite content, though there is less clay content than in argillized rocks. The silicified rocks, which are the most intensely altered of the three alteration types, contain abundant hydrothermal quartz (Abrams et al. 1977). Whereas most of the altered rocks, particularly the silicified ones,

TABLE 7.1   ELEVEN OF 24 SPECTRAL BANDS IN THE 0.4–2.5 μm WAVELENGTH REGION OF THE MSDS AIRBORNE MULTISPECTRAL SCANNER (SPECTRAL BAND NUMBERS NOT ORIGINAL)

Band No.	Wavelength Coverage (μm)
1	0.46–0.50
2	0.53–0.57
3	0.57–0.63
4	0.64–0.68
5	0.71–0.75
6	0.76–0.80
7	0.82–0.87
8	0.97–1.05
9	1.18–1.30
10	1.52–1.73
11	2.10–2.36

have little limonite (hydrated ferric oxide) content, limonite is relatively abundant in the argillized rocks and is spotty in the opalized rocks. Desert varnish is mostly absent in the altered rocks, except for silicified rocks that have developed a vuggy (with holes), pockmarked, weathered surface.

Color Plate CP20 shows a color ratio composite image of the Cuprite mining district, consisting of spectral ratios $R_{3,8}, R_{10,1}, R_{10,11}$ displayed as blue, green, and red, respectively. (See Table 7.1 for spectral band number designations.) Numbers in the image refer to sites where portable field spectrometer measurements were made by the authors (Abrams et al. 1977). Color Plate CP21 shows a color aerial photo of the same Cuprite mining district.

The LANDSAT TM spectral ratio codes of Appendix C can help us get an idea about what colors various minerals would appear as in CP20, even though the MSDS scanner has somewhat different spectral bands than a LANDSAT TM sensor and the ratio codes in Appendix C were calculated from a library of laboratory spectra. Note that the closest LANDSAT TM equivalents to MSDS spectral bands 1, 3, 8, 10, and 11 are the TM bands 1, 3, 4, 5, and 7, respectively. Therefore, the MSDS spectral ratios of $R_{3,8}, R_{10,1}$, and $R_{10,11}$ are approximately equivalent to LANDSAT TM spectral ratios of $R_{3,4}, R_{5,1}$, and, $R_{5,7}$, respectively, where TM band numbers have been employed. The following three paragraphs may be somewhat confusing to follow because the MSDS bands are numbered differently than the TM bands, two of the spectral ratios are reciprocals of the ratios in Appendix C, and the order of colors in CP20 will be addressed as red, green, blue, rather than the opposite order of previous discussions. However, the exercise is worthwhile because it proves that ratio codes of Appendix C are tools that can be trusted, within limits. The predicted color of a mineral in a particular color ratio composite may be slightly mistaken (e.g., yellow-green instead of green), but not greatly mistaken (e.g., green will not be mistaken for red, blue, purple, pink, and so on).

According to Appendix C, the $R_{7,5}$ ratio codes for kaolinite and montmorillonite are 2 and 3, respectively; these values mean that they would appear in the darkest 20% and 30%, respectively, of all the minerals in the database in an $R_{7,5}$ spectral ratio image. (Remember that a ratio code of 0 represents the darkest 10%.) To find the inverse ratio code for $R_{5,7}$, the $R_{7,5}$ ratio code (3 for kaolinite and 2 for montmorillonite) must be subtracted from 9, to yield 6 and 7, respectively, for kaolinite and montmorillonite. Therefore, the $(R_{10,11})$MSDS = $(R_{5,7})$TM ratio code for kaolinite is 6 and for montmorillonite is 7, which is an above-average red component in the $R_{3,8}, R_{10,1}$, and $R_{10,11}$ MSDS color ratio composite image (CP20). Likewise, the $(R_{10,1})$MSDS = $(R_{5,1})$TM ratio code of kaolinite and montmorillonite are 4 and 2, respectively, which means that these clays will have a relatively weak green component in CP20. The $(R_{3,8})$MSDS = $(R_{3,4})$TM ratio codes (after calculating the inverse of $R_{4,3}$) for kaolinite and montmorillonite are 4 and 5, respectively, which means that those two clays will the average amount of blue component in the CP20 color ratio composite. Therefore, kaolinite and montmorillonite could be expected to appear red-to-purple. Alunite will also have a high red display color component in CP20 because it has an $R_{7,5}$ ratio code of 1, or an $(R_{10,11})$MSDS = $(R_{5,7})$TM ratio code of $9 - 1 = 8$, due to sulfate overtone absorption bands in the TM band 7 (MSDS band 11) wavelength region. The $(R_{10,1})$MSDS = $(R_{5,1})$TM ratio code of alunite is 2 and the $(R_{3,8})$MSDS $= (R_{3,4})$TM ratio code is 2, meaning that alunite will have a higher red, but lower green and blue display color components in CP20 than kaolinite or montmorillonite. In CP20, alunite would be expected to be red, redder than the clays.

If hematite is selected from the Appendix C library of mineral ratio codes as representative of ferric oxides, the $(R_{10,11})$MSDS $= (R_{5,7})$TM ratio code would be 1 (very little red display color component), the $(R_{10,1})$MSDS $= (R_{5,1})$TM ratio code would be 9 (high green display color component), and the $(R_{3,8})$MSDS $= (R_{3,4})$TM ratio code would be 0 (almost no blue display color component), for an overall display color of green for hematite. The same ratio codes for goethite, a ferric hydroxide, are 2, 9, and 0, which translate to a green or yellow-green color in CP20 for goethite. Because the mineral population of the library of ratio codes and the scene in CP20 are no doubt different, and the image in CP20 is contrast-stretched, it is appropriate to "stretch" the above color predictions to the point of predicting that alunite and clays will appear red and ferric oxides and hydroxides will appear green in the color ratio image.

Since yellow is a mixture of red and green display colors, yellow in the color ratio image of CP20 would represent a mixture of ferric oxide or hydroxide (green) and alunite or clay (red). From the previous paragraph, green represents ferric oxide or hydroxide without much alunite or clay, and red represents the presence of alunite or clay without much ferric oxide or hydroxide. According to the field descriptions from above, the argillized rocks have the most ferric oxide and clays of the three alteration types, and they appear mostly green and yellow in the color ratio image. The opalized rocks have relatively high clay content and relatively low ferric oxide content, and they appear red in the color ratio image. The silicified rocks contain few clays or ferric oxides, and they appear dark brown in the color ratio image. Thus, the color ratio image of CP20 matches the geologic map of CP19 in a manner consistent with the ratio codes of the laboratory reflectance spectra of the products of alteration. Since the thermal bands in the 8–14 μm wavelength region did not operate correctly in this data set, Abrams et al. (1977) were not able to map silicified rocks on the basis of silica content, but had to rely on other characteristics, such as the absence of ferric oxides and clays from the silicified zone.

Vincent et al. (1984) integrated Airborne Thematic Mapper (ATM) multispectral scanner data (which included the LANDSAT TM bands) with Thermal Infrared Multispectral (TIMS) airborne scanner data for mapping the same Cuprite, Nevada District ore body. Data from an October 5, 1981, 1:30 P.M. flight of the ATM scanner were resampled to match data collected by the TIMS scanner during an August 25, 1982, 12:00 P.M. flight. The resampled spatial resolution (or pixel size) of both data sets was 30 m. Table 7.2 gives the spectral band numbers and wavelength ranges for both the ATM and the TIMS airborne multispectral scanners.

Color Plate CP22 (Vincent et al. 1984) shows a different color ratio image of the same Cuprite mining district as shown in CP20 and CP21. Color Plate CP22, which mixes ratios from ATM and TIMS scanners, is a color ratio image of $(R_{3,2})$ATM, $(R_{5,7})$ATM, and $(R_{1,2})$TIMS displayed in red, green, and blue, respectively. From Appendix C, smoky quartz has ratio codes of 6 and 1 (inverted $R_{5,7}$) for $(R_{3,2})$ATM and $(R_{5,7})$ATM, respectively. If the quartz were not smoky, the visible spectrum of quartz would be flatter, and the $R_{3,2}$ ratio code would be much smaller (1 or 2). From an examination of the quartz thermal infrared spectrum in Figure 3.6 it can be seen that

TABLE 7.2 AIRBORNE THEMATIC MAPPER (ATM) SPECTRAL BANDS, NUMBERED TO COINCIDE WITH MATCHING LANDSAT TM BANDS, AND THERMAL INFRARED MULTISPECTRAL SCANNER (TIMS) SPECTRAL BANDS

ATM		TIMS	
Band No.	**Range (μm)**	**Band No.**	**Range (μm)**
	0.42–0.45	TIMS1	8.20–8.54
ATM1	0.45–0.52	TIMS2	8.60–8.95
ATM2	0.52–0.60	TIMS3	9.01–9.35
	0.60–0.62	TIMS4	9.60–10.15
ATM3	0.63–0.69	TIMS5	10.31–11.12
	0.69–0.75	TIMS6	11.26–11.60
ATM4	0.76–0.90		
	0.91–1.05		
ATM5	1.55–1.75		
ATM7	2.08–2.35		
	8.50–13.0		

quartz has a higher reflectance in the TIMS2 spectral band (Table 7.2) than in TIMS1. As Equation 2.7 (Kirchhoff's law) dictates, TIMS2 would have less emitted energy than TIMS1 from quartz, which implies a higher $(R_{1,2})$TIMS ratio code for in emission. Therefore, quartz (if not smoky) should have little red or green and more blue in CP22. The silicified zone is, indeed, blue in that image, as the ratio codes would predict. Replacement of $(R_{1,2})$TIMS by $(R_{3,6})$TIMS would have probably done a better job of segregating quartz from clays, alunite, or ferric oxides. No multispectral scanner in orbit has yet collected all of these bands from the same platform. Geologists will benefit greatly when such a capability exists in a commercial satellite. NASA plans to orbit the EOS ASTER instrument (Appendix B) in 1998 on the EOS AM-1 platform, which will have multispectral visible-reflective infrared and thermal infrared capability. However, the thermal infrared spectral bands will have a spatial resolution of only 90 m.

The value of multispectral remote sensing with Thematic Mapper Simulator (aircraft scanner with the LANDSAT TM bands) for copper porphyry exploration is summed up well in the Joint NASA/Geosat Test Case Project (Abrams et al. 1984), which resulted from joint collaboration between NASA and commercial petroleum, mining, and geological remote sensing services firms. For three Arizona copper porphyry sites the following results were noted:

Sites	Separable Zones
Silver Bell	Potassic, phyllic, and propylitic alteration zones
Helvetia	Argillic and skarn zones
Safford	Propylitic and phyllic zones

In all three cases, the iron oxide and clay mineral absorption features were critical to success. Iron oxides and hydrous minerals (such as clays, alunite, and pyrophyllite, all of which have absorption bands in the 2.08–2.35 μm area covered by

TM band 7) are key indicators of hydrothermal alteration zones, thus contributing to their success in copper porphyry exploration. The greatest advantage of mapping these minerals from TM bands is that no previous information of the ground is required, only adequate exposures of surface materials and outcrops. The NASA/Geosat report states that all three of the above copper porphyry deposits would have been identified as promising areas for further work solely on the basis of remote sensing. This report supports the idea that multispectral remote sensing is an excellent reconnaissance tool for copper porphyry exploration. A more recent paper by Spatz and Wilson (1994) gives brief case studies of these and other copper porphyry remote sensing experiments and includes data from GERS-63 and GEOSCAN (both commercial airborne imaging spectrometers), as well as from the JERS (Japanese Remote Sensing System) satellite.

Like uranium, however, copper exploration is no longer widely conducted in the United States. World prices of copper, the lower grade of U.S. copper deposits relative to foreign deposits, and environmental considerations have caused copper deposits in foreign countries to have more appeal than domestic copper, even to U.S. mining companies. For both copper and uranium, multispectral remote sensing came too late to play a vital domestic exploration role, although it still is important for exploration in less developed countries, particularly those with semiarid terrains.

Lead, Zinc, Silver, and Tungsten Exploration

Massive sulfide deposits of lead, zinc, and silver were explored in Canada with LANDSAT MSS data in the early days of GeoSpectra's contracted geological remote sensing services, with some degree of success. This type of lead, zinc, and silver deposit, thought to be exhalative deposits hosted in sedimentary rocks, are most likely formed by hydrothermal solutions venting into restricted basins on the seafloor, but in relatively shallow seas. Barite deposits often occur nearby, as do iron-rich and manganese-rich formations from iron and manganese that were exhaled along with the lead, zinc, and silver. Depending on transportation costs from the deposit, the barite may be valuable in itself as an ore, because barite is an important constituent of heavy oil drilling "muds." These are poured into a bore-hole to offset the upward pressure of underground oil and gas. This technique prevents blowouts when a hydrocarbon reservoir is encountered by the drill bit. Massive sulfide deposits in the Northwest Territories of Canada were detected primarily by mapping pyritic gossans that contain large exposures of ferric oxides. The latter could be seen as bright features in an $R_{5,4}$ spectral ratio image of MSS data. Massive sulfides also were often noted to have magnetic highs associated with the edges of a lead-zinc deposit.

The Northwest Territories proved to be very difficult for multispectral remote sensing because of a red flowering vine that mimicked ferric oxides in the $R_{5,4}$ ratio image. This region is one of the few places on Earth that vegetation is known to produce an anomaly that could easily be confused with ferric oxides in a LANDSAT MSS $R_{5,4}$ ratio image. Because of this potential confusion and the coarse, 80 m resolution of MSS data, only the largest gossans (many of which were already known) could be mapped. The effectiveness of satellite remote sensing with that type of data was therefore reduced. Lichen on rocks also greatly reduces the spectral contrast in the Northwest Territories, but lichen seem to avoid the largest pyritic gossans, presumably because of the soil's toxicity to most vegetation.

The best silver prospect resulting from the mid-1970s Northwest Territories exploration venture, called the Mars Claim, was followed up with assays that showed the deposit, because of its wilderness location, would become economic, with a world silver price of approximately $20 per troy ounce. Although the silver commodities market reached a price of $55 per troy ounce shortly thereafter, the silver "bust" quickly brought the price down to a tenth of that price, and $20 silver has never again been attained since that time. The Mars Claim has long since expired.

LANDSAT TM data, which as yet have not been applied to lead-zinc exploration, offer considerably better potential because of the greater spatial resolution (30 m) and the greater number of spectral bands. For instance, the zinc ore mineral smithsonite, which is often associated with Mississippi Valley Type (MVT) lead-zinc deposits, is difficult to identify in the field because of its bland visible color, but would be relatively easy to enhance with TM data because of its absorption band near the 1.0 μm wavelength region (see Figure 3.10). Mapping smithsonite with a multispectral scanner should be considerably easier than mapping it on the ground by human observation. The iron-bearing carbonates ankerite and siderite are also often associated with MVT lead-zinc deposits, and they have distinctive spectral reflectance spectra. A perusal of the LANDSAT TM spectral ratio codes in Appendix C will show that siderite has TM ratio codes of 9, 0, and 8 for spectral ratios $R_{3,1}$, $R_{4,3}$, and $R_{5,4}$, respectively. No other mineral in the spectral library of Appendix C comes close, except for clinozoisite (9,2,8), fayalite (7,0,8), lepidolite (7,1,7), natrojarosite (9,1,8), and serpentine (7,1,6) for those three spectral ratios. A color ratio image of $R_{3,4}$, $R_{1,3}$, and $R_{4,5}$, displayed as red, green, and blue, respectively, would make siderite the reddest of any mineral (of the minerals in Appendix C) in the image. Ankerite is not in the spectral library of Appendix C.

In 1975, GeoSpectra mapped tungsten deposits in the Yukon Territory of Canada with LANDSAT MSS $R_{5,4}$ spectral ratio images. The ferric-oxide-rich skarns, products of contact metamorphism that included tungsten-rich minerals, were easily seen as necklaces of bright dots in almost perfectly circular arrangement around the granitic intrusives, which were displayed in the ratio image as dark, circular features. An example of an MSS $R_{5,4}$ spectral ratio image showing some of these tungsten-rich skarns is shown in Figure 7.2. Most of these skarn deposits were already known at the time of exploration, but some were not.

Gold Exploration

The metal for which multispectral remote sensing has the greatest potential is gold, especially for disseminated gold deposits, where the gold particles are too fine to be seen with the naked eye. Products of hydrothermal alteration, such as ferric oxides, clays, and alunite, have as much significance for gold deposits as they do for other metals that are laid down by hydrothermal processes. Recently, however, a new associated mineral suite has emerged in the form of ammoniated minerals, which are almost uniquely associated with deposits of disseminated gold and some other rela-

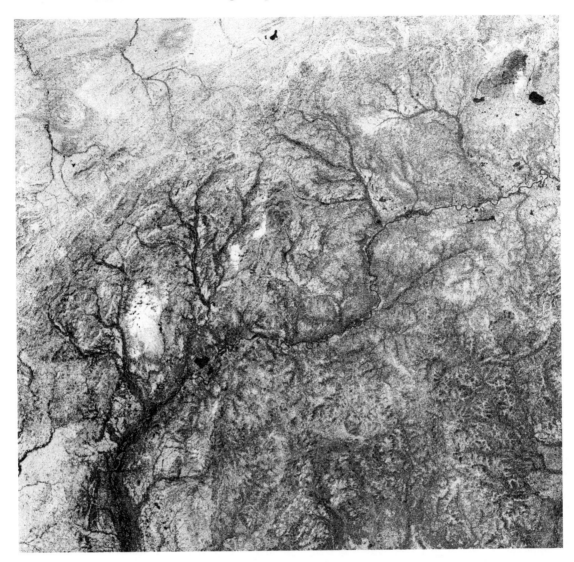

Figure 7.2 LANDSAT MSS $R_{5,4}$ spectral ratio image of the Drenchwater frame in the Yukon Territories, Canada, covering 185 km $\times$ 185 km in area, with North about 15° left of the vertical axis of the image. Ferric oxide is bright. Note the "pearl necklace" around the approximately circular granite intrusive near the middle of the right edge of the image; the "pearls" are ferric oxides of skarns (contact-metamorphosed carbonates) that are exposed around the intrusive. Some of these skarns have been mined for tungsten. *(Courtesy of GeoSpectra Corp.)*

tively rare types of metal deposits associated with organic matter. Interestingly, ammoniated minerals are spectrally discernible in multispectral remote sensing images, yet cannot be distinguished from nonammoniated minerals by the naked eye, because the spectral bands associated with ammoniation occur in the reflective infrared wavelength region.

The first natural ammonium silicate detected in hot spring deposits was buddingtonite, an ammoniated feldspar, which was found in a Hg-bearing hot spring in Sulphur Bank, California (White and Roberson 1962). Twenty years later (Sterne et al. 1982), ammonium-bearing illites were found in the black shales of stratiform Pb-Zn-Ag deposits of Alaska. Shortly thereafter, Von Damm et al. (1985) reported that heated fluids from undersea "black smoker" hydrothermal vents in the Guyamas Basin of the Gulf of California were possible modern analogs to such Pb-Zn-Ag deposits. The heated fluids from such vents contain significant amounts of NH_4^+ that are derived from organic-rich sediments. Several organic-rich sedimentary environments related to hydrocarbons had been observed earlier with associated ammonium minerals, including buddingtonite that was found in midcontinent oil-shale deposits (Gulbrandsen 1974).

Multispectral remote sensing was first used in 1985 when Goetz and others (Goetz and Srivastava 1985; Goetz et al. 1985) reported results of an imaging spectrometer test flight over a hot spring deposit at Cuprite, Nevada. They detected a mineral occurrence that was so unusual, it was not represented in Jet Propulsion Laboratory's large library of mineral reflectance spectra. To compound the mystery,

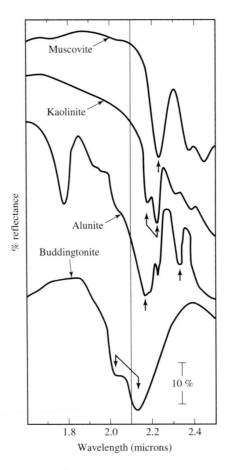

Figure 7.3 Laboratory near-infrared spectra of several common minerals in hydrothermally altered rocks (Lee and Raines 1984) compared to buddingtonite. Note how ammonium absorption features near 2.1 μm in buddingtonite (downward arrows) are distinctive from common OH adsorption features near 2.2 μm in muscovite, kaolinite, and alunite (upward arrows). *(After Krohn et al. 1988.)*

geologists in the field could not see any surface exposure patterns that coincided with the unidentified mineral occurrence. Later x-ray analysis proved that the mystery mineral was buddingtonite, which has a unique spectral reflectance in the 1.8–2.4 μm spectral region. Three years later, Krohn et al. (1988) published a paper presenting reflectance spectra of buddingtonite and three other minerals (muscovite, kaolinite, and alunite) that are commonly found in hydrothermally altered rocks, but that are not ammoniated. A spectral reflectance minimum at 2.1 μm is caused by NH_4 absorption, which is separate and distinct from the common OH absorption features near 2.2 μm in nonammoniated samples of muscovite, kaolinite, and alunite, as is shown in Figure 7.3 (from Krohn et al. 1988; and Lee and Raines 1984). In the same paper, Krohn and others stated that the concentration of ammonium-bearing minerals suggests fracture-controlled deposition. They also state that the persistence of ammonium-bearing minerals in the surface environment and their apparent association with organic-rich host rocks suggest that multispectral remote sensing will have utility in geochemical prospecting for ore deposits associated with hydrothermal fluids circulated through organic matter, including disseminated gold deposits.

In the Cuprite, Nevada deposit, buddingtonite was found near the transition from the argillic to the opaline zones of alteration. Krohn et al. (1988) examined ammoniated minerals from Cuprite, Nevada; Golconda, Nevada; Ivanhoe, Nevada; and Sulphur Bank, California, all of which were sites of hot spring deposits. They found that the ammonium-bearing minerals tend to occur in patches that radiate away from the center, cross-cutting alteration zones, especially in the Ivanhoe case. They were linked to disseminated gold deposits in Golconda, Nevada and Preble, Nevada.

Concurrently, Altaner et al. (1988) published a paper that reported the finding of ammonium in alunites at The Geysers (Sonoma County), California, and near a fossil hot-spring with mercury-gold mineralization in the Ivanhoe district, Elko County, Nevada. When they compared spectral reflectance spectra of synthetic alunite with synthetic ammonioalunite, they showed that the spectral reflectance minimum near 2.2 μm in the nonammonium-bearing alunite is broadened and deepened in the 2.0–2.15 μm region of the ammonium-bearing alunite. The significance of these findings is that there are at least two types, and probably more, of ammoniated minerals that can be exploited with multispectral remote sensing tools for disseminated gold and other types of metal deposits associated with hot springs in hydrothermally altered basalts near organic-rich sedimentary rocks.

More recently, Baugh and Kruse (1994) used field spectral measurements and lab analysis to find a correlation between the depth of the 2.12 μm absorption ammonium (NH_4) absorption band and the concentration of mineral-bound ammonium in buddingtonite that is located in hydrothermally altered volcanic rocks of the southern Cedar Mountains, Esmeralda County, Nevada. They then applied this correlation to data from the Airborne Visible/Infrared Imaging Spectrometer (AVIRIS) by dividing the 0.01 μm wide spectral band that includes the 2.12 μm wavelength by the "continuum" (slowly varying reflectance component of average of reflectance in bands outside the absorption band). Their method was essentially the same as the three-band spectral ratioing method described in Chapter 5 for methane imaging with one spectral band located in the absorption band and two spectral bands located just outside of the absorption band. This "continuum" spectral ratio from the AVIRIS data was then used to predict ppm concentrations of ammonium in surface rocks, with the lab measurements and field spectra providing the linear correlation between the "continuum" spectral ratio and ammonium concentration. Figures 7.4a and b show, respectively, the plot of NH_4 concentration vs. continuum-removed band depth

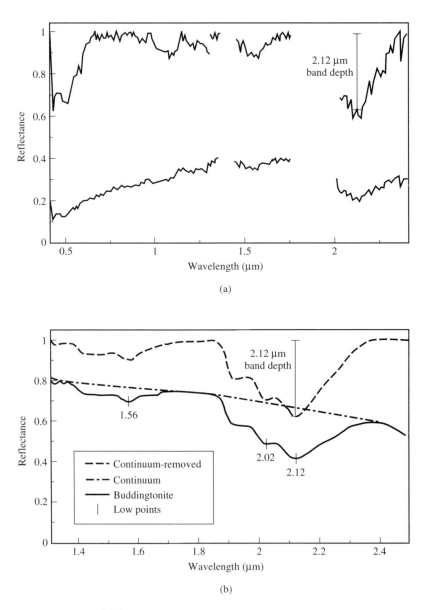

(a)

(b)

Figure 7.4 (a) AVIRIS (Airborne Visible-IR Imaging Spectrometer) spectrum of buddingtonite (lower) and corresponding continuum-removed spectrum (upper). (b) Buddingtonite spectrum from the Cedar Mountains. *(After Baugh and Kruse 1994. Reprinted with permission of the Environmental Research Institute of Michigan.)*

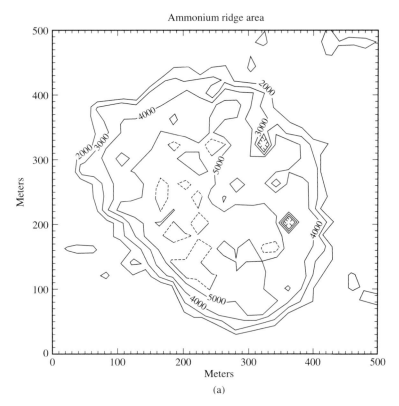

(a)

Figure 7.5 (a) AVIRIS NH_4 concentration map: contour interval is 1000 ppm; dashed contour is 5500 ppm; hatches indicate concentration less than 2000 ppm. *(After Baugh and Kruse 1994. Reprinted with permission of the Environmental Research Institute of Michigan.)*

and a buddingtonite spectrum recorded by AVIRIS, with and without continuum removal. Figures 7.5a and b show the AVIRIS NH_4 concentration map and the ground truth NH_4 concentration map, respectively, of a test area in the Cedar Mountains. They agree moderately well, even though the ground sampling was on a coarser grid (30 m) than the AVIRIS data resolution (20 m). None of these ammonium concentration differences can be detected by the naked eye, because the ammonium absorption band is totally outside the visible wavelength region.

In summary, multispectral remote sensing is proving to be a powerful tool for disseminated gold exploration in sparsely vegetated terrain.

Geobotanical Remote Sensing for Metal Exploration

Even in very vegetated regions, however, there is promise of exploration assistance from remote sensing. There are two basic types of information, structural and geobotanical, that can assist metal exploration in regions completely covered by vegetation. Photo interpretative mapping of curvilinear and linear features in radar or multispectral scanner images can yield useful structural information about hidden calderas,

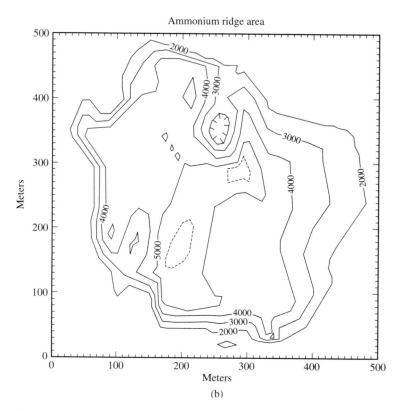

Figure 7.5 *(continued)* (b) Ground truth NH₄ concentration map: contour interval is 1000 ppm; dashed contour is 5500 ppm; hatches indicate less than 2000 ppm.

stocks, or faults associated with metal deposits. There is, however, a more subtle, indirect way of exploring heavily vegetated terrain for metals that exploits botanical associations with localized mineralization of the soils in which the vegetation grows. This is called geobotanical remote sensing, a method that is highly multispectral in nature.

Radiometers with very high spectral resolution have been used to detect geobotanical anomalies related to base-metal mineralization (Collins 1978; Birnie and Francisca 1981). Two types of features in vegetation growing on mineralized pyritic ground were found: a shift in the position of the inflection point of the steep increase in spectral reflectance versus wavelength near 0.7 μm and a change in the ratio of reflectances at 0.565 and 0.465 μm wavelengths. The shift to slightly lower wavelengths of the spectral reflectance of vegetation over mineralized areas (as compared to vegetation elsewhere in the same general region, in the 0.67–0.73 μm portion of the spectrum, where vegetation reflectance increases rapidly with wavelength) is called the blue shift. Examples are given in Figure 7.6, as discussed below.

In British Columbia, Canada, Lourim and Buxton (1988) found a blue shift in yellow cedar, mountain hemlock, and western hemlock that was associated with elevated levels of arsenic, associated with gold deposits. Figure 7.6 shows the average spectral curves of about two dozen samples of each of these three types of vegetation,

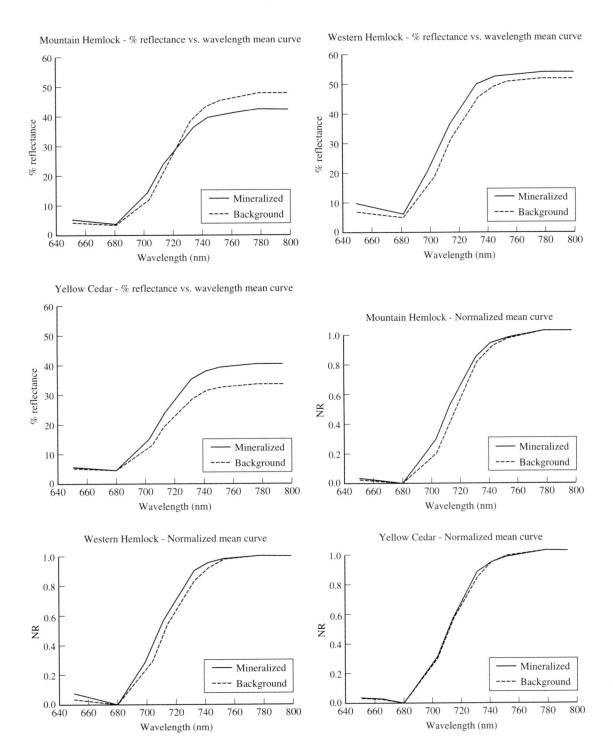

Figure 7.6 Average spectral reflectance curves and normalized spectral curves of mountain hemlock, western hemlock, and yellow cedar both on (solid lines) and off (dashed lines) areas mineralized with arsenic and gold in British Columbia, Canada. The normalized curves were determined by scaling the minimum and maximum of the average spectral reflectance curve between 0.0 and 1.0, respectively, for each of the three types of vegetation. *(After Lourim and Buxton 1988. Reprinted with permission of the Environmental Research Institute of Michigan.)*

approximately evenly split between areas of high arsenic (and high gold) concentrations and control areas that were not mineralized. Since LANDSAT TM bands 3 and 4 cover the spectral regions from 630–690 nm and 760–904 nm, respectively, a study of Figure 7.6 shows that mountain hemlock would display a lower $R_{4,3}$ spectral ratio over mineralized soils than over nonmineralized soils, whereas yellow cedar would show just the opposite behavior (higher $R_{4,3}$ over mineralized soils), and western hemlock would show little difference in $R_{4,3}$ between mineralized versus nonmineralized areas. Because of this complicated response of vegetation to mineralization, a hyperspectral scanner with the same total wavelength coverage as TM bands 3 and 4 would probably not help this particular interpretive problem as much as the inclusion of other spectral regions outside this wavelength range would help. Separation of mountain hemlock and yellow cedar as distinct classes of vegetation with all the reflective LANDSAT TM spectral bands, and the use of those two classes as a mask for the $R_{4,3}$ ratio image would help geologists increase the certainty that the mapped anomalous spectral behavior was due to vegetation stress caused by mineralization.

Torcoletti and Birnie (1988) were able to map major vegetation zones in New Hampshire with principal component imaging of LANDSAT TM data, but they warned that the $R_{5,4}$ spectral ratio, which enhances stressed vegetation, is positively correlated with increasing elevations above sea level. This correlation means that vegetation stress from higher elevations can appear similar to vegetation stress from high metal concentrations. Hornsby et al. (1988) found in the Grenville geological province of Canada that employing LANDSAT images, elevation data (DEM), and geochemical/geological spot measurements in the field permits the identification of well-defined areas of apparent mineral potential with maximum efficiency in vegetated terrains. They also found that radar images were useful to geobotanical prospecting only for slope and aspect information, which could be gained from the elevation data (DEM).

Toxic mineral stress on different types of vegetation can produce spectral anomalies in different spectral regions, some of very narrow spectral bandwidth, as has been reported in a number of the above references. Certainly, interpretation is not easy, as is indicated by Tapper and Dempsey (1988). They reported that stressed vegetation supervised classification maps (produced from 1-meter-spatial-resolution data collected by an airborne linear array with five narrow spectral bands centered at wavelengths of 0.680, 0.713, 0.753, 0.782, and 0.797 μm) required extensive ground-based geochemical information to make them reliable. However, they neither utilized the longer reflective infrared wavelength region between 0.8 μm and 2.5 μm nor neutralized topographic slope and solar illumination effects.

Geobotanical remote sensing is an important area for future research, since many of the unexplored areas of the world for base and precious metals are heavily vegetated, and since commercial hyperspectral sensors will be in orbit in the next few years. As discussed in the first section of Chapter 3 and in the forests and wetlands section in Chapter 9, Curran and Kupiec (1995) found high correlations between reflectances as recorded by an airborne hyperspectral imaging device and foliar biochemical concentrations of water, cellulose, lignin, nitrogen, and chlorophyll. Once the stresses on vegetation caused by unusual local soil mineralization have been quantitatively related to these foliar biochemical concentrations, it should be possible to greatly improve the use of hyperspectral image data for geobotanical remote sensing. With their great, common interests in geobotany, environmental scientists and exploration geologists will both benefit from the type of research that can provide such answers.

THE USE OF REMOTE SENSING IN EXPLORATION FOR DIAMONDS AND RARE EARTH MINERALS

As world market conditions continued to favor higher-priced commodities, satellite remote sensing efforts turned toward exploration for precious metals, diamonds, and rare earth minerals during the second decade of LANDSAT history. This section is dedicated to diamond and rare earth mineral exploration, both of which have a geological association with ultramafic rocks. Since ultramafic rocks come from mantle material, they are rare.

Exploration for Diamonds

There has been considerable exploration for diamonds with remote sensing data, but not much published about the subject. There are several unique properties of diamond occurrences that can be exploited with remote sensing, depending principally on the unusual geological conditions required for diamond production. Diamonds are associated with ultramafic magmas of upper mantle origin that are thought to explosively rise through Earth's crust along deep fractures in the crust. Kimberlites and lamproites, both of ultramafic composition, are the products of similar phenomena and tend to occur in close proximity to one another in diamond provinces. These diamond provinces are usually located in seismically stable interiors of continental plates, called cratons. The oldest prolific production of diamonds, starting about a century ago, was in the southern part of Africa, where they were found in economic quantities only in kimberlites, not in lamproites. Outlines of some of the Lesotho kimberlite pipes in South Africa, with areas in hectares given by the included numbers, are shown in Figure 7.7 (Evans 1993; after Nixon 1980). One LANDSAT TM pixel (of all but the thermal infrared spectral band) covers 900 square meters, or 0.09 hectare, which means that there are 11.11 LANDSAT TM pixels in one hectare. Thus, even the smallest of the kimberlite pipes in Figure 7.7 covers an area the size of several TM pixels.

In the 1980s, significant economic deposits of diamonds were found in lamproites in Western Australia, much to the surprise of most South African geologists. The early exploration of Western Australia for diamonds was carried out by geologists from a South African firm. Although some scattered diamond occurrences had been found in Western Australia, the traditional South African methods used for diamond exploration, involving soil sampling of regularly spaced sites along a two-dimensional grid, failed to show the index mineral "signature" that was the norm for South African deposits. Only one of every two hundred or so kimberlites contain economic diamond deposits anyway, but the kimberlites of Western Australia seemed relatively barren, even by that standard. The geologists that had conducted

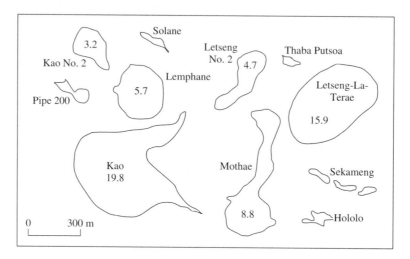

Figure 7.7 Surface outlines and areas (in hectares) of some Lesotho kimberlite pipes from various localities. *(Evans 1993; after Nixon 1980.)*

the exploration were convinced that the soils of Western Australia were less residual (had been carried farther from their source) than soils in the southern part of Africa. Their conjuncture would make the traditional African methods of diamond exploration ineffective in Australia. They also believed that lamproites, rather than kimberlites, would be the more productive of diamonds in Western Australia. However, since traditional exploration methods die hard, these ideas were rejected by the company, and Australian exploration for diamonds was ordered closed. In the spirit of entrepreneurship, the exploration geologists quit the South African company and founded their own diamond exploration company in Australia. They subsequently found economic diamond deposits in several lamproites in Western Australia and became a very substantial diamond mining company. Figure 7.8 shows the occurrences of lamproites, kimberlites, and diamonds in Western Australia (Evans 1993; after Atkinson et al. 1984).

Lamproites and kimberlites share at least two things in common that can be exploited by remote sensing. First, they are both ultramafic in composition, which means that a multispectral scanner with at least two spectral bands in the 8–14 μm wavelength region should be able to detect exposures of either lamproites or kimberlites with relative ease. Olivine, which is a more important constituent of kimberlites than lamproites, has a longer wavelength reststrahlen band than any other silicate mineral. Unfortunately, there are no commercial satellites with more than one spectral band in the 8–14 μm wavelength region, so multispectral remote sensing for uniquely recognizing ultramafic rock exposures currently can only be performed with airborne scanners, such as TIMS (Thermal Infrared Multispectral Scanner), manufactured by Daedalus Enterprises of Ann Arbor, Michigan. Note by comparing the spectral bands of TIMS, given in Appendix B, with the reflectance spectrum of fayalite olivine in Figure 3.6 that the ($R_{6,4}$) TIMS spectral ratio would be an excellent choice for mapping exposed outcrops of kimberlites and lamproites.

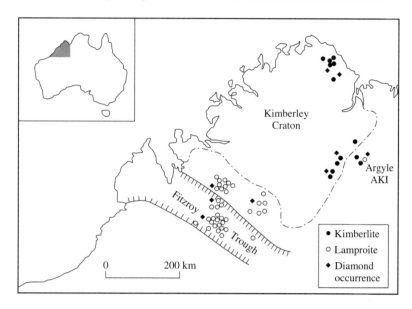

Figure 7.8 Kimberlites, lamproites, and diamond occurrences in the Kimberley region of Western Australia. *(Evans 1993; after Atkinson et al. 1984.)*

However, kimberlites are rarely found exposed as outcrops because olivine has very low resistance to weathering.

The second feature amenable to remote sensing that kimberlites and lamproites have in common is a strong, circular magnetic anomaly that is usually tightly confined to the immediate vicinity of the ultramafic rock occurrences. When geophysical images of aerial magnetometer data are produced, kimberlites and lamproites (whether exposed at Earth's surface or unexposed near the surface) exhibit circular magnetic high anomalies that typically are less than 10 km in diameter. As a further advantage of geophysical images, it is also often possible to observe a magnetic expression of the fault along which the explosive event occurred. This can aid in finding other circular magnetic anomalies along the same fracture. However, such deep-seated fractures are usually quite long, which means that the area surveyed by aerial magnetometer data should be large (covering 50 km × 50 km areas or larger) for best results.

Kimberlites and lamproites do display one difference that is quite important to remote sensing, however. Lamproites tend to be more resistant to erosion than kimberlites. In Australia, lamproites are often seen as circular, pluglike features rising above surrounding sedimentary rocks. Kimberlites, on the other hand, are more easily weathered than most sedimentary rocks. This characteristic makes kimberlite exposures relatively rare, but it also leads to the formation of shallow, circular depressions above the kimberlite. In regions like Western Australia, it creates "pans"; these are circular clay-filled areas that occur because surface water runoff after a rainfall brings in fine-grained material that then settles in these shallow ponds. Upon evaporation of the water, the clays are left behind, usually in bright contrast to the surrounding sandstone or limestone. These clay pans can be seen on false-color composites of two visible and one reflective infrared band as white, circular features about 0.25–1.0 km in diameter. With LANDSAT TM data, it would be possible to recognize these areas as clays. The positive identification of clays is not necessary for success, especially if aeromagnetic data are also available, such that the bright circular features in the false-color composite image that also display well-defined, circular-shaped magnetic anomalies can be singled out as potential underground kimberlite sites. Therefore, the integration of multispectral scanner data with geophysical images of aeromagnetic data is especially helpful for diamond exploration.

Exploration for Rare Earth Minerals

The most important source of rare earth minerals are unusual types of rocks called carbonatites, which are found in association with alkaline igneous rocks. Whereas most carbonate rocks at Earth's surface are sedimentary in origin, carbonatite complexes consist of intrusive magmatic carbonates and their associated alkalic-basic igneous rocks. The latter are generally found in stable cratonic areas associated with regional domes and their attendant faulting and rifting (Evans 1993). Carbonatite complexes can occur as plutonic or volcanic bodies, as hydrothermal solution precipitates in dilatant fractures, and as replacements of earlier carbonatites and silicate rocks. Carbonatites are usually emplaced later than the alkalic rocks with which they are often associated.

There are about 330 carbonatites occurrences worldwide. Of the four types recognized by geologists, the ferrocarbonatites are most productive of rare earth minerals and radioactive minerals. They contain iron-bearing carbonates, such as siderite and ankerite, as well as the more common carbonate minerals, calcite and dolomite. The ferrocarbonatites are becoming increasingly more important because of three factors. First, they are the source of rare earth elements, for which demand is increasing for their use in lasers and high-temperature superconductors. Second, they often contain significant quantities of niobium, an element used in high-temperature alloys required for supersonic aircraft manufacture. Third, they contain relatively large amounts of siderite and ankerite; these have unique multispectral signatures that make them easily separable from ordinary carbonates by multispectral remote sensing methods, as discussed in the subsection on lead, zinc, and silver exploration with regard to Mississippi Valley Type deposits.

The most productive ferrocarbonatite is located at Mountain Pass, California, which is the world's largest ore body of rare earth elements in the world. The primary rare earth minerals that occur there are bastnastite and parisite, with barite and pyrochlore (a niobium mineral) also present. The Mountain Pass carbonatite is a Precambrian intrusion into Precambrian gneisses, which makes its present tectonic environment a function of considerably later deformation, and which may explain why it is not now in an obviously alkaline igneous province. One mass of carbonate rock is about 200 m in maximum width and about 730 m long. Its large exposure and inclusion of siderite and ankerite make it stand out from the surrounding rock in a

LANDSAT TM color composite image of bands 1, 4, and 7, shown in Color Plate CP23. In general, the ferrous oxide absorption band near the wavelength of 1.0 μm is an important remote sensing clue about the occurrence of ferrocarbonatites. Columbite, an iron-niobium oxide found with some ferrocarbonatites, has increasing reflectance with increasing wavelength from 0.4–2.5 μm, and displays one of the highest TM band 7 to TM band 5 spectral ratios in nature. Although the Mountain Pass carbonatite does not have local occurrences of alkalic-basic igneous rocks, such as nepheline syenite, these unusual silicate rocks have unusual multispectral signatures (especially if the 8–14 μm thermal infrared region is included in the scanner data). These signatures that can be generally useful for the mapping of such carbonatites in other places. Carbonatites often also display strong magnetic anomalies, which geophysical images of aeromagnetic data can detect. As with diamond exploration, the integration of geophysical images and multispectral remote sensing images can aid in exploration for carbonatite complexes.

Among carbonatites of all four classes that occur around the world, these rare igneous carbonate occurrences are also mined for phosphorus (from apatite), magnetite, zirconia, fluorite, barite, strontianite, lime, thorium, and some copper.

THE USE OF REMOTE SENSING IN EXPLORATION FOR INDUSTRIAL MINERALS

Industrial minerals can be defined as any naturally occurring substance of economic value that excludes metallic ores, mineral fuels, and gemstones. Between 1900 and 1970, the world use of metallic minerals increased about 1000 times, energy minerals increased about 1100 times, and industrial minerals increased about 1900 times. By 1983, industrial minerals comprised 72% of the world production of all minerals, making up 40% of the value of all minerals. The second largest category, solid fuel minerals, comprised 24% of world mineral production, representing 38% of the total value of all minerals, and metals and ores (the third largest category) comprised 4% of world mineral production amounting to 13% of the total output value (Evans 1993). In fact, one of the best tests of the degree of a nation's industrialization is a ratio of the value of its industrial minerals to the value of the other minerals that are extracted from within its borders. This ratio is greater than 1.0 for all of the world's recognized industrialized nations.

Perhaps the largest difference between industrial minerals and other minerals is that the unit value of the former is sufficiently low as to be exceeded by transportation costs over any appreciable distance. This differential puts a premium on the discovery of industrial minerals nearest the markets for their use, hence creating an excellent opportunity for the combination of remote sensing with geographical information systems (GIS) for finding new deposits and weighting them according to their distance from market. Remote sensing and GIS papers on industrial mineral exploration are relatively scarce in the open literature. Therefore industrial mineral exploration could become an important market opportunity for integrated remote sensing and GIS software and services.

Moreover, this integration would not be very difficult. If the enhanced or classified remote sensing image that identifies the industrial mineral outcrop or soil is resampled to the same map from which the vectorized transportation routes and marketplaces for the minerals have been digitized, the two will be integrated. There are geometric resampling algorithms in all of the commercial image processing software packages that quickly do such tasks. The user identifies in the image 10 or fewer ground control points for which the map coordinates are known. A polygon outline of the deposits can then be used as the GIS import from the remote sensing image. The minimum number of truck route miles from the marketplace (such as an asphalt plant, a road construction site, or a glass factory) to each of the industrial mineral deposits can then be determined. This minimum mileage can then be converted to transportation cost, which is the largest factor that determines the economic value of an industrial mineral deposit.

Instead of listing the industrial minerals one by one and explaining what remote sensing applications may be applicable, this section will take the reverse tack and list remote sensing capabilities, followed by suggestions as to which industrial minerals these capabilities could be applied. It will not be an exhaustive list because of the relative scarcity of such applications in the past. However, it may be helpful during the following discussion to reexamine the rock and mineral spectra of Chapter 3.

The mapping of quartz content in exposed rocks and soils by spectral ratioing of two thermal infrared bands in the 8–14 μm region has already been used in mapping outcrops of sandstones containing high-quality sand that is used for glass making (Vincent et al. 1972). This technique could be combined with spectral ratioing for ferric oxide to eliminate high-silica sands that are contaminated with iron oxides, thus relegating a lower economic value to the sand deposit. It could also be combined with thermal inertia methods (Gillespie and Kahle 1977; Watson 1975) to help locate sand deposits in old river valleys and glacial features. Thermal inertia measurements are performed with a single thermal band collected at two different times of day (usually one at night and one in daytime). This thermal band primarily maps areas that cool or heat up faster or slower than surrounding areas. Even a single nighttime data collection can provide useful information for environmental scientists and geologists. For example, a single nighttime overpass of the Heat Capacity Mapping Mission (HCMM) satellite (no longer in orbit) with a coarse, 600-meter spatial resolution was used in Michigan to map high infiltration rate soils (sandy soils) and peat bogs (Vincent et al. 1981). A thermal infrared multispectral (more than one band) scanner can perform both spectral ratioing and thermal mapping tasks.

The silica-mapping capability may also assist in exploration for perlite, a rhyolitic, glassy rock that is used in heat and sound insulation materials. Distinctive thermal infrared and reflective infrared spectral features of phosphate minerals may help in a similar way (through spectral ratioing of a pair of thermal infrared and a pair of reflective infrared bands) in the mapping of phosphate deposits. Figure 3.13 shows the thermal infrared spectral reflectance of a phosphatic shale, which displays a prominent phosphate reststrahlen band in the 9–10 μm wavelength region.

Clays of all kinds are used as industrial minerals for applications ranging from ceramics to brick-making to catalysts. The hydroxyl ion absorption bands in the

2.0–2.5 μm region can be exploited to map industrial clays, as they have been for mapping clays associated with hydrothermal alteration (discussed earlier in this chapter). Ehmann and Vergo (1986) have showed that zeolites, which are claylike industrial minerals that readily absorb and release water and various-sized cations without structural modification, can be distinguished from dioctahedral clays (such as kaolinite, montmorillonite, and muscovite) by the reflectance spectrum absence in zeolites and the presence in dioctahedral clays of a 2.2 μm absorption feature (reflectance minimum) caused by an Al—OH absorption band. Although this spectral feature is too narrow to be exploited by band 7 of LANDSAT TM (2.08–2.35 μm wavelength coverage), it is possible to map zeolites uniquely with hyperspectral sensors that split the 2.0–2.5 μm wavelength region into a number of spectral bands, such as AVIRIS. Figures 7.9 and 7.10 (Boardman and Kruse 1994), respectively, show spectral reflectance curves of zeolite, calcite, dolomite, and three forms of sericite, as well as AVIRIS-data-derived, high-abundance maps (actually, a spectral similarity map) of each of these materials for a well-mapped region in the North Grapevine Mountains, Nevada.

Spectral ratioing has been used to isolate gypsum outcrops. Their absorption bands are in the 2.0–2.5 μm region and their much darker reflectance is in the 1.55–1.75 μm region, as compared to calcite and clays, which also have absorption bands in the 2.0–2.5 μm region. Gypsum, a hydrated calcium sulfate (called anhydrite if it is not hydrated), is one of the more important industrial evaporite minerals, not only for the value of gypsum itself in construction materials and in the production of sulfuric acid, but also for its association with sulfur deposits. The following case study is about sulfur exploration with satellite data.

A client required the mapping of gypsum outcrops in the late 1980s in west Texas from LANDSAT TM spectral ratio images; field checks proved to be substantially successful. When the client asked for the same spectral ratioing technique, which depended heavily on LANDSAT TM bands 5 and 7, to be applied to an area in Tunisia, the resulting spectral ratio images produced maps of gypsum outcrops that disagreed 100% with geological maps of the Tunisian government. After initial discouragement, the spectral ratio images were taken into the field, anyway, to see why the two disagreed so greatly. Upon completion of the field work, the client exclaimed that the spectral ratio images were "like a road map" to the gypsum outcrops. No explanation was found for the incorrect geological maps. The fact that gypsum could be mapped in Tunisia with a technique developed for west Texas by geologists with no field experience in Tunisia attests to the robustness of spectral ratio images with regard to the suppression of atmospheric, solar illumination, and vegetation cover (different species, although both in semiarid terrain) effects.

Magnesite and bauxite deposits have been mapped by photointerpretation of LANDSAT MSS color composite images and color ratio images in South India (Joshi 1988). Magnesite from hydrothermally altered dunite-pyroxenite rocks were found as bright, white areas in a color composite of MSS bands 4, 5, and 7 displayed as blue, green, and red, respectively, at least partly because there was no vegetation on such sites. Bauxite deposits were located with the help of MSS color ratio images at high elevations in thin, coarse-textured, lateritic soils over flat-topped hills that are sparsely vegetated.

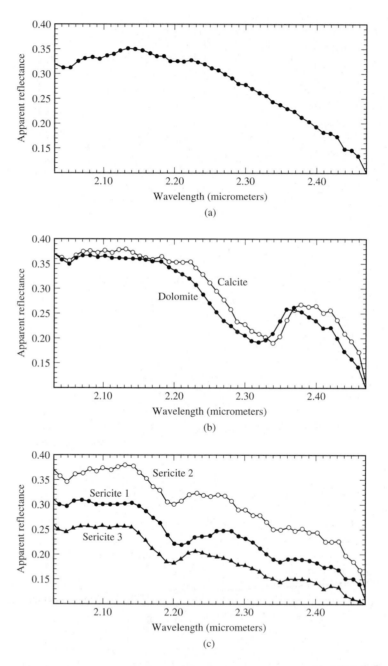

Figure 7.9 Reflectance spectra in the 2.0–2.5 μm wavelength region for (a) zeolite endmember; (b) carbonate (dolomite and calcite) endmembers; (c) phyllosilicate (sericite 1, sericite 2, and sericite 3) endmembers. *(After Boardman and Kruse 1994. Reprinted with permission of the Environmental Research Institute of Michigan.)*

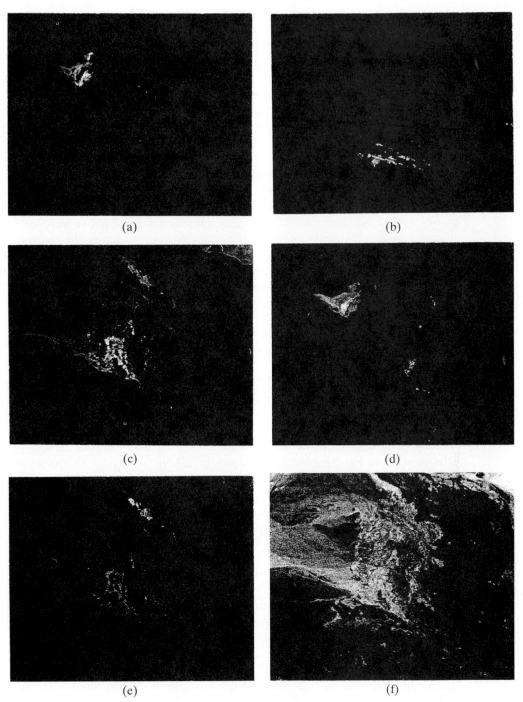

Figure 7.10 Spatial maps of the six endmembers illustrated in Figure 7.9 (a–c): (a) zeoite; (b) dolomite; (c) calcite; (d) sericite 1; (e) sericite 2; (f) sericite 3. *(After Boardman and Kruse 1994. Reprinted with permission of the Environmental Research Institute of Michigan.)*

Besides these multispectral applications, remote sensing applications of digital photogrammetry could help in the exploration for industrial minerals, especially for coarse-grained aggregates (like gravel) in certain glacial deposits, such as eskers. Snakelike curvilinear features with topographic highs that are too slight to be mapped in a government-produced digital elevation model (DEM) or a topographic map might be easily seen in a high-resolution DEM of 2-meter posting intervals produced by an every-pixel-posted digital photogrammetry software package, with input from NHAP (National High Altitude Photographs) photos that are available in stereo pairs for the entire U.S. at 1:80,000 scale. NHAP photos cost only $9 per photo when ordered from the U.S. Geological Survey's EROS Data Center in Sioux Falls, South Dakota.

Of course, similar remote sensing technology could also be applied to the production of industrial minerals. Only two examples will be given here as possibilities. First, silica content in limestones or other materials could be determined in the quarry walls if the same remote sensing techniques described above were used from ground-based instruments. Second, the volume of overburden or the change in the pit volume as a function of time could be measured for any quarry with high accuracy if every-pixel-posted DEM-generation digital photogrammetry software were employed. Digitized stereo photos from low-altitude aircraft or from tethered balloons could be used as source data for such applications.

REVIEW

The earliest application of remote sensing data for geological purposes was in mineral exploration. With multispectral scanners, it is possible to map both visible and invisible colors that identify minerals associated with ore bodies. Many times these minerals are not ore minerals themselves, but are by-products of the same chemical reactions that deposited the ore. Uranium is the earliest example of that type of satellite exploration. Spectral ratio images of LANDSAT MSS data were used to map the occurrence of secondary ferric oxides that had been laid down by geochemical cells in arkosic sandstones, which also caused uranium to be precipitated in the form of uranium oxides and salts. The reductant responsible for uranium ore precipitation is often pyrite that has resulted from ancient hydrocarbon seeps from deeper reservoirs. LANDSAT satellite remote sensing has proven to be a very cost-effective tool for uranium exploration, though the world market price of uranium is too weak to require much exploration at this time.

Copper porphyry deposits are another type of ore deposit that geological remote sensing has been effective at finding. They have up to four different zones of alteration (propylitic, argillic, phyllic, and potassic core) that contain suites of minerals, some of which are readily mapped by remote sensing methods. For instance, kaolinite and montmorillonite clays, ferric oxides and hydroxides, quartz, and other minerals can be used to map copper porphyry deposits, although to do the job thoroughly, multiple thermal spectral bands in the 8–14 μm region are required, along with the visible and reflective infrared bands of LANDSAT TM types of sensors. The multiple thermal infrared bands make possible the mapping of quartz content in exposed rocks and soils. With other types of hydrothermally altered deposits, sulfates like alunite play an even more important role than ferric oxides because they occur exclusively where hydrothermal alteration has taken place.

Mapping spectral reflectance changes in vegetation caused by elevated metal concentrations in residual soils can be used as an exploration tool for base metals and precious metals in vegetated terrain. This use will become an increasingly important, though complex, remote sensing method as hyperspectral sensor data from satellites and aircraft become more available in the future, especially for heavily vegetated terrain.

Massive sulfide deposits containing economic amounts of lead, zinc, and silver can be explored by mapping ferric oxide gossans, barite exposures, and magnetic highs in close proximity. LANDSAT MSS and TM can be used to map the gossans, TM can be used to map barite, and aeromagnetometer data can be turned into geophysical images for massive sulfide exploration. Smithsonite, a zinc carbonate that can occur in both massive sulfide and Mississippi Valley Type (MVT) lead-zinc deposits, could be mapped well with LANDSAT TM data. Ankerite and siderite are other carbonates found in MVT deposits that could be mapped uniquely from LANDSAT TM. Contact metamorphism around circular granitic intrusives that are productive of tungsten can be explored by utilization of spectral ratio images to enhance bright spots of ferric oxides that form a circular-shaped "necklace" pattern.

Multispectral remote sensing has shown more effectiveness for disseminated gold deposits than for almost any other type of metal deposit because of gold's mineral and spectral uniqueness. Recent evidence has supported the hypothesis that disseminated gold deposits are related to geysers and black smokers, where hot water is brought to the surface of the land or sea bottom, respectively, after coming into contact with hydrocarbons underground. Ferric oxides, clays, and alunite are all associated with hydrothermal alteration, and they all can be mapped with LANDSAT TM spectral ratio images. However, the introduction of hydrocarbons to the hydrothermal plume also produces rare, ammoniated feldspar (buddingtonite) and ammoniated alunite. The latter can be discriminated from nonammoniated feldspars and alunites by hyperspectral sensors that include several spectral bands in the 2.0–2.5 μm wavelength region. The spectral effects of ammoniation and the unique spectral features of alunite cannot be observed in the visible wavelength region by the naked eye, just as also the finely disseminated gold cannot be observed. These are "invisible" gold deposits in more ways than one, but multispectral remote sensing renders them detectable, and often identifiable.

Diamonds are found in ultramafic rocks called kimberlites and lamproites, which come from mantle material. If either of these ultramafic rock formations is exposed at the surface, multispectral thermal infrared spectral bands can be used to map them uniquely as ultramafic rocks. However, if they are eroded down beneath the surface, they can still be explored by the remote sensing search for circular depressions of a few kilometers in diameter or smaller. In semiarid terrain, circular clay pans often are associated with the slight depressions above weathered kimberlites. Circular magnetic highs, which can be displayed well in geophysical images, are also associated with kimberlite and lamproite pipes.

Carbonatites, which are volcanic carbonates in which the world's largest quantities of rare earth elements are found, are associated with alkalic basic silicate rocks and, like kimberlite and lamproites, occur in cratonic regions. The use of multispectral thermal infrared spectral bands can identify the alkalic basic silicates and confirm that carbonates are present. Ferrocarbonous carbonatites, which contain the most rare earth minerals, often contain the unusual carbonates siderite and ankerite. These carbonates have such unique spectral properties that LANDSAT TM spectral bands, properly ratioed, can be used to map them. Carbonatites also display high magnetic anomalies.

Industrial minerals have a low price per unit volume and must be found near the sites where they will be used in order for them to be economically valuable. Therefore, remote sensing and GIS methods must be integrated to perform effective exploration for industrial minerals. For sand and gravel deposits, multispectral thermal infrared spectral bands can be used to map silica content in exposed rocks and soils, and thermal infrared broad-band images of night and day overpasses can be used to map thermal inertia anomalies. Surface exposures of clays, gypsum, and sulfur deposits can be mapped with LANDSAT TM data, and phosphates can be mapped with multiple thermal infrared spectral bands because they have spectral features that are sufficiently characteristic for them to be mapped by multispectral methods. Hyperspectral sensors can be used to map surface exposures of zeolites, calcite, dolomite, and at least three different varieties of sericite. Quarrying operations for industrial minerals can benefit from digital photogrammetry by the automated volume estimation of overburden and excavation.

Exercises

1. Name two minerals that can be mapped with LANDSAT Thematic Mapper that are diagnostic of hydrothermal alteration, in that they are not also formed during the weathering of both altered and unaltered rocks.

2. (a) Name four types of deposits that have been cited in this course for which geobotanical indicators have been found by remote sensing methods.

   (b) Describe at least two problems associated with geobotanical prospecting by remote sensing methods and what can be done to minimize these problems.

   (c) If there are problems with geobotanical prospecting methods, why use them at all?

3. Greenbolia is a political entity that covers about 130,000 square kilometers in a domed cratonic area, with a modest seacoast along which beaches rich in nepheline syenite fragments are located. Last year the value of mineral exports exceeded the value of minerals extracted from and used in Greenbolia by about 5%. Magnetic and gravity data are available throughout the area. Your company has made you the project manager of a remote sensing study of Greenbolia that will result in a final report that locates natural resources and advises how the new resources should be exploited. You are instructed to keep exploration costs to a minimum.

   (a) What can you deduce about the state of the Greenbolian economy and how did you arrive at that deduction?

   (b) To what type of exploration target(s) would you give priority and why?

   (c) About how much and what type of data would you order?

   (d) What data processing would you perform and what would you look for with regard to each target(s)?

(e) What natural resource(s) could possibly be produced from the exploration target(s), should you be successful?

Cited References

ABRAMS, M. J., and B. S. SIEGAL. 1976. *Detection of Alteration Associated with a Porphyry Copper Deposit in Southern Arizona.* JPL Technical Memorandum 33-810. Pasadena, Calif.: Jet Propulsion Laboratory,

ABRAMS, M. J., J. E. CONEL, and H. R. LANG. 1984. *The Joint NASA/Geosat Test Case Project Final Report,* part 2, vol. 1, 3–13. available from AAPG Bookstore, Tulsa, OK.

ABRAMS, M. J., L. ASHLEY, L. ROWAN, A. GOETZ, and A. KAHLE. 1977. Mapping of Hydrothermal Alteration in the Cuprite Mining District, Nevada, Using Aircraft Scanner Images for the Spectral Region 0.46 to 2.36 μm. *Geology* 5:713–718.

ALTANER, S. P., J. J. FITZPATRICK, M.D. KROHN, P. M. BETHKE, D. O. HAYBA, J. A. GOSS, and Z. A. BROWN, 1988. Ammonium in Alunites. *American Mineralogist* 73:145–152.

ASHLEY, R. P., and M. L. SILBERMAN. 1976. Direct Dating of Mineralization at Goldfield, Nevada, by Potassium-Argon and Fission-Track Methods. *Economic Geology* 71:904–924.

ATKINSON, W. J., F. E. HUGHES, and C. B. SMITH. 1984. A Review of the Kimberlitic Rocks of Western Australia. In *Kimberlites I: Kimberlites and Related Rocks,* ed. J. Kornprobst, 195–224. Amsterdam: Elsevier Press.

BAUGH, W. M., and F. A. KRUSE. 1994. Quantitative Geochemical Mapping of Ammonium Minerals Using Field and Airborne Spectrometers, Cedar Mountains, Esmeraldo County, Nevada. In *Proceedings of the Tenth Thematic Conference on Geologic Remote Sensing,* vol. 2, 304–312. Ann Arbor: Environmental Research Institute of Michigan.

BIRNIE, R., and J. FRANCISCA. 1981. Remote Detection of Geobotanical Anomalies Related to Porphyry Copper Mineralization. *Economic Geology,* 76(3):637–647.

BOARDMAN, J. W., and F. A. KRUSE. 1994. *Proceedings of the Tenth Thematic Conference on Geologic Remote Sensing,* vol. 1, 407–418. Ann Arbor: Environmental Research Institute of Michigan.

COLLINS, W. 1978. Airborne Spectroradiometer Detection of Heavy-Metal Stress in Vegetation Canopies. 5th IAGOD Symposium, Snowbird, Utah.

CURRAN, P. J., and J. A. KUPIEC. 1995. Imaging Spectrometry: A New Tool for Ecology. In *Advances in Environmental Remote Sensing,* eds. F. M. Danson and S. E. Plummer, 71–88. New York: John Wiley & Sons.

EHMANN, W. J., and N. VERGO. 1986. Spectral Discrimination of Zeolites and Dioctahedral Clays in the Near-Infrared. In *Proceedings of the Fifth Thematic Conference on Remote Sensing for Exploration Geology,* vol. 1, 417–425. Ann Arbor: Environmental Research Institute of Michigan.

EVANS, A. M. 1993. *Ore Geology and Industrial Minerals.* 3rd ed. London: Blackwell Scientific Publications.

GILLESPIE, A. R., and A. B. KAHLE. 1977. Construction and Interpretation of a Digital Thermal Inertia Image. *Photogrammetric Engineering and Remote Sensing* 43, no. 8:983–1000.

GOETZ, A. F. H., and V. SRIVASTAVA. 1985. Mineralogical Mapping in the Cuprite Mining District, Nevada. In *Proceedings of the Airborne Imaging Spectrometer Data Analysis Workshop,* eds. G. Vane and A. F. H. Goetz, 22–31. JPL publication 85-41. Pasadena, Calif.: Jet Propulsion Laboratory.

GOETZ, A. F.H., G. VANE, J. E. SOLOMON, and B. N. ROCK. 1985. Imaging Spectrometry for Earth Remote Sensing. *Science* 228:1147–1153.

GULBRANDSEN, R. A. 1974. Buddingtonite, Ammonium Feldspar, in the Phosphoria Formation, Southeastern Idaho. *Journal of Research, U.S. Geological Survey* 2:693–697.

HORNSBY, J. K., B. BRUCE, J. HARRIS, and A. N. RENCZ. 1988. Implementation of Background
and Target Geobotanical Techniques in Mineral Exploration. In *Proceedings of the Sixth
Thematic Conference on Remote Sensing for Exploration Geology,* vol. 2, 511–521. Ann
Arbor: Environmental Research Institute of Michigan.

HUNT, G. R., J. W. SALISBURY, and C. J. LENHOFF. 1971. Visible and Near-Infrared Spectra of
Minerals and Rocks. IV. Sulphides and Sulphates. *Modern Geology* 3:1–14.

———. 1973. Visible and Near-Infrared Spectra of Minerals. VI. Additional Silicates. *Modern
Geology* 4:85–106.

JOSHI, A. K. 1988. Identification of Magnesite and Bauxite Deposits on LANDSAT Imagery,
South India. In *Proceedings of the Sixth Thematic Conference on Remote Sensing for Ex-
ploration Geology,* vol. 2, 475–483. Ann Arbor: Environmental Research Institute of
Michigan.

KROHN, M. D., S. P. ALTANER, and D. O. HAYBA. 1988. Distribution of Ammonium Minerals at
Hg/Au-Bearing Hot Springs Deposits: Initial Evidence from Near-Infrared Spectral Properties. In *Bulk Mineable Precious Metal Deposits of the Western United States, Symposium
Proceedings,* eds. R. W. Schafer, J. J. Cooper, and P. G. Vikre, 661–679. Reno: Geological Society of Nevada.

LEE, K., and G. L. RAINES. 1984. Reflectance Spectra of Some Alteration Minerals; A Chart
Compiled from Published Data, 0.4 μm–2.5 μm. Open-File Report USGSOFR840096, U.S.
Geological Survey, Reston, Va.

LOURIM, J. and R. A. H. BUXTON. 1988. A Spectral Geobotanical Survey of an Elevated Ar-
senic-Gold Property in Mountainous Terrain in British Columbia, Canada. In *Proceedings
of the Sixth Thematic Conference on Remote Sensing for Exploration Geology,* vol. 2,
613–621. Ann Arbor: Environmental Research Institute of Michigan.

LOWELL, J. D., and D. M. GUILBERT. 1970. Lateral and Vertical Alteration-Mineralization Zoning in Porphyry Ore Deposits. *Economic Geology* 65(4):373–408.

NIXON, P. H. 1980. The Morphology and Mineralogy of Diamond Pipes. *Kimberlites and Diamonds,* eds. J. E. Glover and D. I. Groves, 32–47. Nedlands: Extension Services, University
of Western Australia.

OFFIELD, T. W. 1976. Remote Sensing in Uranium Exploration. In *Uranium Exploration Geology,* 731–744. Panel Proceedings. Vienna: International Atomic Energy Agency.

RAINES, G. L., T. W. OFFIELD, and E. S. SANTOS. 1978. Remote Sensing and Subsurface Definition of Facies and Structure Related to Uranium Deposits, Powder River Basin, Wyoming.
Economic Geology 73(8): 1706–1723.

SALMON, B. C., and W. W. PILLARS. 1975. Multispectral Processing of ERTS-A Data for Uranium Exploration in the Wind River Basin, Wyoming. ERIM Report 110400-2-F. Ann
Arbor: Environmental Research Institute of Michigan.

SPATZ, D. M., and R. T. WILSON. 1994. Exploration Remote Sensing for Porphyry Copper De-
posits, Western America Cordillera. In *Proceedings of the Tenth Thematic Conference on
Geologic Remote Sensing,* vol. 1, 227–240. Ann Arbor: Environmental Research Institute of
Michigan.

SPIRAKIS, C. S., and C. D. CONDIT. 1975. Preliminary Report on the Use of LANDSAT-1
(ERTS-1) Reflectance Data in Locating Alteration Zones Associated with Uranium Mineralization Near Cameron, Arizona, U.S. Geological Survey Open File Report 75-416.

STERNE, E. J., R. C. REYNOLDS, JR., and H. ZANTOP. 1982. Natural Ammonium Illites from Black Shales Hosting a Stratiform Base Metal Deposit, DeLong Mountains, Northern Alaska. *Clays and Clay Minerals* 30, no. 3:161–166.

TAPPER, G. O., and D. A. DEMPSEY. 1988. MEIS II and Surface Data Integration for Detection of Geobotanical Anomalies. In *Proceedings of the Sixth Thematic Conference on Remote Sensing for Exploration Geology,* vol. 2, 499–508. Ann Arbor: Environmental Research Institute of Michigan.

TORCOLETTI, P. J. and R. W. BIRNIE. 1988. A LANDSAT Thematic Mapper Investigation of the Geobotanical Relationships in the Northern Spruce-Fir Forest, Mt. Moosilauke, New Hampshire. In *Proceedings of the Sixth Thematic Conference on Remote Sensing for Exploration Geology,* vol. 2, 541–550. Ann Arbor: Environmental Research Institute of Michigan.

VINCENT, R. K. 1977. Uranium Exploration with Computer-Processed LANDSAT Data. *Geophysics* 42(3):536–541.

VINCENT, R. K., D. H. COUPLAND, and J. B. PARRISH. 1981. HCMM Night-Time Thermal IR Imaging Experiment in Michigan. In *Proceedings of the Fifteenth International Symposium on Remote Sensing of Environment,* vol. 2, 975–984. Ann Arbor: Environmental Research Institute of Michigan.

VINCENT, R. K., P. K. PLEITNER, and M. L. WILSON. 1984. Integration of Airborne Thematic Mapper and Thermal Infrared Multispectral Scanner Data for Lithologic and Hydrothermal Alteration Mapping. In *Proceedings of the International Symposium on Remote Sensing of Environment, Third Thematic Conference, Remote Sensing for Exploration Geology,* vol. 1, 219–226. Ann Arbor: Environmental Research Institute of Michigan.

VINCENT, R. K., F. THOMSON, and K. WATSON. 1972. Recognition of Exposed Quartz Sand and Sandstone by Two-channel Infrared Imagery. *Journal of Geophysical Research* 77: 2473–2477. Later reprinted in *Geophysics Reprint Series No. 3: Remote Sensing,* 1983, eds. K. Watson and R. D. Regan. Society of Exploration Geophysicists.

VON DAMM, K. L., J. M. EDMOND, C. I. MEASURE, and B. GRANT. 1985. Chemistry of Submarine Hydrothermal Solutions at Guaymas Basin, Gulf of California. *Geochim. Cos. Acta* 49: 2221–2237.

WATSON, K. 1975. Goelogic Applications of Thermal Infrared Images. *Proceedings of the IEEE,* 63, no. 1:128–137.

WHITE, D. E., and C. E. ROBERSON. 1962. Sulphur Bank, California, A Major Hotspring Quicksilver Deposit. In *Petrologic Studies: Buddingtonite Volume,* eds. A. E. J. Engel, H. L. James, and B. F. Leonard, 397–428. Geological Society of America.

Additional References (Uncited)

CROWLEY, J. K., and N. VERGO. 1985. Visible and Near-Infrared (0.4–2.5 μm) Reflectance Spectra of Selected Mixed-Layer Clays and Related Minerals. *Proceedings of the Sixth Thematic Conference on Remote Sensing for Exploration Geology,* vol. 2, 597–606. Ann Arbor: Environmental Research Institute of Michigan.

8

Remote Sensing Applications to Petroleum and Ground Water Exploration

▼ ▼ ▼ ▼ ▼ ▼ ▼ ▼ ▼ ▼ ▼ ▼ ▼ ▼ ▼

INTRODUCTION

In the early and mid-1970s, when satellite remote sensing was in its infancy, most geologists were convinced that the primary use of geological remote sensing would be for mineral exploration. With the notable exceptions of petroleum explorationists at Exxon along with a minority of oil companies that had similarly emphasized the use of photogeology from aerial photos, few thought that remote sensing could contribute significantly to petroleum exploration. Because electromagnetic radiation penetrates only the topmost millimeter or so of materials at Earth's surface, multispectral remote sensing was thought by many to be almost useless in the search for oil and gas reservoirs, which occur deep underground. This expectation turned out to be false, but proof that remote sensing could be useful to petroleum exploration took almost a decade of research by a host of geologists from many different petroleum companies and geological remote sensing service firms. This research, which was aided by the 1976 formation of a nonprofit consortium called the Geosat Committee, showed that the probable occurrence of oil and gas reservoirs, some of which were many thousands of feet deep, could be inferred from structural features and compositional variations at Earth's surface, as mapped by remote sensing methods. Such exploration will be the subject of the next discussion, in the longer of the two sections that make up this chapter.

The search for ground water has many similarities with the search for petroleum; this similarity is not surprising because both petroleum and ground water reservoirs usually involve a porous, permeable sedimentary stratum that has been tilted and/or displaced by folding or faulting, thereby providing a trap for the migrating fluids. Ground water reservoirs usually occur at far shallower depths than petroleum reservoirs, and there is a marked difference between the effects that ground water and hydrocarbons have on Earth's surface when they escape along faults that intersect both the reservoir and the surface. However, the mapping of surface expressions of underground geologic structures by remote sensing methods is a common element in the exploration for both petroleum and ground water. The second section of this chapter deals with remote sensing for ground water exploration. Over the next decade, as the world's supply of fresh water continues to fall increasingly short of demand, this exploration will likely become even more important than petroleum exploration.

THE USE OF REMOTE SENSING FOR PETROLEUM EXPLORATION

A commercial deposit of oil and gas requires three basic elements: the presence of hydrocarbons from source rocks that are now or once were connected to the commercial deposit by permeable paths of migration; rock strata that are permeable enough to permit the migration of fluids and gases over a substantial distance and porous enough to store them; and structural or stratigraphic traps that stop the migration and permit the pooling of fluids in the porous reservoir rock (often the same rock stratum through which the fluids have migrated). Figure 8.1 diagrams several types of traps that help form underground oil and gas reservoirs (Holmes 1965). Note that in three of these types, rocks exposed at the surface are different than they would have been in the absence of the structural trap that created the reservoir. Stratigraphic traps are the most difficult to find because there are no structural clues of their presence at the surface and there is nothing to act as a strong reflector of seismic waves, as there would be in the other four types of traps.

However, all oil and gas reservoirs become cracked over the millions of years that they are in existence, and some gas and oil escapes to the surface via these cracks. In the case of fault traps, gas and oil also can escape along the trap-forming faults. However they escape, hydrocarbons interact with rocks and soils above the reservoir and alter their chemical composition. Thus, there are both structural and geochemical clues possible in surface rocks and soils that can indicate the presence subsurface oil and gas reservoirs.

Petroleum exploration with remote sensing data has at least three aspects that will be dealt with in the following subsections: structural mapping, which is the search for geologic structures that could trap commercial quantities of petroleum; ancient seep mapping, which searches for geochemical staining of surface rocks and soils caused by past hydrocarbon escape along faults and other zones of relatively high permeability; and ongoing hydrocarbon seep mapping, which onshore often involves the search for abnormal vegetation that has been altered by hydrocarbons (geobotanical anomalies) and offshore becomes the search for oil slicks and gas bubbles.

Structural mapping primarily involves photo interpretation and spatial processing to far greater extent than multispectral image processing. Contrastingly, the

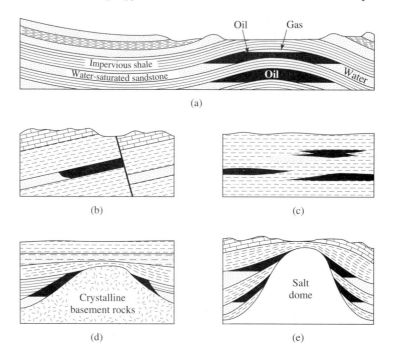

Figure 8.1 Illustrations of various types of traps favorable for the accumulation of oil and gas: (a) anticline; (b) fault; (c) stratigraphic;(d) crystalline basement rocks; and (e) salt dome. Notice that traps of types a, b, and e (sometimes also type d) have an influence on the types of rocks exposed at the surface. Stratigraphic traps (type c) do not influence the types of rocks exposed at the surface. *(After Holmes 1965)*

mapping of geochemical anomalies associated with ancient hydrocarbon seeps, the mapping of vegetative anomalies associated with recent hydrocarbon seeps onshore, and the mapping of oil slicks offshore involve more multispectral image processing than spatial processing. The following subsections deal with these aspects of remote sensing for petroleum exploration and case studies are given for each of them.

Structural Mapping

For decades, photo geologists have used aerial photography for mapping rock out-crops and surface topographic expressions of geologic structures that could provide traps for underground oil and gas reservoirs. It is therefore not surprising that the earliest and most extensive use to date of satellite multispectral scanner data for petroleum exploration has been structural mapping, a natural extension of photo geology.

The LANDSAT series of Earth resources satellites are able to offer two advantages over aerial photographs for structural mapping: a synoptic view and multi-spectral imaging. The synoptic view permits geologists to examine entire basins within just a few images for linear and curvilinear features indicative of underlying geologic structures that would have required hundreds or thousands of aerial photos to cover. Multispectral imaging, even with the oldest LANDSAT MSS data, provides contrasts among surface rock exposures of different chemical compositions that are often difficult or impossible to observe in black and white aerial photographs. This imaging aids the geologist in mapping faults with vertical displacements that are large enough to cause rocks of different compositions and ages to occur on either side of the fault. It is also used for mapping compositional differences between older rocks exposed on the tops of breached (eroded) anticlines and younger rocks on anticlinal flanks. Also, particularly in semiarid terrain, the enhancement of vegetation in false-color images of LANDSAT images permits easier mapping of faults that have created moist zones of ground water along the fault (including strike-slip faults) or perched reservoirs of ground water on the up-thrown side of vertical faults.

The use of LANDSAT MSS data for photogeologic mapping of structural features related to petroleum reservoirs was documented by Halbouty (1976, 1980). In the second of those two papers, Halbouty applied photo interpretation of LANDSAT MSS images to 15 giant oil and gas fields around the world and found that most of these fields displayed structural features in the satellite images that could have led to their discovery, had they been still undrilled. Miller (1975) interpreted LANDSAT MSS images to map major linear features that appeared to form boundaries for a sedimentary basin in eastern Kenya; these features were later confirmed by geophysical and subsurface data. Sabins (1978) showed that photo interpretation of LANDSAT MSS data could add details to existing reconnaissance maps and provide a cartographic base for use in petroleum exploration and development programs. More recently, Berger (1994) has detailed the use of photo interpretation aerial photos and satellite images for structural mapping, and readers wanting more about the subject should seek that reference.

Geophysical images of aeromagnetic or gravity data, an adaptation of remote sensing image processing, have provided a useful adjunct to satellite images for structural mapping by yielding information about deep-seated lithologic boundaries and faults that are related to basement structure, whether or not there may be surface expressions of the structure. As stated in the chapter on spatial image processing, geophysical images provide such cost-effective information on basement structures that they are arguably the most cost-effective preleasing tool for off-shore tracts and for deciding the locations of off-shore seismic lines.

Again with the exception of Exxon geologists and a few other experienced practitioners of photogeology, the value of photo interpretation with either aerial photographs or satellite images was doubted over covered basins (such as glacial terrain) up through the early years of LANDSAT. This doubt was owing to the limitations of aerial photography in the past (especially their small-area coverage) and to a lack of theoretical explanation for the possibility of surface expressions of older faults that were now covered by glacial till. Those doubts were largely removed in the last decade by experimental evidence. Some linear features seen in LANDSAT images of glacially covered basins proved to be traces of faults in the underlying bedrock, as observed in seismic lines. These observations are particularly noted in the two case studies below.

An explanation for why these faults have an observable surface expression, sometimes through hundreds of feet of glacial till that is considerably younger than the latest reactivation dates of the observed faults, was given by Prost (1988). He experimented with glass boxes of uneven wooden blocks (with 6–12 mm of height difference), over which several centimeters of dry and damp sand were poured, respectively, and found that after five minutes of vibrating the box with lateral vibrations (like seismic Love waves) at a frequency of 26 Hz (1550 rpm), the damp sand had formed fault scarps over the edges of the uneven basement blocks. These tests confirm that glacial till or alluvium subjected to repeated vibrations can propagate fractures upward from the margins of irregular basement blocks. Preferential compaction, first described by Barosh (1968), seems to be part of the answer. Draping of sediments over underground structures and the trapping of water along fault zones are other phenomena that can yield surface expressions in covered basins of underground structures (Berger 1984; Berger et al. 1984), with little or no help from seismic shaking.

Part of the proof for the existence of some of these covered basin faults has been the coincidence of LANDSAT linear features with linear features in geophysical images of magnetic and gravity data, sometimes followed by vertical offsets in seismic lines that cross the linear features. The structural interpretation of LANDSAT images and geophysical images and their resultant comparison with seismic data has been especially useful in exploring covered basins in the United States. The exploration has been for deep-seated geologic structures that provide traps that may be, or have been proved to contain, commercial oil and gas deposits. Two examples of the joint use of LANDSAT and geophysical images for photogeologic petroleum exploration in covered basins of the United States will now be given, both of which have yielded seismic evidence for structures mapped by remote sensing, but only one of which has been drilled and found productive of hydrocarbons.

The first case study, which took place in the mid-continental region of the United States, comes from Herman et al. (1984, 1985). Gold and Soter (1980) published a hypothesis that accumulations of mantle-derived, abiogenic methane might occur in commercial deposits deep underground at places for which avenues of escape existed for gas from the mantle into the crust. In 1981, a small, independent oil company contracted for a remote sensing study to find a frontier play with deep gas potential. The Mid-Continent Geophysical Anomaly (MGA), thought to be an old rift zone, was selected for study with the thought that a rift zone would provide escape routes for methane from the mantle. This idea, coupled with the discovery a few years earlier of deep gas along the Mid-Michigan Geophysical Anomaly and the known occurrences of Precambrian oil shale (the Nonesuch Shale) in the Upper Peninsula of Michigan along the northern extension of the MGA, led GeoSpectra geologists to suspect that there could be at least two possible sources of gas in Precambrian sediments along the Iowa portion of the MGA: abiogenic mantle gas and the Nonesuch Shale of the Oronto Group.

Since the MGA in Southwest Iowa forms the northwestern border of a known, hydrocarbon-producing basin (the Forest City Basin), it was decided that the eastern flank of the MGA in Iowa should be the general target of exploration. Linear features were automatically mapped by LIRA (Linear Recognition and Analysis computer program, described in Chapter 6) from LANDSAT MSS data and linear features were photo-interpreted from geophysical images of aeromagnetic and gravity data in the vicinity of the intersection between the Central Iowa Arch and the MGA. Figure 8.2 shows a generalized map of the area, with the major structural features displayed.

After these linear features were studied, priority areas for leasing were recommended, and the oil company proceeded to lease 120,000 acres in Adair, Guthrie, and Audubon Counties in Iowa, the locations of which are shown in Figure 8.3. The locations of holes drilled for oil and gas exploration and for gas storage are also shown in Figure 8.3. The remote sensing geologists picked six three-mile-long seismic lines that were nearly perpendicular to six prominent linear features mapped from LANDSAT and geophysical images. The oil company had those lines collected by Seiscom Delta, with 125 ft spacing between geophones and dynamite as the seismic source. The six seismic line locations that were selected and collected are shown in Figure 8.3.

Good correspondences were found between linear features in geophysical images of Bouger gravity anomalies and several anticlines and faults identified on the six seismic sections. Linear features from LANDSAT and geophysical images of topographic (DEM) data also corresponded to several structures identified on the seismic lines. These included several large anticlinal structures (along two seismic lines) in the Paleozoic section that may be related to the Central Iowa Arch. Large faults that displace Paleozoic and Precambrian units were identified on two of the seismic lines, and a pinch-out of Cambrian or upper Precambrian strata was

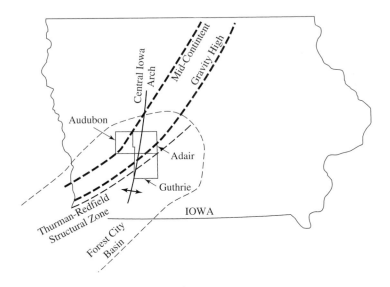

Figure 8.2 Location of the Iowa three-county study area and major geological features in the area, including the Central Iowa Arch, Thurman-Redfield Structural Zone, Forest City Basin, and Mid-Continent Gravity Anomaly (MGA). *(After Herman et al. 1985. Reprinted with permission of the Environmental Research Institute of Michigan.)*

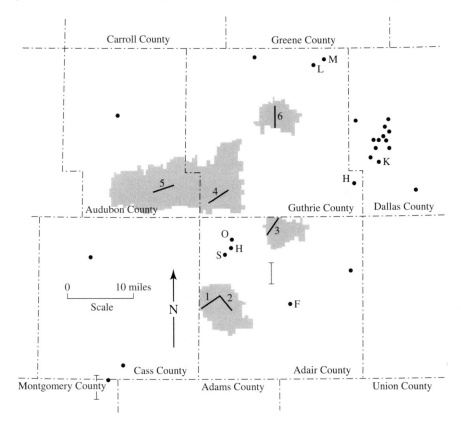

Figure 8.3   Locations of lease blocks, seismic lines 1 through 6, and holes drilled for either oil and gas exploration or for gas storage in the Iowa three-county study area. *(After Herman et al. 1985. Reprinted with permission of the Environmental Research Institute of Michigan.)*

observed along another. The most interesting seismic line, however, was line 6, shown in Figure 8.4. A possible thrust fault is indicated by several strong, coherent reflectors that are apparent along the entire section between 2.0 and 3.4 seconds of two-way travel time, reflectors that are probably located at depths in excess of 14,000 ft. This indicates that a large section of Precambrian sediments and possibly some Precambrian volcanics have been thrust over other Precambrian sediments by a thrust fault in a generally North-South direction. A gravity linear feature that trends NW-SE (approximately perpendicular to the MGA) crosses the center of the seismic line, leading to the hypothesis that the feature in question may be along a NW-SE trending transform fault that has been overthrust toward the south.

Thrusting related to transform faults have been postulated by Freund (1982) near the East African Rift, and Bonatti and Crane (1984) showed that compression along transform faults can cause uplift of slivers of deep crust. For a location closer to this area in Iowa, Craddock (1972) described the Belle Plaine Fault in southeastern Minnesota along the MGA as an apparent right-lateral, strike-slip fault or

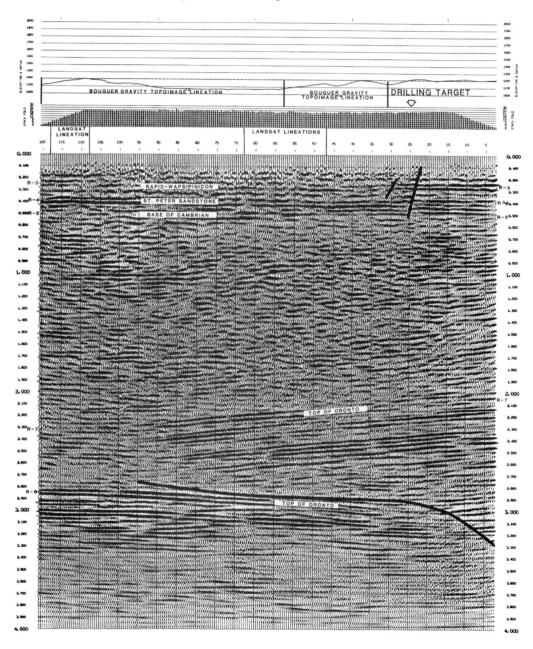

Figure 8.4 Seismic section for line 6, showing overthrusting of the Oronto Group (Precambrian sedimentary rocks) at two-way travel times greater than 20 seconds (a depth of approximately 5000 meters). *(After Herman et al. 1985. Reprinted with permission of the Environmental Research Institute of Michigan.)*

transform fault with a reverse faulting component involving a minimum of 700–1000 feet of throw.

Travel time estimates in Figure 8.4 suggest that the deep reflectors involved in the overthrust may originate from the Oronto Group of Precambrian sediments that are described in Wisconsin and Michigan parts of the MGA. If so, significant oil and gas deposits could occur from the possible presence of the Nonesuch Shale (an oil shale). The anticlinal structures related to the thrust faulting could also provide ideal traps for gas originating from mantle gas, as well as from the Nonesuch Shale. Other trapping mechanisms have also been suggested for this seismic line (Herman et al. 1984, 1985).

Because of the downturn in U.S. petroleum exploration, not a single exploratory well of any depth was drilled in this entire 120,000 acre lease block, although Amoco did drill a deep, dry hole on what may have been an extension of the same transform fault on the opposite side (northwest) of the MGA from the seismic line shown in Figure 8.4. Amoco found porous rocks in the Precambrian and some gas shows in the Paleozoic section, but it is not known whether overthrusting was found in their location. However, the side of the MGA where the lease block from this case study was located may have produced better results because it is the edge of a known productive basin (the Forest City Basin). This lease block played an important role in oil companies starting the Mid-Continent lease play that covered millions of acres (mostly undrilled before the exploration downturn). The lease block has long since expired, but is still a good gas prospect.

The most important lesson to be learned from the project described above is that structural interpretation of LANDSAT together with geophysical images led to the discovery of major subsurface structures with the collection of only 18 line-miles of seismic data. Hundreds of line-miles of seismic data may have been required to find the apparent overthrusting shown in Figure 8.4, had not remote sensing been applied first, to prioritize areas for land leasing and seismic data collection (without tipping off competitors as to the location of the structures). This case study shows that oil companies that fail to use remote sensing as a tool prior to leasing and seismic collection are leaving money on the table.

The second example of the application of structural interpretation to LANDSAT and geophysical images to petroleum exploration in a covered basin, taken from Herman et al. (1991) for the Michigan Basin, has resulted in at least three deep gas wells (the largest producing about 2 million cubic feet of gas per day) on the original lease block, although not for the original lease holders for whom the exploration project was performed. The story begins in 1975, when a linear feature in Bay County, Michigan was mapped from LANDSAT images that appeared to extend the straight, linear southeast edge of Saginaw Bay inland by 20 miles or longer. Later field work showed that this linear feature was a long "moisture stripe" that was slightly lower (by a few feet) than the surrounding terrain. In fact, every time the linear feature crosses a North-South road, the road sags a few feet and standing water accumulates after a rain. It was with this standing water that the linear feature was most easily observed in the field. In light of later developments, it is probable that this linear feature could also have been detected in directional gradient-enhancement images of high-resolution digital elevation models (DEM) of 1-meter (or less) posting interval. A

DEM produced from NHAP stereo photos (1:80,000 scale photos available from the U.S. Geological Survey for the entire United States) by every-pixel-measured digital photogrammetry software, like ATOM, would have resulted in 1-meter postings, if the photos were scanned at 2,000 dpi. Directional gradient-enhancement images of government-produced DEMs, the best of which have 30-meter postings, likely would not reveal such a subtle topographic feature.

Ben Drake (later with Amoco International) was assigned the task of structural interpretation. With the help of LANDSAT images, topographic data, and very sketchy well log data, he hypothesized that a graben (which he called the Saginaw Bay Graben) affecting rocks of Silurian age and older existed under the landward, southwest extension of Saginaw Bay (Drake 1976; Drake and Vincent 1975). About 14,000 acres were leased on both sides of the so-called graben in the late 1970s; these acres were transferred to Wiser Oil with a prospect report that recommended drilling to Ordovician-aged rocks of about 9,000- to 10,000-ft depths. During the 10-year life of the leases, Wiser and Dart Petroleum proceeded to drill three holes to Devonian (maximum depth about 3500 ft). No producing wells resulted, but two of the holes exhibited good shows of oil and gas. No holes were drilled to rocks of Silurian or Ordovician age during that time.

In the late 1980s, after the leases had expired, three deep gas wells were struck in the Lower-Ordovician-aged Prairie du Chien Group (topped by the Glenwood Formation) on those former leases, and several other deep gas wells were drilled by Shell on other acreage in the vicinity of the Saginaw Bay "Graben." Herman et al. (1991) later published more complete remote sensing data for the area, with a new interpretation that the Saginaw Bay "Graben" was actually a more complex structure, possibly controlled or influenced by the Grenville Front. Figure 8.5 shows the location of the study area. Figure 8.6 shows the study area only, denoting the location of the "graben" limits as mapped by Drake (Drake and Vincent 1975) and three transects of well logs taken partially from new deep-gas wells in Fraser Township of Bay County that were drilled subsequent to the original study. The Shell Pross 1–12, Federated Natural Resources Metz 1–15, and the Shell Western LaHar 1–7 wells are on leases that were once acquired as a result of the original study, but they had expired before deep drilling occurred. The seismic line shown in Color Plate CP12 as a multifrequency color image was collected about 1 mile west of the Federated Natural Resources well, which is denoted as well log data site 5 on transect BB' of Figure 8.6.

Geophysical images of magnetic, gravity, and elevation data were produced and interpreted. Whereas directional gradient enhancement (shaded relief) images are first-derivative images, some second-derivative images of magnetic and gravity data were also produced for this study. Figures 8.7, 8.8 (p. 228), and 8.9 (p. 229), respectively, show the top of the Devonian from the A–A' well-log transect, the tops of the Silurian section and the Glenwood Formation (the top of the Lower Ordovician rocks) from the B–B' transect (which is parallel to the "graben"), and the tops of the Silurian section and the Glenwood Formation from the C–C' transect. By Drake's earlier interpretation, the "graben" affected Silurian and older rocks, but not the Devonian, as transect A–A' attests in Figure 8.7. The magnetic and gravity linear features mapped from the geophysical images are shown at the top and bottom,

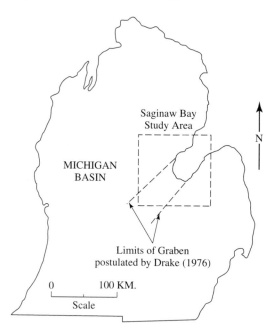

Figure 8.5 Locations of Saginaw Bay "graben" and study area discussed in detail in this chapter, relative to the Michigan Basin, which is centered near the center of the state of Michigan. *(After Herman et al. 1991 in Geological Society of America Special Paper 256. Reprinted with permission.)*

respectively, of Figures 8.7 and 8.9. The LANDSAT linear feature that started the whole study is marked as the "graben" limit closest to A' in Figure 8.7 and closest to C' in Figure 8.9, and it coincides fairly closely with magnetic and gravity linear features in both figures (denoted by M5, M6, G1, and G7).

Gravity images are useful for delineation of density contrasts within the basement, while magnetic images reveal magnetic susceptibility contrasts at or near the surface of the basement. Determination of basement boundaries is important for hydrocarbon exploration. If the basement boundaries are also fault contacts, they could have been reactivated during basin subsidence, causing faulting or folding of younger strata. Also, if basement boundaries separate lithologies of greatly contrasting erosion resistance, faults in the overlying Paleozoic rock units could be caused by their draping (through differential compaction during deposition and diagenesis) over paleotopographic differences produced by erosion during earlier subaerial exposure of the basement rocks.

The complete study (Herman et al. 1991) concludes that important basement structural features occur at the southwest edge of Saginaw Bay and appear to have influenced deep gas reservoirs. The original model appears to be too simplistic, however. Northeast-trending, basement-influenced, monoclinal or anticlinal structures within the Paleozoic section do appear to extend southwest from the northwestern and southeastern sides of Saginaw Bay. The southeastern side (about where the LANDSAT linear occurs) may have been influenced by Precambrian basement faults or contacts that developed during the formation of the Grenville Front, and their intermittent movement along their Northeast trends during Paleozoic times may have controlled sedimentation patterns, at least through Silurian times. The deep deposition center postulated originally by Drake inside the "graben" at the

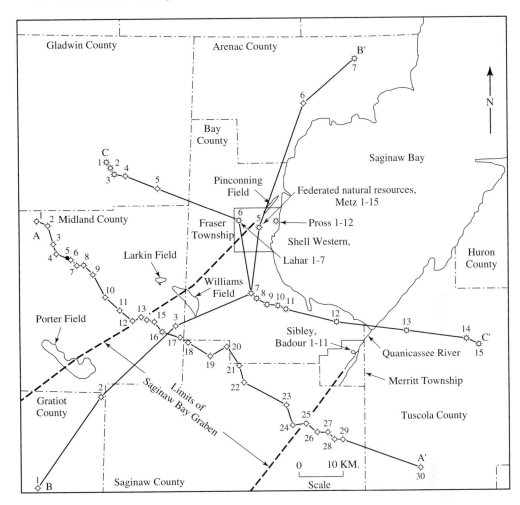

Figure 8.6 Enlarged view of the Saginaw Bay study area, showing boundaries of the Saginaw Bay "graben," cross-section locations, and other features discussed in the text. This area is outlined on succeeding figures and labeled SBSA. (*After Herman et al. 1991 in Geological Society of America Special Paper 256. Reprinted with permission.*)

southwestern end of Saginaw Bay also appears to be part of the reason for the deep gas play in Bay County. Structural interpretation of remote sensing images for petroleum exploration has thus proved useful even for deep (3–5 km) hydrocarbon reservoirs in a basin covered by several hundred feet of glacial till.

Remote Sensing Exploration for Ancient Hydrocarbon Seeps

In the foregoing discussion, multispectral remote sensing played a supporting role to the major roles played by the synoptic view of satellite data and to the utility of geophysical images. However, in this subsection and the following one, the multispectral

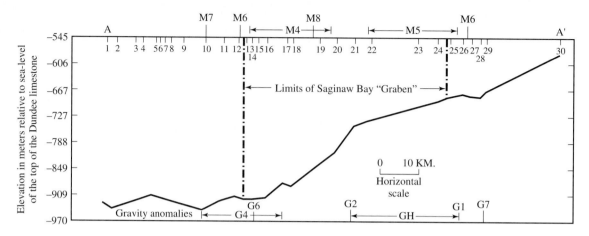

Figure 8.7 Structure section A–A′ showing the elevation of the top of the Devonian Dundee Limestone relative to sea level. The locations where various magnetic and gravity anomalies and linear features (mapped from gravity and magnetic images) and the Saginaw Bay "graben" limits (mapped from LANDSAT images) intersect the section are also shown. Location of the section is shown in Figure 8.6. *(After Herman et al. 1991 in Geological Society of America Special Paper 256. Reprinted with permission.)*

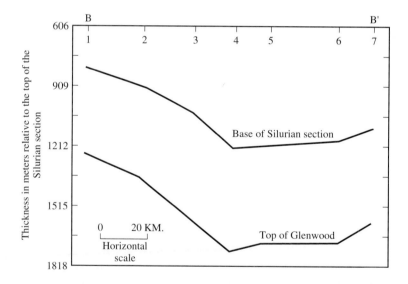

Figure 8.8 Thickness section B–B′ showing the thickness of the Silurian section and the section from the base of the Silurian to the top of the Ordovician Glenwood Formation relative to the top of the Silurian (top of the Silurian datum). Location of the section is shown in Figure 8.6. *(After Herman et al. 1991 in Geological Society of America Special Paper 256. Reprinted with permission.)*

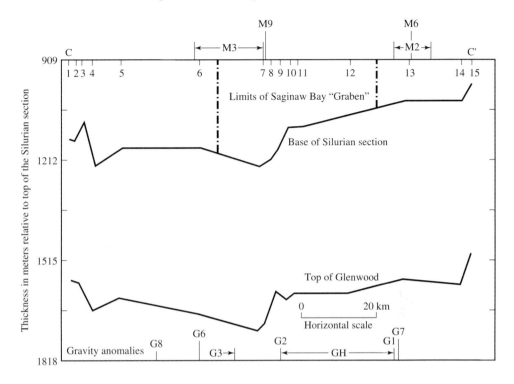

Figure 8.9 Thickness section C–C′ showing the thickness of the same sections depicted on section B–B′. Locations where various magnetic and gravity anomalies and linear features intersect the section are also shown. Location of the section is shown in Figure 8.6. *(After Herman et al. 1991 in Geological Society of America Special Paper 256. Reprinted with permission.)*

aspect of remote sensing becomes vital, for in these discussions the spotlight is on the detection of hydrocarbon seeps, both ancient and recent. The direct visible observation of geologists has thus far found most of the world's major oil and gas fields; these observations were of currently active oil and gas seeps that brought oil or gas to the surface from underlying hydrocarbon reservoirs. If the locations of ancient seeps that are no longer bringing visible amounts of oil and gas to the surface could be detected, the most successful means of discovering commercial deposits of oil and gas could be extended through past ages of seeps.

Because of the lithostatic pressure of the rocks above them, oil and gas will rise from a reservoir if they find an avenue of escape. Some hydrocarbons escape along the very faults that produce the trap for structural reservoirs. If a fault that intersects the initial reservoir also intersects the surface, seeps can occur without having to reside for some time in a higher, secondary reservoir. Sometimes another trapping fault in the secondary reservoir will intersect the surface, and the hydrocarbons will escape along it, or ever higher in the stratigraphic column, escape can occur in a repeated series of faults in tertiary reservoirs. Even stratigraphically trapped reservoirs can become faulted over millions of years; this situation offers an opportunity for some of the trapped oil and gas to escape to the surface. In fact, it is difficult to imagine any oil or gas reservoir that has not been faulted some time later than the time of hydrocarbon genesis or emplacement in that reservoir. Obviously, the escape routes usually permit only small quantities relative to the total amounts of hydrocarbons in the reservoir to escape, else there would be no commercial deposits of oil or gas. But what happens to the rocks along the avenue of escape? Are they not subjected to an unusual sort of geochemistry, and are they not altered or marked by it in some way that endures after those rocks are exposed by erosion? Can chemical alterations associated with ancient hydrocarbon seeps be detected in those rocks by multispectral remote sensing methods?

As a consequence of being a LANDSAT I principal investigator, I spent considerable time studying ferric oxide exposures in the Wind River Basin, primarily for mineral exploration, as discussed in Chapter 7. While in the field, I became aware of what at first seemed a coincidental occurrence of ferric oxides on and adjacent to oil and gas fields in that basin. Since most of the Wind River Basin consists of bland-colored sediments, the presence of ferric oxides were somewhat unusual, except near the basin edges, where Triassic redbeds of the Chugwater Formation were exposed. I made a plot of known field locations (from the 1:250,000 topographic map) occurring in the basin and overlaid the plot on a LANDSAT MSS color spectral ratio image with MSS spectral ratios $R_{5,4}$, $R_{6,5}$, and $R_{7,6}$ displayed as red, green, and blue, respectively, as shown in Color Plate CP5. Petroleum occurrences that coincided with ferric oxide exposures, as mapped from the color ratio image, appeared to fall into two categories. First, several commercially productive, classic anticlines (denoted in CP5 by numbers 1–4), such as Dallas Dome (denoted in the image by the number 1), contain exposures of Triassic Chugwater Formation at their centers. Here primary redbeds, which appear as a light orange color in CP5, have been exposed by the erosion of anticlinal structures. The primary redbeds acted merely as marker beds for older rocks at the centers of exposed anticlines; this process helped prove the value of surface compositional mapping for structural information. However, localized ferric oxide exposures in ordinarily bland-colored Tertiary-aged sediments farther out in the basin were somewhat more intriguing, because many of them (appearing darker orange in CP5) were also on or adjacent to known oil and gas fields. One of those (Vincent 1975) is the Beaver Creek Field, denoted with a number 5 in CP5. It occurs as an erosional remnant in Tertiary-aged rocks that have so much ferric oxide as to have once been mistakenly mapped as part of the Triassic-aged Chugwater Formation. Could those ferric oxides that are dark orange in CP5 be secondary iron oxides caused by ancient seeps of oil and gas?

Terrence Donovan (1974) published a paper that showed that now-exposed reservoir rocks (normally redbeds) of Permian age had been bleached by reducing fluids produced when hydrocarbons seeped through micropores of gypsum, altering calcium sulfate to calcium carbonate, and releasing H_2S. The subsequent H_2S-enriched fluids then reduced Fe^{3+} in the redbed reservoir rocks to Fe^{2+}. The latter is soluble enough in the reducing fluid to be carried away by it. Thus, the reservoir rocks had become white and almost iron-free, containing abundant calcium carbonate. Extending what was learned from that paper, I speculated (Vincent 1975) that if rocks now exposed at the surface had been located along the escape route above the reservoir at the time of the seep, they could have become enriched in iron from the same type of micropore seepage mechanism, via precipitation of pyrite (FeS_2). (The mechanism could work even if the gypsiferous unit were along the escape route, instead of in the reservoir rock.) Subsequently, the FeS_2 could be oxidized into ferric oxides by oxygenated ground water or surface water run-off. Thus, if this extension of Donovan's findings proves correct, oil and gas reservoirs located at or below gypsiferous strata could "stain" younger rocks at or near the surface with ferric oxides through faulted escape routes that provided connectivity between the petroleum reservoir and the surface.

One potential piece of evidence for this extended theory is the coincidence of ferric oxides in otherwise bland-colored Tertiary-aged sediments of the Wind River Basin, Wyoming with known occurrences of petroleum reservoirs, some of which have been discovered since 1975. The large gas field near Lost Cabin (marked as number 6 in CP5) in the Northeastern part of the basin, just south of the Big Horn Mountains, presents an interesting story. The elliptical-shaped light-orange feature in the color ratio image that extends roughly West-to-East in that location was noted in 1972 as being bounded on the north side by a mapped reverse fault and coinciding with a mapped synclinal axis. Field trips to the region revealed extensive ferric oxide (and possibly jarosite) exposures that appeared to be staining the surface sandstones of Eocene age. Mapped sulfur springs were located in a few places within the elliptical ferric oxide anomaly. In 1974, this area was recommended to a major oil company as a good prospect for gas reservoirs, under the assumption that the ferric oxides at the surface were caused by massive hydrocarbon escape along the reverse fault from underground gas reservoirs. Gas was thought more likely than oil because of the great areal extent of the iron oxide anomaly. The major oil company never drilled it, but an independent oil company found deep gas at a depth of about 15,000 ft on the edge of the anomaly a few years later. In the late 1980s, Louisiana Land and Exploration Co. purchased the wells and leases in that area and started drilling for deep gas, partly guided by the elliptical iron oxide anomaly (number 6 in CP5). They have since hit over 50 deep gas wells near that anomaly and are far from finished with it, if natural gas prices behave reasonably. Oil fields numbered 7 and 8 in CP5 were recommended for drilling on the basis of this spectral ratio image and were later productively drilled, though not by the clients who received the recommendations.

One of the greatest case studies concerning the creation by escaping hydrocarbons of large ferric oxide anomalies that are detectable from satellite multispectral scanner data is the Lisbon Valley, Utah study, which was performed as a uranium test case for the NASA/Geosat Test Case Project (Abrams et al. 1984a). Lisbon Valley is a NW-trending subsequent stream valley cut along bleached salt anticlines in the Paradox Basin of the Colorado Plateau (Wood 1968), and it is located in southeastern Utah, near the Colorado state line. Uranium ore is located in bleached sections of Wingate Sandstone of Upper Triassic age. The Wingate sandstone in the area consists of a massively bedded red sandstone lower unit (which the NASA/Geosat report authors call unaltered); a thinly interlayered, alternately-bleached-and-pigmented middle unit called the laminated unit; and an upper unit consisting of predominantly bleached rock (called altered). The uranium ore is found in the older, lower

Triassic rocks of the Chinle Formation (the Moss Back Member) and in the still older, Permian-aged Cutler Formation. However, bleached Wingate Sandstone overlies almost all of the uranium ore deposits, which in turn occur in bleached sections of the Chinle and Cutler Formations.

Even more interesting, oil and gas reservoirs in older Mississippian rocks have production limits that, if extended upward, almost completely enclose the distribution of uranium deposits and the distribution of bleaching. This configuration is demonstrated by Figures 8.10 and 8.11 (both from Abrams et al. 1984a). Figure 8.10 shows the positions of the oil and gas field limits and the mined uranium deposits, as well as the bleached, partially bleached, and unbleached outcrops of the Wingate Sandstone. Most of the uranium deposits are underlain by gas reservoirs and overlain by bleached or partially bleached Wingate Sandstone. Most of the gas fields are overlain by bleached or partially bleached Wingate Sandstone, where Wingate outcrops occur. The gas field in the middle of T30S and R25E in Figure 8.10 has no Wingate outcrops; it would be interesting to see if well cores from the gas wells of that field show bleached underground Wingate Sandstone overlying this gas field, too.

Figure 8.11 is a lithologic cross section of a transect (E–E′ in Figure 8.11), showing the generalized distribution of Wingate Sandstone bleaching and the relations of bleaching to uranium mineralization and oil and gas accumulations. Abrams et al. found that the bleached Wingate Sandstone could be readily discriminated from the unbleached Wingate, as well as from most other rock exposures, with data from the NASA-owned, Bendix 24-channel airborne multispectral scanner, as is shown in Color Plate CP25. Bleached (whitish) Wingate Sandstone is displayed as red, and unbleached (very red) Wingate Sandstone is displayed as white and light yellow in CP25. That figure shows a canonical transform image that resulted from supervised training on the following four types of rock outcrop categories: bleached Wingate Sandstone, unbleached Wingate Sandstone, alluvium, and all other rock types. This image, though of different scale from Figure 8.10, shows that bleached Wingate rocks, which overlie the uranium and gas deposits, are quite well mapped by a canonical transformation image from the Bendix 24-channel airborne multispectral scanner data. It is likely that they could also be well mapped by spectral ratio images produced from LANDSAT TM data. For comparison, Color Plate CP26 shows a color infrared (CIR) photo mosaic of approximately the same area, but at slightly different scale. Bleached and unbleached Wingate Sandstone are barely, if at all, discernible from other areas in the image. This result is further proof that multispectral scanner data can be used to detect compositional differences that are virtually impossible to detect by film cameras or the human eye.

The same NASA/Geosat report (Abrams et al. 1984a) gives a two-step process for the removal of pigment (bleaching) from the originally red Wingate Sandstone. Pyrite is formed (Karstev et al. 1959) from the following reaction:

$$2H_2S + Fe_2O_3 \rightarrow FeS_2 + FeO + 2H_2O \qquad \text{(Eqn. 8.1)}$$

The iron sulfides are then put into solution in oxygenated water by the following stochiometric reaction (Singer and Stumm 1968):

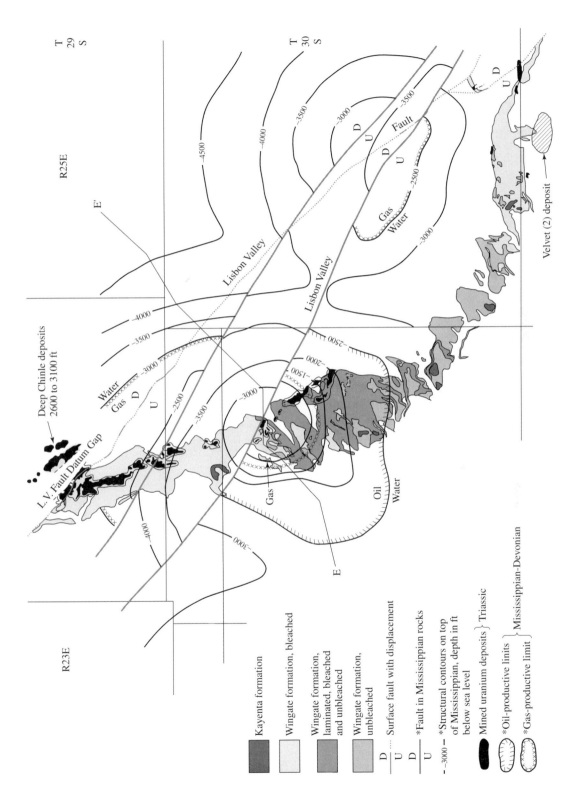

Figure 8.10 Structural contour map of Mississippian-age reservoir rocks (Leadville Dolomite) in Lisbon Anticline and the distribution of Wingate bleaching and Moss Back uranium mineralization. Northeast of Lisbon Valley Fault, the minor normal faults at the surface have been omitted. (*After Abrams et al. 1984a. Reprinted with the permission of the American Association of Geologists.*)

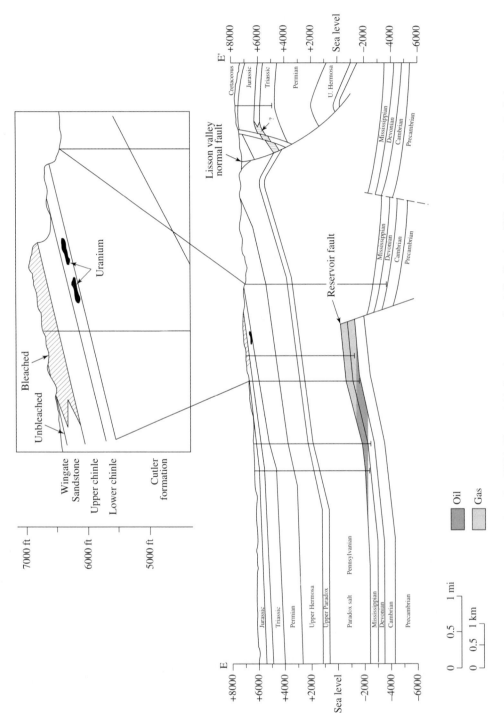

Figure 8.11 Cross section along line EE' in Figure 8.11 showing generalized distribution of Wingate bleaching and relations of bleaching to uranium mineralization and oil and gas accumulations. Principal section was modified from Parker (1966); enlarged portion from Weir and Dodson (1958a, 1958b, 1958c). Minor normal faulting northeast of Lisbon Valley Fault not shown. (*After Abrams et al. 1984a. Reprinted with the permission of the American Association of Petroleum Geologists.*)

$$FeS_2 + \frac{7}{2}O_2 + H_2O \rightarrow Fe^{+2} + 2SO_4^{-2} + 2H^{+1}$$

and

$$Fe^{+2} + \frac{1}{4}O_2 + H^{+1} \rightarrow Fe^{+3} + \frac{1}{2}H_2O \qquad \text{(Eqn. 8.2)}$$

If there is some pyrite still available in solution, as would be likely while the solution is still in the rock being bleached, the following reaction can occur (Abrams et al. 1984a):

$$FeS_2 + 14Fe^{+3} + 8H_2O \rightarrow 15Fe^{+2} + 2SO_4^{-2} + 16H^{+1} \qquad \text{(Eqn. 8.3)}$$

There are two noteworthy items in this reaction. First, very acidic waters are produced (because sulfuric acid results in a following step). This acidity will often dissolve carbonate cement completely, as it has in the bleached Wingate Sandstone, where no carbonate cement currently exists. Second, the ferrous (Fe^{+2}) iron produced will be changed back to ferric (Fe^{+3}) by the second reaction in Equation 8.2.

This series of reactions probably continues, making the solution increasingly acidic, until there is no more pyrite (FeS_2) left, and one of the two reactions below occurs:

$$Fe^{+3} + 3H_2O \rightarrow Fe(OH)_3 + 3H^{+1}$$

or   (Eqn. 8.4)

$$2Fe^{+3} + 3H_2O \rightarrow Fe_2O_3 + 6H^{+1}$$

The first reaction in Equation 8.4 yields ferric hydroxide (goethite) and the second reaction produces ferric oxide (hematite). If the decrease in amount of pyrite was caused by the movement of solutions away from the place along the hydrocarbon escape route where pyrite is most readily formed, assumedly the region of bleaching, goethite and hematite will start to form some distance from where the bleaching is located. If the motion is along a vertical fault, this would imply that the regions of goethitic and hematitic precipitation are above the bleached areas.

In Lisbon Valley, the bleached Permian redbeds are exposed at the surface, such that regions of goethitic and hematitic occurrences caused by hydrocarbon escape have been removed by erosion. However, in the Wind River Basin, the Eocene sandstones that are locally enriched in hematite and goethite and are near producing oil and gas fields may be examples of a version of the above chemical process, as applied to the movement of hydrocarbon/water mixtures upward along a fault or joint. Once permeable sandstones are intercepted by the upwardly migrating fluid, the fluid could invade the aquifer and move in the direction of the hydrostatic gradient. If this aquifer is at the surface or is later exposed by erosion, a large lateral area stained by hematite and goethite could result, as appears to have happened near some of the oil and gas fields in the Wind River Basin that are producing from strata below the Triassic-aged Chugwater Formation. The best test for this hypothesis would be the examination of well cores from the Chugwater Formation below those fields that have a localized ferric oxide anomaly in the exposed Eocene sandstones on or adjacent to the field. If the wells near a fault that defines the edge of the field (and perhaps controls the ferric oxide anomaly at the surface) show bleaching in the Chugwater Formation, the hypothesis would be strengthened considerably.

The case studies given above for the Wind River Basin and Lisbon Valley, plus bleached Permian redbeds over the anticlinal crests of Oklahoma oil fields of Velma, Eola, and Chickasha (Ferguson 1979a, 1979b, 1981), provide some evidence for the hypothesis that ferric oxides (enriched, as in the case of the Eocene-aged arkosic sandstones of the Wind River Basin, or bleached, as in the Upper-Triassic Wingate Sandstone of Lisbon Valley) can be indicators of ancient hydrocarbon seeps. Well-core and field studies are needed to further test the hypothesis. However, it is entirely possible that further work with hyperspectral data, or even with LANDSAT TM data, will yield information on mineral expressions of hydrocarbon escape other than ferric oxides.

Calcium carbonate content in sandstones, or its inverse (silica content), may bear fruitful information about hydrocarbon seeps because, at least in the Lisbon Valley case, the lowest carbonate content (highest silica content) in the Wingate Sandstone occurs where the greatest bleaching by escaping hydrocarbons took place. That is where the escaping fluids were most acidic. Where could the dissolved carbonate go? Eh and Ph conditions might finally be reached at or near the surface where the iron carbonates siderite and ankerite might precipitate from the escaping fluids. In fact, as mentioned in the previous chapter, both of those carbonates are found in Mississippi Valley Type lead-zinc deposits, which often occur near oil and gas reservoirs. Both carbonates have characteristic ferrous absorption bands in the reflective infrared wavelength region and are highly susceptible to detection by spectral ratioing (see the siderite ratio codes in Appendix C), if they cover enough of the exposed surface to make up a significant portion of an individual pixel. The amount of siderite that could be detected in various sandstones by spectral ratioing could be determined by a linear mixing rule, using laboratory spectra as input.

There is also the chance that more exotic minerals with distinctive infrared spectral features, yet with bland visible spectral characteristics, will be found to be associated with geochemical alteration caused by hydrocarbon escape. Ammoniated feldspars and sulfates, which in the previous chapter were discussed because of their association with disseminated gold deposits, are formed when hydrothermal fluids intercept hydrocarbons or other organic material. These minerals may be useful for hydrocarbon exploration, as well as for gold exploration. However, hydrocarbon reservoirs found by the occurrence of ammoniated minerals would likely have been subjected to elevated temperatures, since the ammoniated minerals are associated with hydrothermal alteration.

Geobotanical Remote Sensing and Oil Slick Mapping for Petroleum Exploration

Where vegetation covers the ground, remote sensing methods cannot directly observe spatial patterns of variations in soil geochemistry because most of the light entering the sensor is reflected or emitted from the plants, not the soil. However, as with metal exploration, geobotanical remote sensing can sometimes link spatial patterns in plant species or in vegetation stress within plant species to soil concentrations of metals. These concentrations are selectively enriched or removed by geochemical phenomena associated with hydrocarbon escape. Once again, as with geobotanical remote sensing for metal exploration, vegetation can be used as an indicator for indirect detection of hydrocarbon escape. Since vegetation is short-lived, geologically speaking, most of the geobotanical clues are suspected to yield information about ongoing or recent seeps, rather than ancient ones.

A case study on the control of vegetation species occurrences by escaping hydrocarbons is given in Section 12 of the joint NASA/Geosat test case project (Abrams et al. 1984b), which is about the Lost River, West Virginia petroleum test site. The Lost River test site, which is heavily covered with vegetation, is located above a simple anticlinal gas reservoir in a structurally complex thrust sheet. Seventy soil samples, from both on and off the gas field, were collected and measured for manganese concentration. These samples revealed enhanced concentrations of soil manganese in the vicinity of the gas field. However, no relationship between manganese concentration and vegetation species cover was noted.

The investigators did, however, find two areas of anomalous stands of maple trees, one a circular stand of sugar maples near a productive gas well and the other a linear stand of red maples parallel to a narrow, NNE-trending ravine between a second gas well and a dry hole. The soils at both of these anomalous maple stands were found to have higher than average soil gas (methane through butane) values for the area (Rock et al. 1986). The sugar maples at the first site (the circular stand) are on an East-facing slope at an elevation of about 650 m. The red maples (linear stand) at the second site cover the North side of the ravine, above the wet areas of the ravine bottom, and about 100 ft up the East side of the ravine, on a WNW-facing slope, and at an elevation of approximately 585 m. Both sites were judged anomalous because obvious factors, such as recent cultural disturbance, topography, slope aspect, soil moisture, and soil type, do not appear as likely causes for the maples at these two stands. In this area, the maple grows in places that are unsuitable for the predominant oak trees. In general, maples prefer East-facing slopes and their large stands usually occur at higher ridge tops (above 740 m) in this part of Appalachia.

At the ravine site, extensive development of white fungal mycelia was observed in the soil, and the area is considered by local residents as being excellent for collecting mushrooms. Mycorrhizal fungi apparently thrive at the site. Red maples are known to form mycorrhizal fungus associations very differently to the mycorrhizal fungus associations in oaks, in that the fungus associated with red maple roots is a nonseptate, filamentous version belonging to phycomycetes (Medve 1971), which is similar to the fungus noted by Davis (1967) as capable of oxidizing gaseous hydrocarbons. Oak root systems are instead associated with the basidiomycetes form of mycorrhizal fungus.

It has also been found that pin oaks (Flower et al. 1981) and black oaks (Leone et al. 1977) are less tolerant of landfill gases (up to 60% methane and 40% carbon dioxide, with a few trace gases) than are red maples. Those same two references found that whereas methane itself does not have a direct toxic effect on tree species, the development of an anaerobic soil atmosphere in the root zone does have a toxic effect on many tree species. The phycomycetes form of mycorrhizal fungus is more tolerant than basidiomycetes of anaerobic soil conditions. Another factor that must be considered is root depth. Flower et al. (1981) found that shallow-rooted vegetation can tolerate sites near landfills from which deep-rooted species have been excluded. Raven et al. (1976) have shown that oaks produce deep taproot systems. The NASA/Geosat study (Abrams et al. 1984b) found that maples in the Lost River, West Virginia area tend to be shallow rooted, as evidenced by many maples that had been uprooted by a 1977 ice storm, while oaks had not. Therefore, from both the aspects of root fungal systems and from root depth, maples are more tolerant of anaerobic soil conditions than oaks.

Leone et al. (1977) also found that the tree species most tolerant of landfill gases is black gum. Black gum is normally a low shrub thicket that occurs throughout the Lost River area with no preference to the gas field. However, a difference occurs with the large trees (trunks exceeding 10 cm in diameter at breast height and that constitute only 5% of the black gums in the area). The large trees of that species occur primarily (about 80% of the large black gums) at the two maple anomaly sites, along the gas pipeline, or near gas wells.

The investigators of the Lost River gas field (Abrams et al. 1984b) were able to employ supervised multispectral classification processing to airborne multispectral scanner data that included the LANDSAT TM bands plus several other spectral bands. They were able to differentiate maples, oaks, and black gum from one another, but there were difficulties in separating different members of the same species. Since slope direction and total elevation above sea level were important discriminators between the aforementioned anomalous and normal maple stands, it would have been helpful for them to employ elevations and slopes derived from a digital elevation model (DEM) as separate "spectral bands" of input to the multispectral classification scheme. Without the addition of DEM information to the multispectral data, it would appear from the Lost River study that the identification of geobotanical species indicators of hydrocarbon escape from multispectral data alone is difficult, if not impossible. This result appears to be in agreement with Price et al. (1985), who used principal components of LANDSAT TM data of an area near Harrisburg, Pennsylvania to show that dividing the first principal component into four parts (an example of canonical analysis) resulted in a classification of four vegetative classes that represented the sunlit and shaded sides of each of the upper-elevation sandstone and the lower-elevation shale. They showed that most of the vegetation information in LANDSAT TM data is controlled by elevation and slope aspect, which are in turn controlled by the lithology (sandstone vs. shale). Therefore, whatever tree species differences there are that are related to oil and gas escape must be winnowed out of a strong tree species dependency on topographic elevation and slope aspect; these in turn are controlled by the gross properties of lithologic units underlying the vegetation. Relationships between the distribution of natural tree species and hydrocarbon escape are discussed by Rock et al. (1986).

Within-species geobotanical indicators of hydrocarbons have also been noted in the literature. Apparently the first such notation was by Vostokova et al. (1961), who reported that vegetation as a hydrocarbon indicator near Siberian oil fields was suggested as early as 1949 by Y. A. Vostokova and S. V. Viktorov. When bitumen was present in the soils, those investigators noted abnormalities in the Soviet vegetation (mostly arid bushes) that included gigantism (oversized), increased branching, and second blooming (two-cycle development).

An interesting master's degree thesis by B. J. Harding (1989) studied tree morphology in and around the Albion-Scipio Oil-Trend in south-central Michigan. He used field observations and field photographs to count the number of twigs per 3 ft length of branch on 58 trees of four different species, some of which were on the linear oil and gas field and some of which were off it. Harding used the Student's T-test to determine significance of the difference in "twigginess" of trees of the same species on and off the field and established the significance level at a maximum of 5%. Harding found that three species passed the significance test: bur oak (1% significance level), red oak (1–2.5%), and sugar maple (less than 0.1%, a finding that made the sugar maple test the most significant of the three). All three species showed more twigging on the field than off the field. White oak (5–10%) failed the test. Soil gas testing was also performed on different transect sites, but the different data collection patterns for the tree sites made it impossible to directly correlate soil gas concentrations of light hydrocarbons with rates of twigging. He did note a C_2/C_1 carbon ratio low over the field on most of his transects, however.

Harding's results were verified by GeoSpectra geologists, who had been using the twigging phenomenon (they called it "hairy trees") as a field exploration tool in Michigan since the mid-1980s. No clear-cut remote sensing method has been yet found for mapping this phenomenon from an airplane or satellite. Passive airborne and spaceborne visible and infrared sensors clearly do not have the spatial resolution required to map the number of twigs per foot of branch extension. However, there may exist other phenomena directly related to twigging, such as fuller leafing or chlorosis level, that passive and infrared sensors have some possibility of mapping, but that have not been fully explored to date. Harding did note in his thesis that the sugar maples on the field appeared quite vigorous, fully leafed, and healthy, with dense bright-green foliage. There is also the possibility that multifrequency radar in the microwave wavelength region or lasers operating in the visible or infrared wavelength regions might measure twigging directly. In any case, classification of tree species is a necessary prerequisite before twigging or related phenomena can be employed as a prospecting tool for hydrocarbons. Different tree species can exhibit markedly different between-species rates of twigging, unrelated to the presence of hydrocarbons in the soil upon which the trees grow.

The search for geobotanical indicators of oil and gas is made more difficult by the fact that causes for vegetation stress other than hydrocarbon escape can produce similar multispectral patterns. One of the most studied sites for geobotanical indicators of hydrocarbon escape has been Patrick Draw, Wyoming, where several investigators (Abrams et al. 1984c; Feder 1985; Marrs and Paylor 1987) have suggested or claimed that a tonal anomaly on the sagebrush/grassland community above the Patrick Draw oil and gas field, as observed in one or more LANDSAT images, was evidence for vegetative stress caused by gas escape from the underground oil and gas reservoirs. A more recent study (Scott et al. 1988) showed that manganese and iron content of the soil was greatest in the "anomalous" blighted sagebrush area (which was over oil and gas reservoirs), second greatest over gas but not on the sagebrush anomaly, third greatest over oil but not on the sagebrush anomaly, and fourth greatest over regions of no production where only dry holes had been drilled. The amounts of iron and manganese in the sagebrush showed a consistent decrease with distance from leaking gases. They found that dead sagebrush dominated in swales of the anomalously blighted area, but not elsewhere. Scott et al. (1988) also found that the salt content of blighted sagebrush was about four times higher than that of background sagebrush away from the blighted region and that airphotos and LANDSAT MSS images collected from July 22, 1974 through October 19, 1978, fail to show the vegetation anomaly (blighted sagebrush). Later airborne and satellite images and photos showed the vegetative anomaly from the period of October 29, 1978 (enhanced airborne TM Simulator images) through August 17, 1985. Finally, they found from BLM records that annual rainfall averaged 20.27 cm per year from 1966–1986, but that rainfall was 1.6–4.72 cm above average during the 1980–1983 period. This finding reinforces the possibility that the death of sagebrush and grasses in swales of the "anomalous" area may have been caused by a higher-than-average water table, or even flooding, for four years in a row.

Several recent papers (Scott and McCoy 1993; Bammel and Birnie 1993; and Bammel et al. 1994) have noted subtle spectral features that appear to be related to soil geochemistry anomalies that are associated with hydrocarbon seeps, but were not mapped in remote sensing images. Cwick et al. (1993) have reported correlation coefficients for Pollard Oil Field in southwestern Alabama as high as -0.675 between the soil Fe/Mn ratio and a video camera NIR/Red ratio, where NIR is a 0.815–0.827 μm band and Red is a 0.656–0.664 μm band of an aerial video camera owned by the U.S. Dept. of Agriculture's Remote Sensing Research Unit in Weslaco, Texas, with 2 m resolution at a flight altitude of 5500 ft.

All of the foregoing discussion has been about the detection of ongoing seeps on land via geobotanical remote sensing. However, seeps occur offshore, also, as evidenced by oil slicks that emerge from the sediment/water boundary, rise to surface, and float on the water. Color Plate CP24 is a LANDSAT II high-contrast image of a region in the Aegean Sea, showing oil slicks on the water in a contrast-stretched image of LANDSAT MSS bands 4, 5, and 7, shown in blue, green, and red, respectively. These oil slicks, seen as darker than the background water, are likely caused by the flushing of oil tanks on tankers that move regularly along this shipping route, but they would appear as similar features if they were caused by natural oil seeps.

Satellite synthetic arperture radar (SAR) images are even better than LANDSAT images for mapping oil slicks, because radar images can be collected at any time, regardless of cloud cover or sun position. Figure 8.12 (Berry 1995) shows an ERS-1 SAR image acquired on September 9, 1993 in the Green Canyon Area of the Gulf of Mexico, near the edge of the continental shelf. The many oil slicks show up as black in the radar image. The gray outlines are positions of oil slicks traced from a LANDSAT TM image from 1984; the outlines indicate that these are natural seeps occurring from stationary positions. In both cases, the source ends of the slicks are the southern ends (North is towards the top of the image). White objects are oil platforms.

If a geophysical imaging study of an offshore region is accompanied up by a multitemporal satellite imaging study that searches for oil slicks, particularly those

Figure 8.12 Oil slicks in an ERS-1 SAR image (T-1304) of the Green Canyon area, Gulf of Mexico, September 9, 1993. (North is toward the top.) Gray outlines are oil slicks that appeared in a 1984 LANDSAT TM image. The appearance of oils slicks in approximately the same locations (the source ends of the slicks are the south ends) at two different times indicate that they are caused by natural seeps. White objects are oil platforms. *(After Berry 1995; reprinted with permission of the Geosat Committee, Inc./GOSAP project, courtesy Dr. E. K. Biegert; image copyright owned by the European Space Agency. Copyright © 1993 ESA, data distributed by RSI. Image processing courtesy of Pecten International Company.)*

that appear to coincide with significant linear and closed structural features in the geophysical images, it is possible to plan the placement of offshore seismic lines for optimal recording of subsea structures related to hydrocarbon reservoirs. The use of geophysical images and satellite imaging of seeps together can pay for themselves many times over in the increased number of potential hydrocarbon reservoirs discovered per line-km of seismic data collected.

Off-shore gas seeps have also been detected underwater in the same Green Canyon area of the Gulf of Mexico, though by more unusual remote sensing methods. Lee et al. (1993) reported on an offshore gas seep called Bush Hill, which was imaged by sonar from a nuclear-powered submarine (USS NR-1) on loan from the U.S. Navy. Figure 8.13 shows four types of acoustical anomalies that were encountered, and Figure 8.14 shows where they occur on Bush Hill (contours represent depth below the water surface). Type I shows strong bottom echoes from authigenic carbonate outcrops on the sea bottom that supports thriving tube worm communities and displays the strongest gas seeps. Type II is acoustically turbid with no subbottom echoes, a profile that is characteristic of gas in seafloor sediment. Type III is acoustically turbid with strong subbottom echoes and is found all over the mound called Bush Hill. It is likely caused by a gas hydrate layer. Type IV is acoustically transparent with parallel, continuous subbottom reflectors and is found surrounding

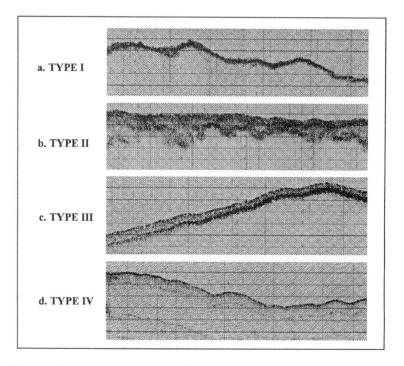

Figure 8.13 Example of four types of acoustic anomalies from 25-Khz, high-resolution, subbottom profiles, imaged by sonar from a nuclear-powered submarine (USS NR-1). These profiles were collected around an ocean bottom feature called Bush Hill, in the Green Canyon area of the Gulf of Mexico. *(After Lee et al. 1993. Abrams et al. 1984a. Reprinted with the permission of the Environmental Research Institute of Michigan.)*

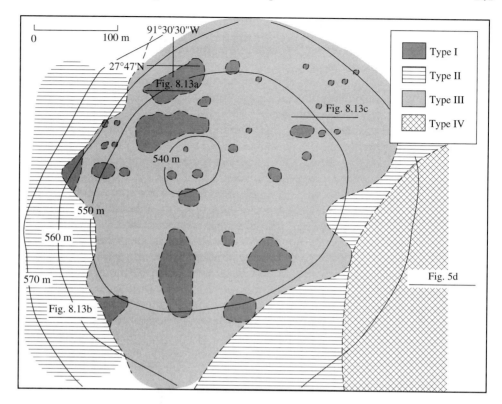

Figure 8.14 Distribution of the four types of acoustic anomalies shown in Figure 8.17 from 25-Khz, high-resolution, subbottom profiles, imaged by sonar from a nuclear-powered submarine (USS NR-1) for Bush Hill, in the Green Canyon area of the Gulf of Mexico. *(After Lee et al. 1993. Reprinted with the permission of the Environmental Research Institute of Michigan.)*

the Bush Hill mound. It is considered to consist of muddy sediments unaffected by hydrocarbon seepage.

THE USE OF REMOTE SENSING FOR GROUND WATER EXPLORATION

During the first decade and a half of satellite remote sensing history, oil was more precious than ground water almost everywhere. Since then, however, the value per barrel of potable ground water has outpaced the value of a barrel of oil in many areas of the world, particularly in the Middle East. Although there are nowhere near the same number of ground water exploration case histories that there are for petroleum exploration, it is appropriate that ground water be given equal status in this chapter with petroleum, even though this section will necessarily be shorter than the previous one.

Ground water exploration has several aspects of geologic structure in common with hydrocarbon exploration. As with petroleum reservoirs, confined ground water aquifers (which have aquitards above and below the aquifer) require sufficient permeability for long-distance transport, sufficient porosity for storage, and a trap (structural or stratigraphic) that will stop fluid migration, in order for liquids to be accumulated in economic quantities. Most confined aquifers, which usually occur at greater depth than unconfined aquifers, and petroleum reservoirs will leak upward if fractured because they are under pressure (hydrostatic if shallow and lithostatic if deeper). Unconfined aquifers (which have aquitards below but not above them) require porosity sufficient for storage, permeability sufficient for pumping, and rainfall sufficient for surface recharge; the rainfall occurs over a broad area above the relatively shallow, unconfined aquifer. There are no upward seeps from unconfined aquifers, which are under no more pressure than one atmosphere.

In regions of Earth that are now very arid, there are few unconfined reservoirs with significant water still stored in them for two reasons: (1) there has not been enough rainfall in recent years to recharge them and (2) they have been tapped out by heavy human consumption. Only the deeper, confined reservoirs that hold ancient water that has been preserved from wetter times past are still useful in those presently arid areas. In presently humid areas, ground water has become important for agricultural irrigation, primarily as a water management tool. An example of that is northwest Ohio, where green beans are irrigated from mostly unconfined aquifers during a critical two-week period of their growth cycle.

The previous two paragraphs pertain to terrain that is underlain by sedimentary rocks. However, ground water is in demand everywhere for human and animal consumption, even in rocky areas, where the lack of topsoil precludes agriculture. In igneous and metamorphic terrain, the only available aquifers are porous zones of brecciation along faults and joints in the exposed bedrock. The porous rock debris and soil in the cracks of the rocks become natural reservoirs (unconfined aquifers) for the storage of water that does not run off the surface after periods of precipitation. In such terrain, the search for ground water is synonymous with the search for faults and joints in the bedrock. To a large extent, the same can be said of the search for ground water in karst terrain, where solution zones along joints and fractures in underground carbonate rocks become conduits for the migration and storage of ground water.

From a geochemical aspect, hydrocarbons and water are quite different. Hydrocarbons tend to produce reducing environments and are unnecessary for the growth of vegetation, whereas ground water often occurs in oxygenated environments and is absolutely essential for vegetation growth. In places where ground water seeps to the surface from a confined aquifer, the growth of vegetation will be significantly enhanced, compared with its surroundings, in arid and semiarid terrain. An oasis in the desert is an example of this enhancement. In humid regions, such a seep may produce standing water and/or a wetland. Such an environment supports vegetation communities that can be quite different from the prevailing vegetation cover. Wetlands are often maintained by ground water from confined aquifers that are breached by faults, joints, or intersection by Earth's surface.

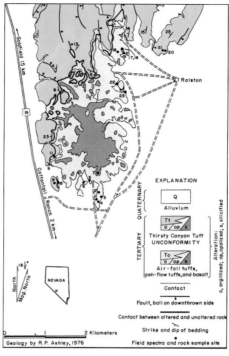

Note: Captions for
Color Plates located
on p. 369

Color Plate 19

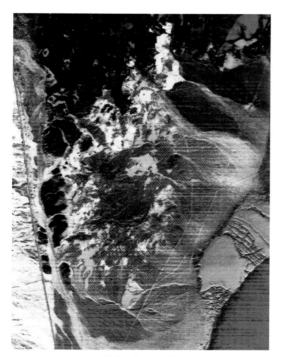

Color Plate 20

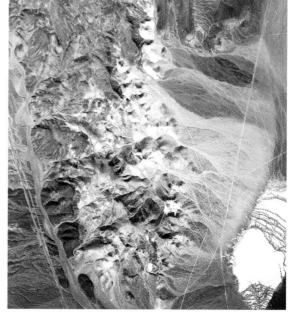

Color Plate 21

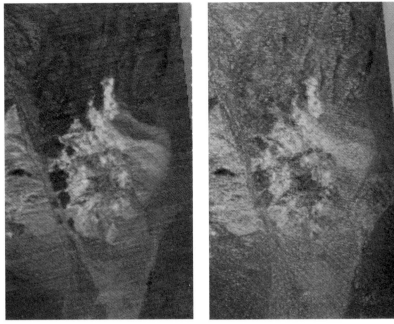

Color Plate 22a   **Color Plate 22b**

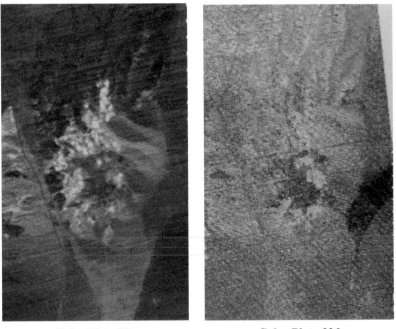

Color Plate 22c   **Color Plate 22d**

Color Plate 23

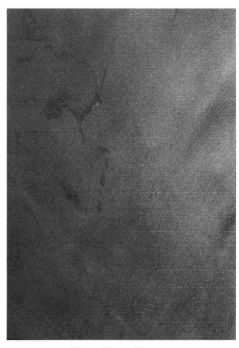

Color Plate 24

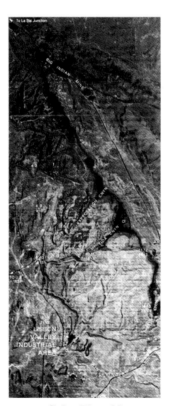

Color Plate 25

Color Plate 26

Color Plate 27a

Color Plate 27b

Color Plate 28a

Color Plate 28b

Color Plate 29a

Color Plate 29b

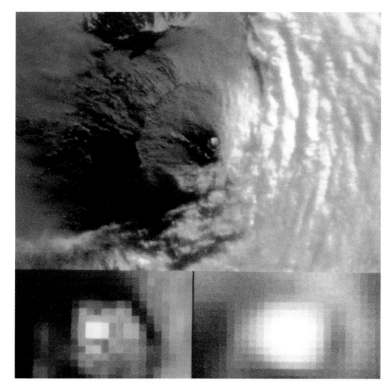

Color Plate 30

Color Plate 31

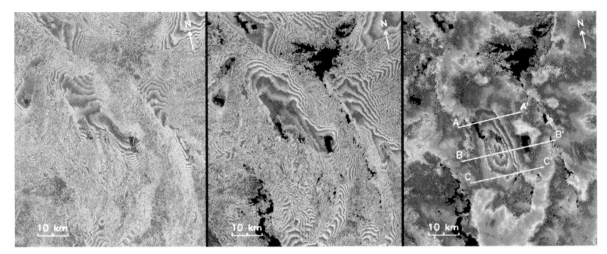

Color Plate 32

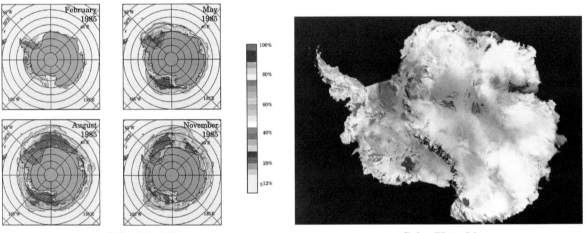

Color Plate 33

Color Plate 34

(a) 8/Jun/83 – 22/Jun/83

(b) 20/Jun/84 – 04/Jul/84

Color Plate 35a & 35b

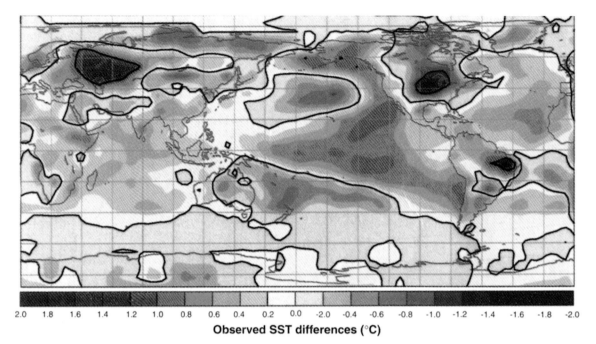

Observed SST differences (°C)

Color Plate 36

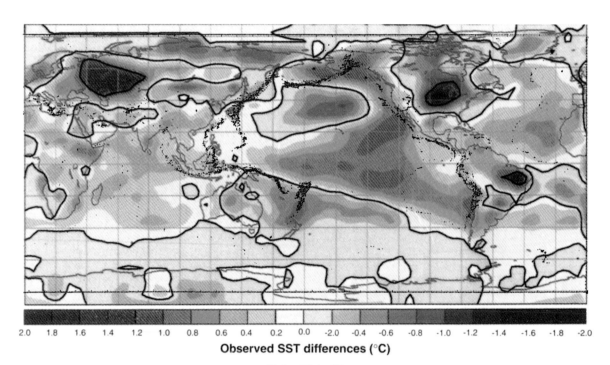

Observed SST differences (°C)

Color Plate 37

Ground Water Exploration in Igneous, Metamorphic, and Sedimentary Karst Terrain

The mapping of linear features associated with fractures (faults and joints) can be performed from images of almost any wavelength region. This type of mapping is an important ground water exploration tool for metamorphic and igneous terrain, as well as for sedimentary karst terrain, because the greatest amount of water will be found near fractures, where the only porosity and permeability will be located. One of the assumptions made in the past about igneous and metamorphic terrain is that the greatest amounts of water could be found where the greatest fracture length density (total line miles of fractures per square mile) occurs (Vincent et al. 1978). However, that early reasoning has proved to be somewhat simplistic, in that later evidence has shown that some fractures are more productive of ground water than others, and in some metamorphic terrain, folds can be as important as fractures.

Li (1994) recently found that in granitic rocks, fractures that were caused by tensional forces often contained more water than compressional fractures. The reason given by Li (1994) for this was that tensional fractures are wider than compressional fractures. However, tensional forces often cause en echelon fractures (sets of parallel fractures). This process provides yet another reason for more ground water to be found along tensional fractures, as was found in a recent experiment that related photo-interpreted linear features to hydraulic anisotropy (Cheema and Islam 1994). Invoking a single porosity model, which assumes that almost all flow is along fractures and not in the igneous and metamorphic bedrock, Cheema and Islam measured the hydraulic outflow of an enclosed, impermeable vinyl sheet that had been cut to match a fracture pattern drawn from photo-interpreted stereo SPOT images, LANDSAT images, and aerial photos of a region in the Black Hills near Rapid City, South Dakota. They found that the direction of maximum hydraulic conductivity coincided with the most prominent orientation of fracture traces. Both the greater fracture width and the dominant fracture trend caused by tensional fractures likely contribute to the greater ground water potential for tensional than for compressional fractures mapped from remote sensing images of igneous and metamorphic terrain. From an image interpretation standpoint, this conclusion would favor en echelon fractures over other fracture types observed from the interpretation of remote sensing images collected over igneous and metamorphic terrain.

Ramasamy and Jayakumar (1993) have shown that in complexly folded and fractured Precambrian rocks of the Southern Indian peninsula, some aquifer systems were controlled by total linear density and total linear intersection density, some were controlled by linear intersections and release fractures (thin, tight fractures parallel to the fold axes and restricted mostly to the axial portion of the anticlines and synclines), and some were controlled by extension and shear fractures. The movement of ground water from anticlines to synclines makes folds an important additional factor in ground water exploration of complexly folded metamorphic terrain.

The mapping of fractures from remote sensing data is also important for ground water exploration in karst terrain that is underlain by carbonate sedimentary rocks (limestones and dolomites). A good example of that is shown in Figure 8.15 for a karst area of Eastern Herzegovina called the Dinarides (Kresic 1994). The LAND-SAT I MSS band 7 image in Figure 8.15a was photo-interpreted for linear features shown in Figure 8.15b, along with known positions of sinks (where surface water flows into the solution zone) and springs (where ground water flows out of the solution zone). Each of the known sinks is connected to a spring by a fracture that was

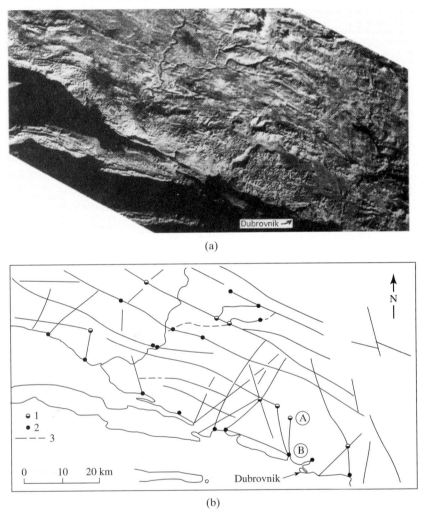

(a)

(b)

Figure 8.15   (a) Part of a LANDSAT I image covering the karst area of Eastern Herzegovina, the Dinarides (acquisition date October 31, 1973, MSS band 7). (b) Visual interpretation of the linear features (3; dashed when inferred) from the LANDSAT MSS band 7 image in (a), and the locations of major sinks (1) and springs (2) of the area. *(After Kresic 1994. Reprinted with permission of the Environmental Research Institute of Michigan.)*

mapped from the LANDSAT image. Figure 8.16a shows an enlarged portion of the previous LANDSAT image. When a binary slice-density enhancement of the part of this image was enclosed by the drawn parallelogram, Figure 8.16b resulted. The dark dimpled features in Figure 8.16a, which are shown in Figure 8.16b as gray areas outlined in black, are interpreted by Kresic (1994) to be karst depressions (uvallas and dolines). A N-S alignment of dolines, which can be seen in both Figure 8.16a and b, connects the sink with the spring. Figure 8.16c shows a rose diagram of the orientation of the outlined karst depressions in Figure 8.16b. Kresic (1994) interprets the rose diagram maxima as the most probable general directions of ground water flow within the aquifer. He also recommends the study of stereo aerial photos for more detailed information about the karsts. It could be added that an every-pixel DEM produced from such aerial photos would also be of assistance for mapping subtle karst depressions. If stereo photos collected decades apart were used to calculate such high-resolution DEM, it is possible that the rate of subsidence over sinkholes could be measured.

Although Kresic (1994) concedes that airborne thermal infrared imagery would be the most useful for mapping karst terrain, he used a LANDSAT I image for the above study because of its relatively low cost. The LANDSAT TM thermal band may also have been useful, but the thermal TM band has 120 m spatial resolution, as compared with the 79 m resolution of the LANDSAT MSS image used by Kresic (1994), and it would have been more expensive.

Ground Water Exploration in Nonkarst, Sedimentary Terrain

The mapping of linear features in remote sensing images is also important for ground water exploration in nonkarst sedimentary terrains. In some cases, linear features may delineate faults with vertical throw that place the aquifer against an impermeable barrier, in which case the linear feature can be a barrier to ground water flow. Taylor et al. (1992) gives an example of a linear feature that acts as barrier to ground water.

In other cases, however, linear features mapped from remote sensing images of sedimentary terrains can delineate regions of higher ground water flow, as they do in metamorphic and igneous terrain. One definitive example of this is given by Minor et al. (1994) for the Tease area in Ghana, which includes the southern part of the Voltaian Sedimentary Basin, a synclinal structure of relatively flat-lying Paleozoic sandstones, shales, and conglomerates. The sandstones are compacted quartz arenites, with ductile grains that have been deformed into concave-convex shapes. Much of the primary porosity and permeability of the arenites have been destroyed by the compaction process, and there is little prospect for secondary porosity being formed by weathering processes because the sandstones are so clean and quartz-rich. The ground water flow is controlled by vertical flow, recharging fissures at depth, and a more subdued subhorizontal flow parallel with stream gradients.

Minor et al. (1994) used LANDSAT TM data, SPOT data, and digitally scanned Russian MK-4 space photography for mapping linear features. They report that the most useful form of the TM data was a principal component image of TM bands 5, 4, and 3. No TM thermal band use was reported. All of the above data were

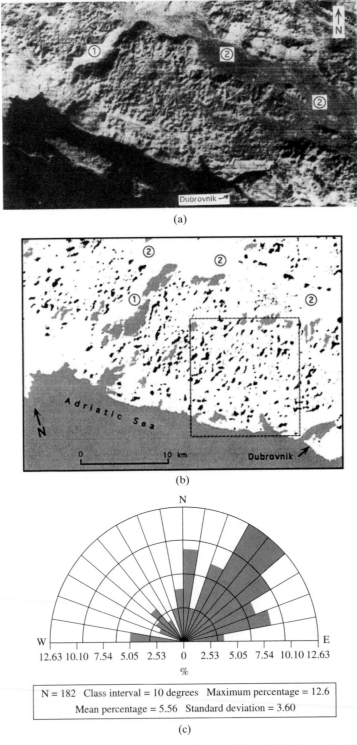

(a)

(b)

(c)

Figure 8.16 (a) Another part of the LANDSAT I MSS band 7 image of Figure 8.19 (a) covering the area of the karst polje of Popovo in Eastern Herzegovina, where(1) denotes a karstified inactive valley (the black tone is the shadow on the SE side of the valley) and (2) denotes flat Quaternary deposits of the polje of Popovo. (b) The result of a binary slice-density enhancement of the area shown in (a), where (1) and (2) denote the same as in (a). The small, black areas ("particles") are interpreted as karst depressions—uvallas and dolines. (c) Rose diagram of the orientation of the karst depressions for the outlined area in (b). *(After Kresic, 1994. Reprinted with permission of the Environmental Research Institute of Michigan.)*

carefully geometrically corrected, with the aid of GPS-measured (Global Positioning System) ground control points, and placed into a GIS database, so that the interpreted linear features could be accurately located on the ground. A GPS survey was then conducted to locate the positions of water well locations to within 10–20 meter accuracy, and these were also added to the GIS database. Figure 8.17 shows the resulting distribution of wet and dry water wells versus distance from the nearest linear feature or fracture trace (a linear feature less than one mile in length). It shows that 100% of all wells located within 100 m of identified linear features and fracture traces were wet. It also shows that the fraction of dry wells increases with increased distance from a fracture. The geometric accuracy of the GIS database, owing to accurate GPS surveys and accurate image processing, is probably the most significant factor contributing to the project's definitive results. Remote sensing results and the location of the drilling rig on the ground must be geometrically accurate to be of much value for ground water exploration, if a few hundred meters is the margin between economic success and failure.

Thermal infrared imagery has special relevance for any type of reservoir that brings ground water near enough to the surface to affect surface temperatures. Since ground water is normally about 55°F year-round, thermal infrared images will often show areas of near-surface ground water as anomalously cool (if the ground is warmer than that, such as during a summer day) or anomalously warm (if the ground is cooler than 55°F, such as at night or in colder seasons). Passive microwave would show virtually the same information as thermal infrared, with less signal-to-noise, because the peak of the black-body curve for the ground under most observing conditions occurs in the 8–14 μm wavelength region. Long wavelength radar (such as L-band, with approximately 29 cm wavelength) images of arid and semiarid terrain can

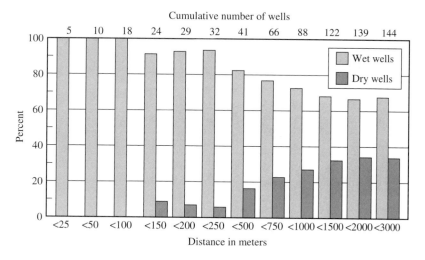

Figure 8.17 Distribution of wet and dry water wells versus distance from the nearest linear feature or fracture trace. *(After Minor et al. 1994. Reprinted with permission of the Environmental Research Institute of Michigan.)*

also highlight areas of near-surface ground water as having a different radar return than drier ground. (The detection can be made from the great difference in the dielectric constants of water and dry soil.) Radar and thermal infrared images will both reveal near-surface ground water in arid and semiarid terrain, the former because of the higher dielectric constant of wet soil and the latter because of the lower temperature (during most daytimes) of near-surface ground water. Standing water appears dark in most radar images because it has a specular surface and a high dielectric constant, which together cause high reflection in a direction away from the radar receiver. Moist soil has higher dielectric constant than dry soil, but can appear either dark or bright in a radar image, depending on whether the diffuse or specular reflectance dominates for the particular radar geometry employed during data collection. Radar remote sensing can be performed at any time, not just when the surface temperatures of the ground are appreciably different from 55°F. However, unlike thermal infrared images collected at night, topographic shadowing will make slopes of hills facing away from the radar system appear dark in the resulting radar images at any time of day.

Figure 8.18 shows a side-looking, synthetic aperture (SAR) airborne radar image of an area south of Reno, Nevada, along the eastern flank of Mount Rose (Brown 1994). Brown (1994) states that the dark area between the two small, dark arrows in the center of the image (also apparent in "other types of imagery") is possibly caused by "ponding" of stream underflow due to bedrock morphology beneath the alluvium and/or by discharge into the alluvium of bedrock aquifers along the fracture that borders the saturated area on the west (white arrows). The west half of a hill of stacked volcanic units, the crest of which is denoted by the white dashed curved line, is in radar shadow and also appears dark.

Figures 8.19a and b show aerial multispectral scanner data (a Daedalus Enterprises ATM scanner) of TM-equivalent spectral bands 1 (visible blue) and 6 (thermal infrared) for a region near Tonopah, Nevada. Note that the dark (cool) region in the thermal infrared image is not obvious in the visible image. It is likely a perched reservoir of near-surface ground water, though field work has not confirmed this interpretation. This example brings to mind that relatively little use of band 6 LANDSAT TM imagery has been reported in the technical literature, even though thermal imagery has a lot to offer ground water exploration, particularly in sedimentary terrain. Even the 120 m resolution of TM band 6 should be useful for structural traps of near-surface ground water that are caused by faults along which vertical motion has occurred. If the fault zone is manifested by ground water moving up along the fault plane only, 120 m resolution may be inadequate. However, if a near-surface aquifer is being dammed by the fault or recharged by upward migration of water along the fault plane, there may be a large-area thermal anomaly bounded by a linear feature that could be easily seen with 120 m spatial resolution.

Thus far, arid and semiarid areas have been the subject of discussion. However, remote sensing can make contributions in ground water exploration even in wetter climes. In humid terrain, where vigorous vegetation covers most of the ground (except for fallow agricultural fields), the mapping of vegetative species associated with wetlands and multitemporal imaging of wetlands becomes important for ground water exploration. An example of the latter is given by Maktav and Kapdash (1994)

Figure 8.18 A side-looking, synthetic aperture airborne radar image (SAR). Some prominent faults and fractures are indicated by thick, black arrows. A major fracture (white arrows) borders the up-gradient side of an area of saturated alluvium (short, black arrows). *(After Brown, 1994. Reprinted with permission of the Environmental Research Institute of Michigan.)*

for the Dalyan wetland in Turkey, where a channel connects Lake Koycegiz to the Mediterranean Sea. A supervised classification of the Dalyan wetland region was produced from TM bands 4, 2, and 1 of three different dates: August, 1984; August, 1988; and July, 1991. The water-covered areas within the reeds of the wetland were much larger in July, 1991 than those of the two earlier dates. However, there was more recorded rainfall in the region in August, 1984 than was recorded in the two later dates. There is also greater surface flow than the rainfall rate (by a factor of 2 or 3) in the months of July and August. The stability of the channel mouth, where it flows into the Mediterranean Sea, indicates that the discharge rate of the channel is very low, which increases the importance of ground water sources for feeding of the Dalyan wetlands. These facts led Maktav and Kapdash (1994) to the conclusion that the Dalyan wetland is fed by ground water and seawater rather than surface flow. There were no significant waves affecting the coast in August, 1984, whereas there were quite significant waves in the direction of the coast that includes the Dalyan wetland. Thus, a seawater intrusion induced by the local wave action could have caused some of the observed increased standing water in the wetlands. However, the increased standing water observed in the July, 1991 LANDSAT TM data also occurred in an apparently unconnected (in the classification map) area at some distance (up to 2 km) landward from the seashore. This event may favor the ground water contribution over the contribution of seawater intrusion. Regardless of the

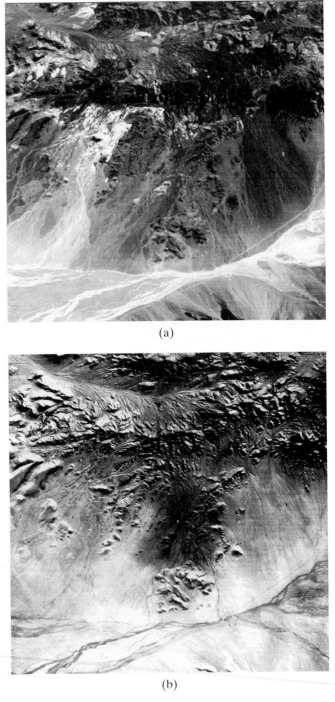

(a)

(b)

Figure 8.19 (a) A visible blue image (band 1) collected by the Daedalus Enterprises ATM scanner of a region near Tonopah, Nevada. (b) A thermal infrared image (ATM band 7) of the same area. Note the dark (cool) area above the rock outcrop, which is suspected of being a perched reservoir of groundwater. *(Courtesy of GeoSpectra Corporation.)*

size of the ground water contribution to the Dalyan wetland, this experiment demonstrated the potential of multitemporal satellite data for monitoring ground water fluctuations in vegetated wetland regions.

REVIEW

Geological remote sensing for petroleum exploration has three aspects: structural mapping, geochemical mapping of ancient hydrocarbon seeps, and geobotanical mapping (onshore) and oil slick mapping (offshore) of ongoing hydrocarbon seeps. Structural mapping, which involves both spatial and multispectral processing, is aided by the integration of satellite multispectral images with geophysical (magnetic, gravimetric, and seismic) data that have been imaged.

Examples of petroleum exploration remote sensing case histories that involve images from LANDSAT, magnetic, and gravity data integrated into the same study were given in two regions covered by glacial soils, one over the Mid-Continent Gravity Anomaly in Iowa and the other over the part of the Michigan Basin located in Bay County, Michigan, near the southwest end of Saginaw Bay. The Mid-Continent study proved that such an integrated study can focus seismic exploration efforts to obtain more pertinent exploration information with lower costs. In Iowa, an overthrust in Precambrian sediments that could contain oil and gas reservoirs was discovered with just 18 line miles of seismic data, because six, three-mile-long lines were collected perpendicular to major linear features that were interpreted from LANDSAT and geophysical images. The Bay County, Michigan study employed LANDSAT and geophysical images to guide the collection of several short seismic lines: all together they provided evidence for major structural features, one of which harbored a deep gas reservoir that is now productive.

The search for ancient hydrocarbon seeps thus far has primarily involved either anomalous absence (bleaching) or anomalous presence (staining) of ferric oxides. A Lisbon Valley, Utah case study has shown than primarily red sandstones have been bleached by hydrocarbons escaping from breached reservoirs in older rocks beneath them. The bleached areas also overlie uranium ore, which is deposited under the Eh and Ph conditions associated with some of the fluids escaping from a hydrocarbon reservoir. In the Wind River Basin, Wyoming, several coincidences have been demonstrated between surface exposures of ferric oxides in otherwise bland-colored Eocene sediments and underground hydrocarbon reservoirs. These coincidences provide circumstantial evidence for a connection between ferric-enriched anomalies now at the surface and ancient hydrocarbon seeps. Direct evidence for or against this hypothesized connection between localized, ferric-enriched anomalies and hydrocarbon escape from deep reservoirs could be found by searching drill cores of older primary redbeds (such as occur in the Chugwater Formation) below fault traces that bound ferric-enriched, surface anomalies to see if bleaching has occurred in the older, underground redbeds. Anomalous concentrations of carbonates, silica, and ammoniated minerals may also prove useful in the search for ancient hydrocarbon seeps.

The search for ongoing hydrocarbon seeps by geobotanical methods onshore and oil slicks offshore presents another method of hydrocarbon exploration by remote sensing. There are subtle relationships between hydrocarbon escape and the occurrence of vegetation species differences and the occurrence of within-species changes. There is evidence that some of these geobotanical anomalies can be imaged by multispectral remote sensing methods, but it is still unclear as to how useful geobotanical remote sensing will be as an exploration tool for hydrocarbons. The reason is that vegetation stress from phenomena other than hydrocarbon escape can yield similar geobotanical anomalies. From an offshore standpoint, it has been demonstrated that oil slicks can be mapped from LANDSAT images (at least under calm sea conditions). Ongoing oil seep mapping thus is possible with satellite remote sensing. Gas seeps have been mapped on the seafloor by a nuclear submarine, creating the possibility that gas seeps could also be mapped by multispectral sensors.

The value of ground water has increased substantially with respect to the price of oil or gas in recent years, and will likely continue to do so in the future. As a result, ground water exploration by remote sensing methods will also increase in importance. Over igneous and metamorphic terrain, the only porosity available for ground water reservoirs is in fracture zones, particularly those caused by tensional stress. Therefore, the mapping of fractures, both manually and automatically, is a useful ground water exploration method. Fractures are also important in karst terrain because water is often stored in carbonate rock solution zones along the fractures. Even in nonkarst sedimentary terrain, fractures can also be important, as a test case in Ghana, where 100% of the water wells dug within 100 m of linear features mapped in LANDSAT images were wet (productive). In ground water exploration, as in other types of exploration, it is important that accurate ground control points and accurate GIS or image processing software are used for geometrical correction of the satellite images. It is also important that GPS devices are used during field-checking and image processing follow-up.

Exercises

1. **(a)** What geobotanical clues that would be helpful in oil and gas exploration could you observe from airborne or satellite remote sensing data or field data in a very vegetated terrain such as the Appalachian foothills?

   (b) How would this be similar to looking for vegetative indicators of ground water contamination by effluents from a solid waste landfill in a forested area, and what is the underlying cause for this similarity?

   (c) If the area were underlain by nonpermeable bedrock, describe at least two types of remote sensing clues that would help you find where the most prolific water wells could be drilled.

2. Your major oil company employer has gained a concession to explore for oil and gas in a 50 km × 50 km region of Russia that is completely covered by young glacial till, and you are put in complete charge of planning the seismic data collection with a budget of $500,000. Seismic data costs $5000 per line-km to collect and process in this region, and satellite study costs approximately $25 per square km, counting the cost of data (in the range of $0.10–$0.70 per square km). As part of the concession, the Russian government has supplied free use of airborne magnetic and gravity data that they collected in lines

with 5 km spacings for the entire region, but it will cost about $5 per square km to process and study that data. If oil and gas are not discovered, you and the people you manage will likely be victims of the next company downsizing. Describe what you would do to maximize the chances of discovering economic deposits of oil and gas in this region.

Cited References

ABRAMS, M. J., J. E. CONEL, and H. R. LANG. 1984a. *The Joint NASA/Geosat Test Case Project Final Report,* Part 2, vol. 1, 8-1–8-101. Tulsa, Okla.: AAPG publisher.

———. 1984b. *The Joint NASA/Geosat Test Case Project Final Report,* Part 2, vol. 11, 12-1–12-96. Tulsa, Okla.: AAPG publisher.

———. 1984c. *The Joint NASA/Geosat Test Case Project Final Report,* Part 2, vol. 11, 11-1–11-112. Tulsa, Okla.: AAPG publisher.

BAMMEL, B. H., and R. W. BIRNIE. 1993. Spectral Reflectance Response of Big Sagebrush to Hydrocarbon-Induced Stress in the Bighorn Basin, Wyoming. In *Proceedings of the Ninth Thematic Conference on Geologic Remote Sensing,* vol. 2, 891–902. Ann Arbor: Environmental Research Institute of Michigan.

BAMMEL, B. H., R. McCoy, G. NASH, and Y. SOHN. 1994. Mapping Field Spectra and Biogeochemical Data from Sagebrush in the Little Buffalo Basin Oil Field, Wyoming. In *Proceedings of the Tenth Thematic Conference on Geologic Remote Sensing,* vol. 1, 481–496. Ann Arbor: Environmental Research Institute of Michigan.

BAROSH, P. J. 1968. Relationships of Explosion-Produced Fracture Patterns to Geologic Structure in Yucca Flat, Nevada Test Site. *Geological Society of America Memoir* 110:199–217.

BERGER, Z. 1984. Structural Analysis of Low-Relief Basins Using LANDSAT Data. In *Proceedings of the Third Thematic Conference on Geologic Remote Sensing,* vol. 1, 251–271. Ann Arbor: Environmental Research Institute of Michigan.

———. 1994. *Satellite Hydrocarbon Exploration—Interpretation and Integration Techniques.* Berlin: Springer-Verlag.

BERGER, Z., R. L. DODGE, and R. L. BROVEY. 1984. Evaluation of Thematic Mapper Data for Hydrocarbon Exploration in Low-Relief Basins. In *Proceedings of the Third Thematic Conference on Geologic Remote Sensing,* vol. 2, 827–836. Ann Arbor: Environmental Research Institute of Michigan.

BERRY, J. L. 1995. Detecting and Evaluating Oil Slicks on the Sea Surface. In *Remote Sensing for Oil Exploration and Environment, Proceedings of the Space Congress,* 90–110. Munich, Germany: European Space Report.

BONATTI, E., and K. CRANE. 1984. Oceanic Fracture Zones. *Scientific American* 250, no. 5:40–52.

BROWN, N. N. 1994. Integrating Structural Geology with Remote Sensing in Hydrogeological Resource Evaluation and Exploration. In *Proceedings of the Tenth Thematic Conference on Geologic Remote Sensing,* vol. 2, 158–169. Ann Arbor: Environmental Research Institute of Michigan.

CHEEMA, T. J., and M. R. ISLAM. 1994. Determination of Hydraulic Anisotropy of the Black Hills Region of South Dakota Using Fracture Trace Analysis. In *Proceedings of the Tenth Thematic Conference on Geologic Remote Sensing,* vol. 2, 158–169. Ann Arbor: Environmental Research Institute of Michigan.

CRADDOCK, C. 1972. *Keweenawan Geology of East-Central and Southeastern Minnesota in Geology of Minnesota: A Centennial Volume.* Minnesota Geological Survey.

CWICK, G. J., M. P. BISHOP, R. C. HOWE, P. W. MAUSEL, J. H. EVERITT, and D. E. ESCOBAR. 1993. Integration of Multispectral Video and Biogeochemical Data for the Assessment of Hydrocarbon Microseepage Conditions in Southwestern Alabama. In *Proceedings of the Ninth Thematic Conference on Geologic Remote Sensing,* vol. 2, 871–881. Ann Arbor: Environmental Research Institute of Michigan.

DAVIS, J. B. 1967. *Petroleum Microbiology.* New York: American Elsevier Scientific Publishing Co.

DONOVAN, T. J. 1974. Petroleum Microseepage at Cement, Oklahoma: Evidence and Mechanism. *AAPG Bulletin,* 58:429–466.

DRAKE, B. 1976. Saginaw Bay Graben and Its Implications for the Origin of the Michigan Basin and Pleistocene Glaciation (abstract and oral paper). Ann Arbor: Midwestern Regional American Geophysical Union Meeting.

DRAKE, B., and R. K. VINCENT. 1975. Geologic Interpretation of LANDSAT I Imagery of the Greater Part of the Michigan Basin. In *Proceedings of the Tenth International Symposium on Remote Sensing of the Environment,* 933–947. Ann Arbor: Environmental Research Institute of Michigan.

FEDER, A. M. 1985. Contemporary Remote Sensing for Hydrocarbon Exploration, Development—With Case Histories. *Oil and Gas Journal* 83:160–171.

FERGUSON, J. D. 1979a. Subsurface Alteration and Mineralization of Permian Redbeds Overlying Several Oil Fields in Southern Oklahoma, Part 1. *Shale Shaker Digest, Journal of the Oklahoma City Geological Society* 29, no. 8:172–178.

———. 1979b. Subsurface Alteration and Mineralization of Permian Redbeds Overlying Several Oil Fields in Southern Oklahoma, Part 2. *Shale Shaker Digest, Journal of the Oklahoma City Geological Society* 29, no. 9:200–208.

———. 1981. Formation of Diagenetic Alteration Zones by Leaking Reservoir Hydrocarbons Over Three Oil Fields in Oklahoma. *AAPG Bulletin* 65, no. 34:924. (Abstract only.)

FLOWER, F. B., E. F. GILMAN, and I. A. LEONE. 1981. Landfill Gas, What It Does to Trees and How Its Injurious Effects May be Prevented. *Journal Aboric.,* vol. 7, no. 2:43–51.

FREUND, R. 1982. The Role of Shear in Rifting, in Continental and Oceanic Rifts. In *Geodynamics Series,* vol. 8, ed. G. Palmason. Boulder, Co.: American Geophysical Union and Geological Society of America.

GOLD, T., and S. SOTER. 1980. The Deep-Earth-Gas Hypothesis. *Scientific American* 242, no. 6: 154–161.

HALBOUTY, M. T. 1976. Application of LANDSAT Imagery to Petroleum and Mineral Exploration. *AAPG Bulletin* 60:745–793.

———. 1980. Geologic Significance of LANDSAT Data for 15 Giant Oil and Gas Fields. *AAPG Bulletin* 64:8–36.

HARDING, B. J. 1989. *Effects of Light Hydrocarbons on Tree Morphology: Albion-Scipio Oil-Trend, South-Central Michigan.* Master's Thesis, Bowling Green State University, Bowling Green, Ohio.

HERMAN, J. D., R. K. VINCENT, and B. DRAKE. 1991. Geological and Geophysical Evaluation of the Region Around Saginaw Bay, Michigan (Central Michigan Basin) with Image Processing Techniques. Boulder, Co.: *Geological Society of America,* Special Paper 256:221–240.

HERMAN, J. D., P. J. ETZLER, M. L. WILSON, and R. K. VINCENT. 1984. Geoscience Finds Possible Iowa Overthrusting. *Oil and Gas Journal* Nov. 5:129–134.

———. 1985. The Mid-Continent Rift Frontier Hydrocarbon Play: A Case Study Based Upon an Economical Approach to Prospect Generation. In *Proceedings of the International Symposium on Remote Sensing of the Environment, Fourth Thematic Conference, Remote Sens-*

ing for Exploration Geology, 1:21–30. Ann Arbor: Environmental Research Institute of Michigan.

HOLMES, A. 1965. *Principles of Physical Geology.* New York: Ronald Press Co.

KARSTEV, A. A., S. A. TOBASARANSKII, M. I. SUBBOTA, and B. A. MOGLIEVSKII. 1959. *Geochemical Methods of Prospecting and Exploration for Petroleum and Natural Gas.* Berkeley: University of California Press.

KRESIC, N. 1994. Remote Sensing of Tectonic Fabric Controlling Groundwater Flow in Dinaric Karst. In *Proceedings of the Tenth Thematic Conference on Geologic Remote Sensing,* vol. 1, 161–167. Ann Arbor: Environmental Research Institute of Michigan.

LEE, C. S., I. R. MacDONALD, and W. W. SAGER. 1993. Seafloor Mapping Using Geophysical Records Obtained by the Submarine NR-1. In *Proceedings of the Ninth Thematic Conference on Geologic Remote Sensing,* vol. 2, 653–660. Ann Arbor: Environmental Research Institute of Michigan.

LEONE, I. A., F. B. FLOWER, J. J. ARTHUR, and E. F. GILMAN. 1977. Damage to Woody Species By Anaerobic Landfill Gases. *Journal Arboric.,* vol. 3, no. 12:221–225.

LI, D. 1994. Fracture Water Network Analysis and Its Applications to Watersupply and Underground Engineerings. In *Proceedings of the Tenth Thematic Conference on Geologic Remote Sensing,* vol. 2, 568–575. Ann Arbor: Environmental Research Institute of Michigan.

MAKTAV, D., and S. KAPDASH. 1994. An Engineering Application of LANDSAT Data to Wetland Investigations in Turkey. In *Proceedings of the Tenth Thematic Conference on Geologic Remote Sensing,* vol. 1, 515–520. Ann Arbor: Environmental Research Institute of Michigan.

MARRS, R. W., and E. D. PAYLOR. 1987. Investigation of a Surface Spectral Anomaly at Table Rock Gas Field, Wyoming. *Geophysics* 52:841–857.

MEDVE, R. J. 1971. Anatomical Study of the Endotrophic Mycorrhizae of Acer Rubrum. *Bulletin Torrey Botanical Club,* vol. 98, no. 1:41–45.

MILLER, J. B. 1975. LANDSAT Images as Applied to Petroleum Exploration in Kenya, NASA Earth Resources Survey Symposium. NASA TM X-58168, 1-B. Houston: Lyndon B. Johnson Space Center.

MINOR, T. B., J. A. CARTER, M. M. CHESLEY, R. B. KNOWLES, and P. GUSTAFSSON. 1994. The Use of GIS and Remote Sensing in Groundwater Exploration for Developing Countries. In *Proceedings of the Tenth Thematic Conference on Geologic Remote Sensing,* vol. 1, 168–179. Ann Arbor: Environmental Research Institute of Michigan.

PARKER, J. M. 1966. Lisbon Field Area, San Juan County, Utah. *AAPG Mem.* 9, no. 2: 1371–1388.

PRICE, C. V., R. W. BIRNIE, T. L. LOGAN, B. N. ROCK, and J. B. PARRISH. 1985. Discrimination of Lithologic Units on the Basis of Botanical Associations and LANDSAT TM Spectral Data in the Ridge and Valley Province, Pennsylvania. In *International Symposium on Remote Sensing of the Environment, Fourth Thematic Conference: Remote Sensing for Exploration Geology,* 125–140. Ann Arbor: Environmental Research Institute of Michigan.

PROST, G. L. 1988. Predicting Subsurface Joint Trends in Undeformed Strata. In *Proceedings of the Sixth Thematic Conference on Remote Sensing for Exploration Geology,* vol. 2, 423–436. Ann Arbor: Environmental Research Institute of Michigan.

RAMASAMY, S., and R. JAYAKUMAR. 1993. Behavior of the Aquifer Systems in Complexly Folded and Fractured Precambrian Regimes of Southern Indian Peninsula. In *Proceedings of the Ninth Thematic Conference on Geologic Remote Sensing,* vol. 1, 597–607. Ann Arbor: Environmental Research Institute of Michigan.

RAVEN, P., R. EVERT, and H. CURTIS. 1976. *Biology of Plants.* 2d ed. New York: Worth Publishing.

ROCK, B. N., H. R. LANG, J. B. PARRISH, and C. J. LAVINE. 1986. Remote Detection and Correlation of Distribution of Natural Tree Species and Soil Concentrations of Low Molecular Weight Hydrocarbons. In *Unconventional Methods in Exploration for Petroleum and Natural Gas, IV,* ed. M. Davidson, 163–171. Dallas: Southern Methodist University Press.

SABINS, F. F., JR. 1978. *Remote Sensing: Principles and Interpretation.* San Francisco: W.H. Freeman.

SCOTT, L. F., R. M. MCCOY, and L. H. WULLSTEIN. 1988. A Closer Look at the Patrick Draw Oil Field Vegetation Anomaly. In *Proceedings of the Sixth Thematic Conference on Remote Sensing for Exploration Geology,* vol. 2, 529–538. Ann Arbor: Environmental Research Institute of Michigan.

SCOTT, L. F., and R. M. MCCOY. 1993. Near Surface Soils: The Interface Between Deep Reservoir Hydrocarbons and Metals in Sagebrush, Lodgepole Oil Field, Utah. *Proceedings of the Ninth Thematic Conference on Geologic Remote Sensing,* vol. 2, 861–867. Ann Arbor: Environmental Research Institute of Michigan.

SINGER, P. C., and W. STUMM. 1968. Kinetics of the Oxidation of Ferrous Iron. Paper presented at the *Second Symposium on Coal Mine Drainage Research,* the Mellon Institute, Pittsburgh.

TAYLOR, K., M. WIDMER, and M. CHESLEY. 1992. Use of Transient Electromagnetics to Define Local Hydrogeology in an Arid Alluvial Environment. *Society of Exploration Geophysics* 47, no. 2:343–352.

VINCENT, R. K. 1975. Oil, Gas Exploration Tool—Composite Mapping of Earth from Satellite Information. *Oil and Gas Journal* Feb. 17:141–142.

VINCENT, R. K., G. N. SCOTT, and S. THILLAIGOVINDARAJAN. 1978. Groundwater Exploration in Northwestern Tamil Nadu, India with LANDSAT Data. In *Proceedings of the Twelfth Symposium on Remote Sensing of Environment* 2: 1053–1062. Ann Arbor: The Environmental Research Institute of Michigan.

VOSTOKOVA, Y. A., D. D. VYSHIVKIN, M. S. KASYNOVA, N. G. NESVETAYOLAVA, and A. M. SHVYRYAYEVA. 1961. Geobotanical Indicators of Bitumen. *International Geological Review* 3, no. 7:598–608.

WEIR, G. W., and C. L. DODSON. 1958a. Preliminary Geologic Map of the Mount Peale 3 NE Quadrangle, San Juan County, Utah. *U.S. Geol. Surv. Invest. Field Studies Map MF-145.*

———. 1958b. Preliminary Geologic Map of the Mount Peale 3 SE Quadrangle, San Juan County, Utah. *U.S. Geol. Surv. Invest. Field Studies Map MF-147.*

———. 1958c. Preliminary Geologic Map of the Mount Peale 4 SW Quadrangle, San Juan County, Utah. *U.S. Geol. Surv. Invest. Field Studies Map MF-148.*

WOOD, H. B. 1968. Geology and Exploitation of Uranium Deposits in the Lisbon Valley Area, Utah. In *Ore Deposits of the United States, 1933–1967,* ed. J. D. Ridge, vol. 1, 770–789. New York: American Institute of Mining, Metallurgical, and Petroleum Engineers.

Additional References (Uncited)

MCCOY, R. M., and L. H. WILLSTEIN. 1988. Mapping Iron and Manganese Concentrations in Sagebrush Over the Blackburn Oil Field, Nevada. In *Proceedings of the Sixth Thematic Conference on Remote Sensing for Exploration Geology,* vol. 2, 523–528. Ann Arbor: Environmental Research Institute of Michigan.

9

Remote Sensing Applications to Environmental Studies and Geological Engineering

▼ ▼ ▼ ▼ ▼ ▼ ▼ ▼ ▼ ▼ ▼ ▼ ▼ ▼ ▼

INTRODUCTION

Whereas Chapters 7 and 8 were about the use of geological remote sensing for natural resource exploration, Chapters 9 and 10 contain mixtures of environmental and geological remote sensing applications. This chapter is about remote sensing applications to environmental studies and geological engineering, both of which nonexclusively include interactions between manmade constructs and nature. Unlike mineral and petroleum exploration, which employ remote sensing on a one-time-only basis (usually up-front), environmental and geological engineering problems often call for remote sensing information on an ongoing basis. This multitemporal coverage requirement increases the commercial market potential for environmental and geological applications relative to remote sensing for exploration purposes.

Some applications in this chapter, such as site selection for all types of manmade features, require only one-time data collection of remote sensing data. There are many other applications, however, that require the monitoring of environmental or geological engineering changes over time. Government regulations already require monitoring for changes in volume, leachate run-off patterns, and methane content of solid waste landfills. Open pit mines require monitoring for volume changes, slumping (a safety issue), reconstruction of topographic contours, and the determination of toxic waste proximities. Wetland delineation requires estimation of the average number of days of high water table during the growing season. Studies of farm land erosion require accurate modeling of surface water run-off after precipitation events, while coastal erosion studies require measurement of volumes of land eroded in the past and estimates of future erosion. Pipelines require routine monitoring for leaks, and highways require roadbed and pavement monitoring for subsidence, surface water channeling, and pavement breakup. All of these monitoring tasks either could or already do benefit from remote sensing technology.

The development of remote sensing methods for the monitoring of environmental and geological engineering changes has lagged behind the development of methods for mineral and petroleum exploration. However, the increasing importance of environmental and infrastructure issues and decreased levels of petroleum and mineral exploration in the post–Cold War era have tipped the scales of remote sensing development toward environmental and geological engineering applications. Even the development of new remote sensing methods in major petroleum companies has progressed from exploration to applications for logistics, pipeline routing, seismic line planning, and facility siting (refineries, marine terminals, and so on). Consequently, this chapter is less extensive than the previous two because of the exploration bias of past applied research goals, but it will be of greater current interest for many students now seeking jobs in environmental or geological engineering companies.

The chapter will be divided into four sections, each concerned with the application of remote sensing to pairs of categories that share similar remote sensing needs. The first section is devoted to solid waste landfills and open pit mines, the second to forests and wetlands, the third to farm land and coastal erosion, and the third to pipelines and highways.

REMOTE SENSING APPLICATIONS FOR SOLID WASTE LANDFILLS AND OPEN PIT MINES

Solid waste landfills are the final resting places for our garbage. Although much of what we place there is biodegradable, some of it is not. Toxic waste landfills are simply solid waste landfills to which toxic wastes have been added.

Open pit mines are surface excavations of metallic ore. In the last half-century, the methods for mining many metals by open pit mines have proved to be far less expensive per ounce of metal produced than underground mining methods. Toxic materials, such as cyanide, are used in the separation of the desired metal from the rocks and soils that have been excavated.

Despite the fact that landfills involve the insertion of disposable wastes into the ground and open pit mines involve extraction of ore from the ground, these two geological engineering applications have at least two similarities. Both require the monitoring of volume changes associated with holes dug into the ground and their partial or complete refilling, and both can benefit from the monitoring of changes in surface composition in and around the excavation sites. The following two subsections deal with these fundamentally different remote sensing tasks that are needed for both landfill and open pit mining operations.

High-Resolution Topographic Mapping for Landfills and Open Pit Mines

For over half a century, photogrammetry (Avery 1977) has been employed for the production of elevation contour maps from a pair of stereo photos. These are two downward-looking photos that overlap (usually around 60%). One-foot contour maps take a long time to produce by traditional, primarily manual, methods. Typically, it takes weeks or months to get a conventionally produced one-foot contour map delivered, once the aerial photos are available for stereo-plotting. Even then, only the contour maps are available; these are insufficient by themselves for measurement of temporal changes in landfill or open pit mine volumes. If those maps are digitized and elevations are interpolated between the contour lines, then the resulting DEM of day 1 can be subtracted from a similarly created DEM for day 2, and volume change can be calculated (sum of the elevation changes for all elevation grid cells multiplied by the area of one grid cell). The image data from the photos, however, remain separate from the contour map and DEM, if traditional photogrammetric methods are employed.

Over the last decade automation has greatly changed photogrammetric technology. As described in Chapter 6, software now exists that is capable of automatically extracting elevation data for every pixel in the overlapped region of two digitized stereo images, while keeping this high-resolution elevation data in perfect co-registration with the image data from one of the two stereo images. Stereo aerial photos can be digitized by scanning them with a digital scanner, of which many types are available commercially. Two digitized stereo images (the source images) and about a half-dozen ground control points, for which x, y, and z locations are known, are the input to DEM-generating software. Output usually consists of two computer files, one a high-resolution DEM and the other a digital orthophoto (a digital image from which parallax has been removed). The two perfectly overlay one another because they were both produced from the same source images.

In the case of the software described in Chapter 6, the resulting DEM and digital orthophoto have the same spatial resolution as the digitized stereo source images. For example, 1:4000 scale photos that are scanned with a 25 μm spot size (1000 dots per inch) yield an overlapped region (an area on the ground of about 3000 ft $\times$ 1500 ft) that contains about 41 million pixels per photo, with each pixel covering about 4 in. square on the ground. This is a suitable spatial resolution for either a landfill or an open pit mine. ATOM can extract an elevation for each of these 4 in. pixels with a root-mean-square error in elevation of about 5 or 6 in. (Vincent et al. 1987), in less than 2 hours on a workstation computer.

For solid waste landfills, the resulting high-resolution DEM can be employed in several useful ways (Vincent 1994). First, it can be used as input to a surface run-off model for determining where rainfall or flood water run-off will transport surface waste fluids and gases from different parts of the landfill to outlying areas and where depressions in the landfill cover may collect and concentrate run-off. Second, after the extraction of elevations from stereo images taken at two different times, it is possible to calculate very accurately what volume change has been caused by subsidence or additional filling between the two data collection times over the solid waste land-fill. Third, numerous computer programs are available that can transform the DEM and digital orthophoto into simulated perspective views, or even into a "fly-around," of an existing solid waste landfill or a potential new landfill site. The combined tech-nology involved with the generation of every-pixel-posted DEM and digital or-thophotos and the subsequent calculation and display of simulated perspective views from any user-selected position of the observer has been called digital holography (Vincent 1989). Perspective views and "fly-arounds" produced by digital holography could be useful for briefing local and state officials about the true appearance of a solid waste landfill or a toxic waste landfill.

In the next chapter, an example will be shown that applies interferrometric synthetic aperture radar (IFSAR) from two-pass, ERS-1 satellite radar data to the mapping of very small topographic elevation changes (28 mm) caused by an earth-quake. Although this technique could be applied to landfill topographic changes, the elevation changes are averaged over one pixel, which is 25 m $\times$ 25 m in area. This as-sessment is far broader than aerial stereo photos can yield, and would be unsuitable for digital holographic "fly-arounds."

Open pit mines also require considerable elevation information because the topography quickly changes with time. Volume estimations of the amount of overbur-den that must be removed or the amount of material that has been removed in the past are currently supplied to open pit operators by conventional civil engineering methods. These methods are slow and inaccurate by the standards of digital photogrammetry. It is now possible for an open pit mine to employ digitized aerial or tethered balloon stereo photographs of the mine in a digital photogrammetry workstation at the mine site to make estimation of volume changes over time periods as short as 24 hours. This data can then be used with three-dimensional mine modeling software to estimate how much overburden must be moved to recover a certain volume of ore, as well as to esti-mate how much material has been removed in the last time interval between stereo photo collections. On a longer time interval, before and after stereo photo pairs of the mining area can be used to create high-resolution DEMS that can document the return of land contours to within regulated deviances from their original shapes. As with land-fills, open pit mines can also benefit from simulated perspective views and "fly-arounds" of the mine for the briefing of company managers or regulatory agency personnel.

Mapping Chemical Indicators Around Landfills and Open Pit Mines

The great expense of acquiring point-measurement data is an impediment toward denser areal coverage than the minimum requirements. This data often require labor-intensive, equipment-intensive, and laboratory-intensive methods for their collection. For instance, one lab analysis of a single water sample can cost between $1200 and $5000, depending on the level of detail in the analysis. Field and lab meth-ods usually require long response times, sometimes too long to permit the informa-tion obtained to be useful as a management planning tool.

Multispectral remote sensing data offer opportunities for the areal extrapo-lation of point measurements made in the field. This advantage results in either an increase in the area for which information can be gleaned from a given set of point measurements, or a decrease in the number of point measurements required to obtain a specified area of coverage. For example, soil coring and surface soil chemical analysis are point-measuring methods that can yield the content of contaminants in soil at the surface and in the shallow underground. However, both methods greatly undersample the area of interest and tend to be slow and expensive for large areas. Remote sensing data can be used to extrapolate a few point measurements over large areas whenever a remotely sensed surface property (such as a subtle topographic expression or a surface geochemical alteration) is found to be correlated with important lab-measured surface or subsurface information at sites of point data collection. The resulting information contains much denser information than could be afforded, both in time and dollars, by point measurements alone.

What information can multispectral remote sensing data provide about subsurface conditions? This is the same question that faced remote sensing companies in the 1970s and 1980s concerning petroleum and mineral exploration. In that case, remote sensing geologists found that geochemical alteration of surface soils had been caused by ancient hydrocarbon seeps that had risen to the surface from ruptured reservoirs thousands of feet deep (Vincent 1977; Abrams et al. 1984). It was also found that detailed observation of surface topography made it possible to detect and locate subtle topographic expressions of some of the structural traps that help define underground petroleum reservoirs (Abrams et al. 1984). In summary, remote sensing geologists were able to infer what was beneath the surface by mapping compositional and topographic indicators that were present at the surface.

The application of multispectral remote sensing to solid waste landfills will be discussed first, followed by a similar discussion for open pit mines. Remote sensing through photo interpretation of historical photos has already been applied to landfills (Erb et al. 1981; Stohr et al. 1987; Stohr et al. 1988) and water resource management (Salomonson 1983). There have also been some applications to landfill problems of remote sensing involving electronic imaging (Lyon 1982; Stohr et al. 1994). However, far more applied research is required before multispectral remote sensing can be used operationally for solid waste landfill monitoring.

The applications of multispectral remote sensing for exploration geology offer several precedents that can be useful for the mapping of surface soil compositions and geochemical alterations caused by elevated concentrations of toxic chemicals that have interacted with surface soils. First, clays have been distinguished (Abrams et al. 1977) with multispectral scanners, and clays are an important constituent of solid waste liners and caps. It would be possible, for instance, to use multispectral data from the 1.55–1.75 μm and the 2.08–2.35 μm region (LANDSAT TM bands 5 and 7, respectively, which are duplicated in the Daedalus Enterprise's DS-1268 Airborne Thematic Mapper) to produce a spectral ratio image (Vincent et al. 1984b) that shows clays in sharp contrast to other materials (except, perhaps, for some carbonates and sulfates). It is likely that soil consisting of at least one-third clay could be recognized as clay-rich by having a band 5 to band 7 spectral ratio above an empirically determined threshold ratio value. Large percentages of landfill surface area covered by recognized clay-rich soil would indicate effective cover of the landfill by a clay cap.

Ferric oxides and hydroxides (such as hematite and goethite) can also be mapped with multispectral scanners (e.g., Rowan et al. 1983; Vincent 1983), especially with spectral ratios of visible red to visible green bands (like LANDSAT MSS bands 5 and 4), and they are often found where ground water has discharged through gravel and sand. Under some circumstances the oxidation-reduction phenomena associated with contaminant leakage into ground water will deposit iron oxides in a manner similar to ferric and ferrous oxide exposures related to geochemical cells, used for mineral exploration. When this happens, it is often a by-product of oxidation-reduction phenomena, rather than the compound of principal interest itself, that produces a telltale sign that such chemical reactions have occurred. Leachate seeps from a landfill can stain surface soils with an orangish coating of limonite, a hydrated ferric oxide. This use of a multispectral scanner is an indirect approach similar to that of exploration geologists who map clay and sulfate exposures when looking for gold deposits (Abrams et al. 1977) that contain gold in far too small quantities to permit direct detection.

It is also likely that surface exposures of exotic compounds with unique spectral features, such as cyanide and chromium compounds, will result from some leachate or contaminant interactions with surface soils and water. An example would be "blue soil," a cyanide compound that I have observed to be associated with coal gas production wastes. Many of these exotic compounds will be more easily observed in the reflective infrared and thermal infrared wavelength regions than in the visible wavelength region. This ability gives the multispectral scanner a distinct advantage over the human eye. Just as in mineral exploration, a properly filtered multispectral sensor can map some chemical composition differences that a human can walk across, but overlook.

It is instructive to examine what kind of contaminants can be expected to emerge from landfills in contaminated ground water. Figure 9.1 shows a ranking of inorganic ground water constituents based on their contamination factor, which is defined as the ratio of the mean concentrations between the contaminated and the uncontaminated ground water (Arneth et al. 1989). Figure 9.2 shows a ranking of the 15 most frequently detected organic contaminants in ground waters downstream of waste sites in Germany and in the United States (Plumb and Pitchford 1985). Aliphatic (meaning that carbon atoms are linked in open chains rather than in rings) chlorinated hydrocarbons are the most common organic contaminants, as shown by the black bars (9 in Germany and 10 in the United States). They include the Dense Non-Aqueous Phase Liquids (DNAPLs), which are denser than water, thereby causing them to quickly sink to the bottom of unconfined aquifers if introduced on or near the natural ground surface.

There are two important areas of research that need to be performed concerning the remote sensing of solid waste landfills. First, a determination must be made of what chemical compounds might be produced by the interaction of these inorganic and organic materials with ground water. Second, spectral reflectance measurements of those telltale compounds and the contaminants themselves in the

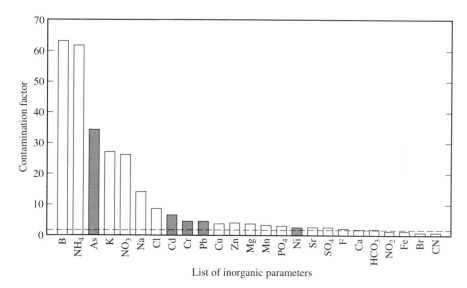

Figure 9.1 Ranking of inorganic ground water constituents based on their contamination factor (the ratio of the mean concentrations between the contaminated and the uncontaminated ground water). Dark gray indicates toxic metals. *(After Arneth et al. 1989. Reprinted courtesy of Springer-Verlag New York, Inc.)*

visible, reflective infrared, and thermal infrared wavelength regions must be made. These two research steps will tell us what chemical compounds we are looking for, which ones of those are likely to be detectable by multispectral remote sensing, and what image processing methods can be used to find them.

Remote sensing data have already proved useful for monitoring solid waste landfills and toxic waste sites by mapping vegetative stress in the vicinity of waste leakage. Herman et al. (1994), reports a study of a waste site in Michigan where escaping toxic liquids had actually killed trees that were part of a nearby swamp. Vegetation stress that ranges from chlorosis to death can be mapped from multispectral data that are sensitive to relative amounts of chlorophyll in vegetation.

Restriction of multispectral efforts to the mapping of solids or liquids associated with landfills would be shortsighted, because gases can be a problem, too. When solid waste decomposes, gases are given off as products of decomposition, with the most prevalent one being methane. Municipal solid waste landfill emissions consist primarily of methane and carbon dioxide, with trace amounts of more than 100 different nonmethane organic compounds (NMOCs) such as ethane, toluene, and benzene (U.S. Environmental Protection Agency 1991). Methane in landfills acts as a stripping (or transport) gas, moving the NMOCs present in the landfill through the landfill to the atmosphere. These landfill gas emissions have adverse health and welfare effects resulting from NMOCs, some of which are known or suspected carcinogens and others that have various noncancer health effects (U.S. Environmental Protection Agency 1991). NMOCs also contribute to low-altitude formation of ozone that can cause lung irritations. Besides contributing to adverse health effects, the NMOCs in landfill gas emissions can cause odor

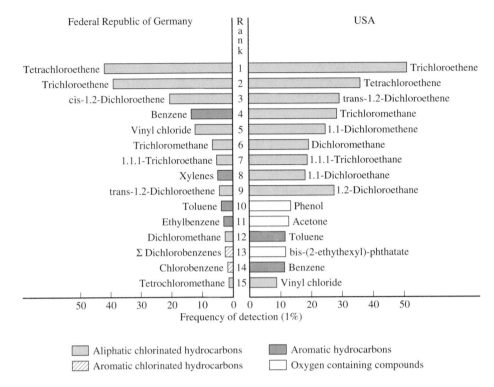

Figure 9.2   The 15 most frequently detected organic contaminants in ground waters downstream of waste sites in the Federal Republic of Germany (92 sites) and the United States (358 sites). *(After Plumb, Jr. and Pitchford 1985.)*

nuisances, and methane itself has caused explosions and fires resulting from its migration to on- and off-site structures or enclosures (U.S. Environmental Protection Agency 1991).

A recent study (Carman 1996) of the methane content in soil gas and in the atmosphere above three Wood County, Ohio solid waste landfills found methane concentrations ranging from 5% to 95% of the soil gas in two inactive landfills and from 10% to 45% in the active Wood County Landfill. Figure 9.3 shows a plot of soil gas methane content versus atmospheric barometric pressure for the six most methane-rich subsurface measuring stations at the active Wood County Landfill. These plots have been translationally normalized by algebraic addition, such that they all show the same soil gas methane content at a barometric pressure of 29.6 in. of mercury. There is an increase of soil gas methane content with an increase in barometric pressure, but this increase is not linear. Figure 9.4 shows plots of the percentage of atmospheric methane as measured by a handheld methane "spotter" at 1 meter above the landfill near a vent (that extends down into the landfill) and the soil gas methane content for the nearest station (#9) where soil gas was measured, versus barometric pressure for the same Wood County Landfill. Two notable items related to this figure are that the atmospheric methane content, as a percentage by volume in air, was

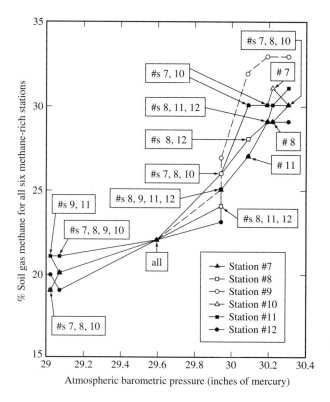

Figure 9.3   Soil gas methane content versus atmospheric barometric pressure from six methane-rich data collection stations on the Wood County Landfill (active) near Bowling Green, Ohio, for 10 different days in the Feb. 27–Mar. 13, 1996 time period. These plots have been translationally normalized by algebraic addition, such that they all show the same soil gas methane content at a barometric pressure of 29.6 in. of mercury. *(After Carman 1996. Reprinted with permission.)*

recorded between 0.5% and 0.8% (5000–8000 ppm), and the atmospheric methane increases with both low and high barometric pressure. The latter item indicates that two different types of physical mechanisms are probably responsible for the methane escape. No measurement was made for NMOCs because the appropriate instrumentation was not available. For methane to be flammable, it must be in the 5% to 15% range of methane in air (50,000–150,000 ppm).

As discussed in Chapter 5, it should be possible to use remote sensing methods to image differences in the chemical composition of gases, provided that the instruments used are properly filtered and the gas concentrations are sufficiently high. Figure 9.5 shows a spectral absorption coefficient (Vincent and Singleton 1994) of methane gas at a pressure of 702 mm of mercury in the 3.0–4.0 μm wavelength region, which is within an atmospheric window. The principal methane absorption band could be exploited by a down-looking, three-channel infrared imaging device that compared the radiance in the 3.298–3.330 μm wavelength region to radiances in

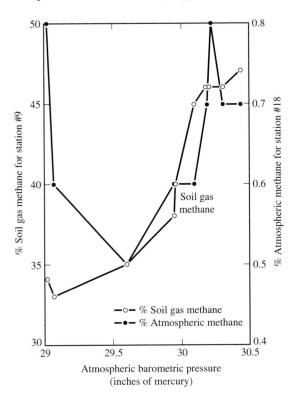

Figure 9.4   Soil gas methane content from station #9 and atmospheric methane content from station #18 versus atmospheric barometric pressure on the Wood County Landfill (active) near Bowling Green, Ohio, for 10 different days in the Feb. 27–Mar. 13, 1996 time period. *(After Carman 1996. Reprinted with permission.)*

the 3.100–3.132 μm and 3.500–3.532 μm wavelength regions by means of spectral ratio imaging. In very humid conditions, it may be necessary to use only the two longest wavelength bands of the three, since water vapor may absorb too much in the 3.100–3.132 μm band.

Such imaging would permit easier location of methane plume exit sites than would be afforded by nonimaging detection methods and would help separate landfill methane from other sources of methane, such as natural gas wells or swamps, by virtue of the plume location. Since methane is the most common landfill gas, it is the primary indicator of those places where landfill gas is escaping to the atmosphere. It is also possible that ground water contamination from the landfill could be traced by the imaging of methane escape from contaminated near-surface aquifers, because methane is dissolved in leachate-contaminated springs. Existing multispectral sensors might also be modified to image other landfill gases such as ozone, ethane, toluene, and benzene with thermal infrared absorption bands from airborne (aircraft or tethered balloon) multispectral platforms, though the concentration of these gases will be nowhere near as high as methane.

The use of multispectral remote sensing for mineral exploration was the subject of Chapter 7. The application of multispectral techniques to open pit mining is a newer field of endeavor, but the early results are significant, primarily because minerals tend to differ from one another by chemical composition and hence have dif-

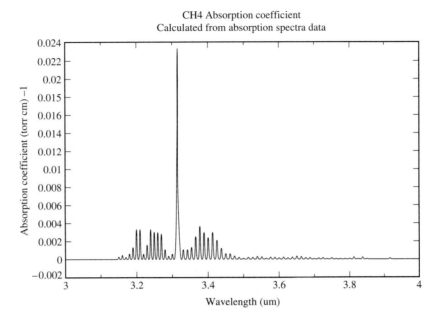

Figure 9.5 Absorption coefficient of methane in units of $(torr\text{-}cm)^{-1}$ versus wavelength in the 3.0–4.0 μm wavelength region. *(After Vincent and Singleton 1994. Reprinted with permission.)*

ferent multispectral "signatures." Examples are presented below that represent several stages of open pit mining, from an estimation of the "diggability" of the rocks, through the spectral analysis of wall rock and underground core samples, to the spectral analysis of tailings piles.

Ayday and Goktan (1993) used GIS technology, field geology, and a digitized topographic map to produce a digitized geologic map that identified the types of rock outcrops and classified them according to their ease of excavation. To do this, they employed two types of schemes for selecting the size of tractor (or blasting) required to excavate the rocks: a diggability classification system (Muftuoglu and Scoble 1985) and a rippability rating chart (Singh et al. 1986). Ayday and Goktan accepted as correct any result that the two schemes agreed upon. The diggability classification scheme was determined from ratings of the following factors for each rock outcrop: degree of weathering, uniaxial compressive strength, joint spacing, and bedding spacing. The rippability rating chart rated according to degree of weathering, tensile strength, seismic velocity, abrasiveness, and discontinuity spacing. Only one rock unit, a claystone, was judged by both schemes to require the same size tractor (designated as a D10 Caterpillar ripper-tractor) for excavation. Since agreement between the two schemes occurred for only one rock unit, it is clear that more work is needed to unify the two excavation assessment schemes, with an attempt to rely solely on surface properties of the outcrop, if possible.

Even though field work, not remote sensing, was used to perform the ratings of each scheme, remote sensing could be employed for the assessment of three out of four ratings in the diggability classification ratings (degree of weathering, joint spacing, and bedding spacing) and two of the five ratings in the rippability ratings (degree of weathering and discontinuity spacing), especially if aerial photography and satellite data were both available. With more research, it appears likely that a method involving GIS and RS (remote sensing) together may one day be able to predict the tractor size required to excavate the rock outcrops in an area that has been selected for mining. This would be tantamount to estimating the cost of mining by GIS/RS methods before mining has begun.

Great strides have been made in the use of field spectrometers for determination of the direction of excavation that will intercept the highest grade ore bodies. Yamaguchi et al. (1994) married a palmtop computer (with expert-system software) and a portable infrared spectroradiometer that operates in the 1.3–2.4 μm wavelength region to produce a system that is capable of identifying 40 different types of minerals. The spectroradiometer is directed toward an exposed soil, rock outcrop, or rock fragment and the names of alteration minerals present will be displayed on the computer screen. Measurement time takes about 8 seconds and identification time takes about 5–20 seconds. The computer converts from radiance to reflectance, smoothes the reflectance spectrum, enhances absorption features via the Kubelka (1948) method, and corrects by the Hull-quotient method (Green and Craig 1985). The choices of the expert-system software are displayed on the screen with the probabilities listed for the presence of each. The authors claim that the system has proved good enough for field use, particularly for the field mapping of hydrothermal alteration zones. However, they do not report any mixing rule algorithms that would attempt to determine the relative percentages of each of the minerals present in a mixture.

A system developed by Kruse et al. (1994) was designed primarily for the automatic scanning and classification of split drill core—cylindrical pieces of rock extracted by a drill that are cut lengthwise into half-cylinders. They used a PIMA II field spectrometer (manufactured by Integrated Spectronics, Sydney, Australia) that has 600 spectral bands in the 1.3–2.5 μm wavelength region. The spectrometer is placed in physical contact with the flat surface of the split core sample and records a reflectance spectrum for every 1 square cm of the sample surface. Hyperspectral imaging software used for airborne imaging spectrometer work was employed, along with an unpublished library of an unreported number of mineral spectra. Spectral classification was used to produce images showing the location of specific minerals in each split core sample. However, they did not stop there. The recognized minerals were taken to be endmembers of a mixture and a linear mixing algorithm was employed to estimate the abundance of each mineral. Though the authors claimed the results were reasonable, based on the appearance of endmember spectral mixes and visual observation of the split core samples, they offered no supporting analytical measurements to verify the mapped mineral distributions.

The combined efforts of the investigators described in the previous two examples indicate that field spectrometers and portable computers with software can automatically identify the presence of pure minerals and possibly even estimate the abundance of minerals present, and they will soon be available for use with soil, rock, or split core samples from open pit mining areas. Field spectrometers will assist open pit mining not only by directing excavation activity, but also by determining the proximity of toxic metals left over as waste products of mining, as the next example demonstrates.

Munts et al. (1993) found that field spectra of tailings at a lead-zinc base metal facility in Colorado showed spectral reflectance characteristics that could be related to lead content of the soil, although the lead produced no reflectance feature by itself. Figure 9.6 shows field spectra of six samples collected at different depths (depth increases from sample A to sample F) in the tailing pile on the side of a hill. The strong reflectance minimum near the 1.9 μm wavelength region is caused by water. This value indicates that the wastes have a high water content. The sharp spectral feature (reflectance minimum) near 2.2 μm in the deeper samples C–F is caused by kaolinite, a well-ordered clay. This sharpness is dulled in the surface sample (A) and the next two shallow samples (B was collected at a depth of 6.3 cm and C at 19 cm), owing to the presence of disordered clay, such as smectite. Smectite decreases with depth, replaced by more ordered clay species (kaolinite).

X-ray diffraction analysis of the samples showed that lead was found in the jarosite, an iron sulfate, which in turn was associated with smectite. Postemplacement oxidation of the tailings creates acidic fluids, which helps free up the lead. Lead appears to concentrate near the surface, hosted in the jarosite, as a result of upward migration of fluids driven by surface evaporation in dry climates. The alteration activity is manifested on the surface through the presence of the disordered smectites and iron oxides. Thus, soils with the highest lead content also have the highest smectite content, and it is the smectite that produces the subtle spectral feature that can be identified by multispectral not-so-remote sensing. This environmentally impor-

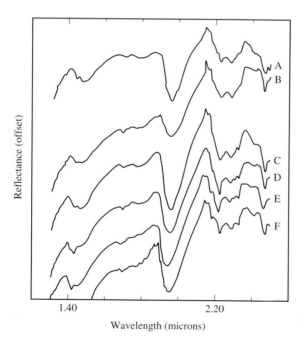

Figure 9.6 Selected short wave infrared (SWIR) spectra from trench 1 in a tailings pile at a lead-zinc open-pit mine in Colorado, collected from different depths: (a) surface; (b) 6.3 cm beneath surface; (c) 19 cm beneath surface; (d) 25 cm beneath surface; (e) 63.5 cm beneath surface; and (f) 76.2 cm beneath surface. SWIR data was collected using a PIMA-II spectrometer built by Integrated Spectronics Pty. Ltd, Sydney, Australia. The increase in sharpness of the 2.2 μm wavelength reflectance minimum with increased depth beneath the surface indicates that disordered clays (like smectite) decrease with depth, replaced by more ordered clay species (kaolinite). *(After Munts et al. 1993. Reprinted with permission of the Environmental Research Institute of Michigan.)*

Reflectance (offset)

1.40   2.20

Wavelength (microns)

tant example for lead-zinc mine tailings may have a corollary with toxic waste land-fills in semiarid areas, since the lead-smectite correlation could also hold there.

A consideration of all the above examples leads to the conclusion that open pit mining likely will become one of the most sophisticated applications of multispectral remote sensing.

REMOTE SENSING APPLICATIONS FOR FORESTS AND WETLANDS

Mapping and monitoring forests and wetlands is an increasingly interesting and valuable application of remote sensing, especially since the advent of hyperspectral sensors. As briefly discussed at the beginning of Chapter 3, Curran and Kupiec (1995) have used data from 181 spectral bands of the AVIRIS (Airborne Visible/Infrared Imaging Spectrometer) sensor of a 60 hectare slash pine (*Pinus elliottii*) plantation in Alachua County, Florida, 20 km northeast of Gainsville, Florida (Gholz et al. 1991). The pine trees were planted in 1965 on a flat, well-drained site. Slash pine retains its needles for two years, so there were only two classes of needles to investigate, the current year's growth and last year's growth. Palmetto (*Serenoa repens*) and gallberry (*Ilex glabra*) were the predominant understory species. Half of the trees were in fertilized plots and the other half were not fertilized. In July, 1992 (the same month as the AVIRIS data collection), 180 branches, each carrying new and old needles, were collected from the upper canopy, half each from the fertilized and unfertilized plots.

Laboratory measurement was made of the foliar biochemical concentrations of chlorophyll, nitrogen, lignin, cellulose, and water as a percentage (%) of the dry sample weight, with the exception of chlorophyll concentration, which was given in milligrams per gram of dry sample weight. Table 9.1 summarizes those lab measurements (Curran and Kupiec 1995). There was a significant difference at the 99% confidence level between fertilized and control plot needles for four of the five biochemical concentrations, excepting lignin.

The AVIRIS data from 181 spectral bands in the 0.4–2.5 μm wavelength range were collected, then first derivatives of the spectral band data as a function of wavelength were calculated. This calculation can be performed by subtracting the DN (digital numbers) of two adjacent spectral bands and dividing that difference by the

TABLE 9.1   BIOCHEMICAL ASSAY DATA FOR SLASH PINE NEEDLES, JULY 1992 (360 SAMPLES).

Biochemical	Mean	Standard Deviation	Coefficient of Variance (%)	Min.	Max.
Chlorophyll (mg/g)	1.71	0.77	45.00	0.45	5.59
Nitrogen (%)	0.98	0.24	24.50	0.52	1.97
Lignen(%)	22.50	1.20	5.30	19.50	25.40
Cellulose (%)	35.50	1.80	5.10	31.20	40.40
Water (%)	57.90	3.60	6.20	48.80	66.30

Source: After Curran and Kupiec 1995.

difference of the center wavelengths for the same two bands. As discussed earlier in Chapter 5, derivative spectra are not as robust a method as calculation of the spectral ratios of two dark-object-corrected spectral bands would be. The results of a stepwise regression analysis between AVIRIS first-derivative reflectances and the same five biochemical concentrations are summarized in Table 9.2. Curran and Kupiec (1995) have given the nearest absorption bands to the location of the spectral band selected per step of the multiple regression as their interpretation of what caused that particular spectral band to be mathematically selected by the multiple regression procedure. This table is very useful because it gives the absorption bands of the five foliar biochemicals in the 0.4–2.5 μm wavelength region. What Curran and Kupiec (1995) call the Red Edge is caused by chlorophyll absorption in the visible red wavelength region. They also state that the 2.2222 μm spectral band (the only one of the 22 AVIRIS spectral bands that was used more than once by the multiple regression procedure) is near a protein absorption feature that occurs at 2.240 μm.

There are two important conclusions to be made from their results. First, it is possible to determine the chlorophyll, nitrogen, lignin, and water contents of slash pine within a high degree of correlation with hyperspectral sensor data. Cellulose content can be determined to a lesser degree than the other four foliar biochemicals. Second, only 21 of the narrow spectral bands of AVIRIS are required to determine those concentrations. To repeat what was said earlier in the geobotanical subsection

TABLE 9.2 STEPWISE REGRESSION ANALYSIS BETWEEN AVIRIS FIRST-DERIVATIVE REFLECTANCE AND FIVE FOLIAR BIOCHEMICAL CONCENTRATIONS IN A SLASH PINE CANOPY.

Biochemical	Step No.	Waveband (μm)	R^2	Interpretation	
Chlorophyll	1	0.7320	0.73	Red Edge	
	2	2.3707	0.87	Nitrogen	(2.350 μm)
	3	1.5515	0.96	Nitrogen	(1.510 μm)
	4	2.0730	0.99	Nitrogen	(2.060 μm)
	5	2.1526	0.99	Nitrogen	(2.180 μm)
Nitrogen	1	1.0308	0.68	Nitrogen	(1.020 μm)
	2	0.9251	0.85	Nitrogen	(0.910 μm)
	3	1.2226	0.94	Lignin	(1.200 μm)
	4	2.3410	0.97	Nitrogen	(2.350 μm)
	5	0.9155	0.98	Nitrogen	(0.910 μm)
Lignin	1	0.9828	0.58	Water	(0.970 μm)
	2	1.9732	0.84	Nitrogen	(1.980 μm)
	3	1.5416	0.92	Cellulose	(1.540 μm)
	4	1.9932	0.96	Nitrogen	(1.980 μm)
	5	1.1939	0.98	Lignin	(1.200 μm)
Cellulose	1	2.2222	0.36	Unattributable	
	2	1.7696	0.61	Cellulose	(1.780 μm)
Water	1	2.2222	0.40	Unattributable	
	2	1.2620	0.59	Water	(1.200 μm)
	3	1.0116	0.79	Water	(0.970 μm)
	4	1.6706	0.96	Lignin	(1.690 μm)
	5	1.2130	0.99	Water	(1.200 μm)

Source: After Curran and Kupiec 1995.

of Chapter 7, the missing research link is the correlation of these five foliar bio-chemicals with various kinds of botanical stress, including stress caused by the localization of toxic metals in the soil. When repeated for other types of forests, this correlation would give multispectral remote sensing methods the capability of mapping toxic metals in the soil from spectral characteristics of the tree tops. This capability would be desirable from both the standpoints of environmental monitoring and mineral exploration. It may also become possible to distinguish various types of stress from one another with multispectral remote sensing data.

If it were possible to uniquely identify the stress on tree foliage caused by water tables that were within 0.3 m of the surface for 12% (consecutive days) or more of the growing season (the nonmarginal wetlands definition given in the U.S. Army Corps of Engineers 1987 manual for wetlands delineation), then remote sensing with hyperspectral sensors could be used to map nonmarginal wetlands with one or a few data collections during the growing season. This would be a great improvement over current attempts to delineate wetlands with color infrared (CIR) photographs, which have three layers of film emulsion that detect reflected sunlight in three wavelength regions (when used with a blue-blocking filter): 0.50–0.58 μm, 0.58–0.68 μm, and 0.68–0.90 μm. These three spectral bands are displayed as blue, green, and red, respectively. Thus, green vegetation appears red in CIR photos, owing to the much higher reflective infrared reflectance (compared to visible reflectance) of green leaves and needles.

There are many misconceptions about what CIR photos reveal about wetlands, the most common of which is the assumption that connected, blue-toned, dendritic patterns in a CIR photo collected during leaf-off conditions (outside the growing season) are wetlands. One-time coverage of the United States by CIR has been used by the U.S. Fish and Wildlife Service (1981) to create a National Wetlands Inventory (NWI) Map, assisted by spot-checked field data of some of the "wetlands" that government agency photo-interpreters attempted to delineate from the CIR photos. Because the NWI map is being used by the U.S. government in criminal cases against businesses and private citizens for illegally changing wetlands, this is an important scientific/public issue.

To demonstrate some of the problems associated with the use of CIR photography for wetlands delineation, I performed a simple CIR photographic experiment with water and fallen tree leaves under natural illumination on my sun porch in Bowling Green, Ohio. Kodak Ektachrome IE135-36 CIR film in a 35 mm format was placed in a 35 mm camera by William Butcher (a technical photographer in the Bowling Green State University Dept. of Geology). The camera (with a MicroNikkor 55 mm focal-length lens) was filtered with a Y52 glass, deep yellow (minus-blue) filter that is similar to the filter normally used in aerial cameras with this same make of film, though in 9-in. format. On a snowy, overcast day (January 9, 1996), a series of photos were taken with 1-second exposures of fallen leaves (ranging from light brown to dark brown in color) from swamp maple and oak trees that were collected in my yard. Natural color photos were also taken. The background included two house plants, a red-flowering poinsettia and a nonblooming green prayer lily, with dry fallen maple and oak leaves (just as they had been collected in the yard a month earlier) in between the two house plants. The foreground changed with every picture, but always included two side-by-side clear pyrex dishes under which was located a black-and-white sign that contained the word MARYLAND (twice written).

Color Plate CP27a and b shows natural color and CIR photos of the same scene, with both pyrex dishes empty. The red flowers of the poinsettia (on the left) appear yellow in the CIR photo, and the green leaves of both the poinsettia and the prayer lily appear purplish red, because of their chlorophyll content. The red flowers of the poinsettia appear yellow in the CIR photo, which displays green, red, and reflective infrared regions as blue, green, and red, respectively, because the red flowers have high red and reflective infrared reflectances (red and green make yellow in the photo), but a low green reflectance (not much blue in the photo). The dry fallen leaves (between the house plants), gray carpet, white paper and towel underneath the pyrex dishes, and the white-painted wood and aluminum frames behind the house plants all appear bluish in the CIR photo. This demonstrates that the blue color in CIR film indicates the absence of chlorophyll, but not much else.

Color Plate CP28a is a CIR photo of the same scene, except that clear tap water has been added to a 2 cm depth in the left pyrex dish and to a 4 cm depth (the deepest that could be handled in the pyrex dishes without getting the carpet wet) in the right one. Color Plate CP28b is a CIR photo of fallen maple leaves in the left dish that have been submerged for 16 days and oak leaves in the right dish that have been submerged in water for 8 days. The water in both dishes is tea-colored from leaf tannins, and the approximate water depth is 4 cm in both dishes. This pair of photos demonstrates that it takes more than a 4 cm depth of clear water to turn a CIR photo blue. However, if the water is tea-colored in the visible wavelength region from tannins, it will be slightly blue in the CIR photo. In neither case is the photo black, which is the color of deep (much greater than 4 cm) water in aerial CIR photos. Black is also the color of shadows in aerial CIR photos.

Color Plate CP29a shows a CIR photo of the same leaves as CP28b, except that they were removed from the water one hour before this photo was taken. Notice that some of these leaves are as light blue as the dry, fallen leaves sitting between the two house plants, and others are darker blue to black. This photo, along with the two in the previous figure, demonstrates that the absorption of sunlight by up to 4 cm of water does not change the appearance of the fallen leaves much in a CIR photo. It may be true that water chemically makes the leaves darker, but research is needed to determine how long a period of soaking in water is required to make an already dark fallen leaf appear dark blue or black in a CIR photo. However, the same leaves in or out of shallow (4 cm or less) water appear nearly the same in a CIR photo.

Color Plate CP29b shows a CIR photo of the front of my home, taken on the same winter day as the previous photos. Note that the snow is bright white, not dark, just like the white aluminum siding on the house. This result demonstrates that CIR photography is primarily detecting reflected sunlight, not thermal infrared radiation. If it were otherwise, the cold snow would be dark, not bright, and the glass front door through which I am peering in CP29b would be bright with heat, not dark.

There are two conclusions that can be drawn from this photographic experiment. One is that the connected, blue-toned, dendritic patterns in CIR photos over deciduously forested areas during times outside the growing season are mostly dead leaves on the ground that are left behind by flowing water after rains. This would be true if there were no standing water, or if there were standing water up to 4 cm. (Perhaps the water levels might be higher, but our experiment was limited to this maximum depth.) Thus, the CIR photos represent drainage patterns that may or may not have been wet at the time the photos were collected. Second, the blue patterns are not temperature patterns, and they offer no information about whether the water table is from 4 cm above the surface to many meters below the surface at the time the CIR photos were collected. Given this last statement, logic would dictate that there is no way to extrapolate leaf-off coverage of a CIR photo to the behavior of the water table during the growing season. The bottom line is that CIR photos collected over deciduous forests outside the growing season, though helpful for mapping drainage patterns, provide insufficient evidence for wetland delineation unless the wetland is almost continually covered by standing water greater than 4 cm in depth. In the case of deep standing water, wetlands in deciduous forests during leaf-off conditions would look like a lake, which usually appears black in CIR photos.

REMOTE SENSING APPLICATIONS FOR FARM LAND AND COASTAL EROSION

Erosion studies for both farm land and coastal land involve determination of the volume and rate of soil loss by erosion in the past, which can be used to help predict rates and volumes of soil that are likely to be eroded in the future. Though both are forms of mass wasting, coastal erosion rates are usually much higher than farm land erosion rates because wave action is added to the other causes of erosion that affect both types of land.

The characterization of erosion in farm land requires detailed knowledge of the topography, farm density, and vegetative cover. In a study of the Hou Gou Men Basin, a loess plateau in China, Jian (1993) produced a GIS classification map of the study area showing five degrees of soil erosion (strongest to lightest) on four different types of soil erosion. The types of soil erosion were labeled scale, sheet, gully, and ravine erosion. To do this, Jian first found principal components of each soil erosion type, based on linear combinations of nine factors, six of which were topographical (relative elevation, mean gradient, maximum gradient, sunlight exposure rate, slope of nonfarm land, and slope of farm land) and three others that include vegetation coverage, cultivation index, and gully density. The topographic factors came from DEM that were produced by the digitization of elevation contours from 1:10,000 scale topographic maps. The vegetation cover, cultivation index, and gully density came from digitized, 1:10,000 scale color infrared (CIR) photography. Based on the principal components, a cluster analysis that minimized the Euclidean distance between soil erosion classes was used to classify the rasterized color CIR photo data. Polygons were drawn around each class, and ARC-INFO (a GIS software package by Environmental Systems Research Institute, or ESRI) was used to plot the soil erosion map of Hou Gou Men Basin shown in Figure 9.7.

The common characteristic between this example and later efforts by other researchers to determine degree of soil erosion is the importance of the combined use

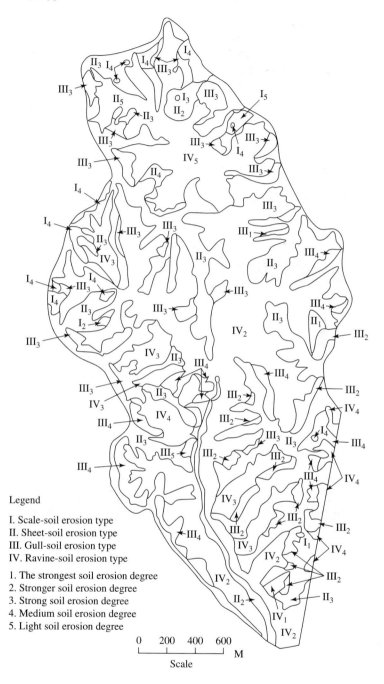

Figure 9.7 A GIS soil erosion classification map of the Hou Gou Men Basin, People's Republic of China, showing five degrees of soil erosion (strongest to lightest) on four different types of soil erosion (scale, sheet, gully, and ravine erosion). *(After Jian 1993. Reprinted with permission of the Environmental Research Institute of Michigan.)*

of high-resolution DEM and CIR photography for the establishment of a GIS data-base from which erosion rates can be calculated or classified. This GIS database must include topographic and vegetative cover parameters that contain the necessary information for determining rates of surface water run-off and erosion, which are closely related. In the Chinese example given above, the DEM and CIR photos came from two different sources. However, it would be possible to produce DEM from stereo pairs of color stereo photos by digital photogrammetry, such that all of the information can come from one source. Besides the obvious advantage of perfect co-registration of the resulting DEM and digitized CIR photos, the DEM would likely be of much higher resolution and greater accuracy than DEM produced from digitization of an existing topographic contour map or sparsely posted elevation data set. These high-resolution DEM will yield much more accurate slope information than current maps or government-produced DEM, and slope is an important parameter for the mapping of soil types and the prediction of surface water run-off directions and velocities. It would appear a safe bet that digitized CIR stereo imagery will be greatly utilized as a common source of both multispectral and elevation data required for GIS database generation in future soil erosion and surface water run-off studies.

Coastal erosion studies have long employed aerial photography, but in a low-tech manner. The traditional method for determining the amount of land lost to coastal erosion along Lake Erie is demonstrated in Figure 9.8 (Mackey 1995). Transects perpendicular to the coastline are visually measured in aerial photos that were

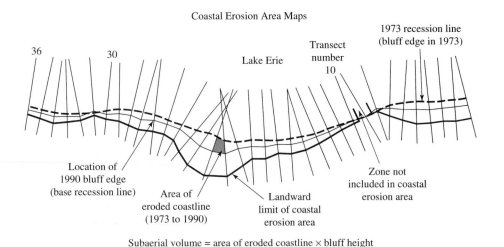

Figure 9.8   Transects perpendicular to the Lake Erie coastline in a traditional method for estimating coastal erosion losses. The annual loss from coastal erosion in Lorain County, which contains the Helen Drive area, is 0.9 hectare per year in area, corresponding to a volume loss of 61,116 cubic meters per year. This loss corresponds to a linear loss rate on the order of 1 m per year. The highest rate of coastal erosion in Ohio is in Ashtabula County, where the annual losses are 1.4 hectares and 229,389 cubic meters per year. *(After Vincent et al. 1995. Reprinted with permission.)*

collected many years (such as a decade) apart in time. The number of meters that the transect length was shortened during that time interval is determined for each transect and averaged for the photo, then multiplied times the coastal length to get the area lost in the intervening time. This result is then multiplied times the average height of the coastal cliff to obtain volume of soil lost to coastal erosion during the given time interval. Division of volume by the time interval yields the volumetric rate of erosion. This result is then used to estimate when a given structure on the coast might be expected to be lost into the water. The answers receive much more than academic interest. Along the Lake Erie coast of northern Ohio, 25% of houses are within 8.3 m of eroded cliff edge, and 47% of the houses are within 16.7 m of the cliff (Mackey 1995).

With digital photogrammetry, it seems possible to calculate the volume of soil lost far more accurately than the above procedure permits. A recent study of the southern coast of Lake Erie (Vincent et al. 1995) employed digital photogrammetry of two stereo pairs collected 10 years apart to estimate the volume rate of erosion on about 600 m of coastline. Figures 9.9a and b show ground photos (originals in color) of eroded cliffs along the Lake Erie coast, with (a) in the vicinity of Helen Drive, just East of Vermillion, Ohio, where the investigation took place. The results of this first attempt at digital photography did not meet expectations, but were instructive in dealing with the problem, as will be shown below.

The two stereo photo pairs of the Helen Drive area were collected at different scales, but one stereo pair was accessible only as hard-copy prints. This was the first of three chief causes for less-than-optimum results, because the hard-copy photo (about 1:6000 scale) could not be scanned with less than a 100 μm spot size by the Optronics P-1000 scanner used for this investigation. Although the second stereo pair was in transparency form and could be scanned with a 25 μm spot size, it had to be resampled to the pixel size of the other stereo pair, which was approximately 0.6 m. This resulted in a minimum detectable elevation difference of 1.0 m, which is too large for this problem by an estimated factor of 3; 0.33 m, or about 1 ft, is estimated to be the maximum resolution in elevation required for this task. The following steps were then taken: DEM were produced with ATOM for the 1980 and 1990 stereo pairs; the 1990 DEM was co-registered to the 1980 DEM; and the difference DEM was calculated. When multiplied times the area of one pixel (0.36 square m), the difference DEM should become the volume of material lost along the coastline.

Figure 9.10a, for 1980, and b, for 1993, show the right photo of each of the two stereo pairs, after the 1980 photo had been resampled and rotated to the 1993 photo, using the same ground control points off a 1:24,000 scale topographic map. This was the second of three chief causes for less-than-optimum results, because the map was much older than the oldest of the photos and it was not possible to select more than four ground control points in the overlap region of the two stereo pairs. Future investigators should employ global positioning satellite (GPS) instruments for determining the accurate positions of points on the ground that have not been changed between the two aerial photo data collection events.

Figure 9.11 shows the resampled DEM generated by ATOM from the 1980 stereo pair (a), the 1993 stereo pair (b), and the difference image of the two DEM

(a)

(b)

Figure 9.9   Ground photos (originals in color) of eroded cliffs (a) in the vicinity of Helen Drive, just east of Vermillion, Ohio, and (b) elsewhere along the Lake Erie coast (Lake County, Ohio). (*After Vincent et al. 1995. Reprinted with permission. Photos courtesy of Ohio Department of Natural Resources—Division of Geological Survey.*)

(a)   (b)

Figure 9.10 Right photos of aerial stereo pairs collected (a) in 1980 and (b) in 1993 for the region around Helen Drive, just east of Vermillion, Ohio on the Lake Erie coast. The left photo has been resampled to the right photo. North is toward the top and the right photo width is approximately 6 km. *(After Vincent et al. 1995. Reprinted with permission.)*

(c). In the (a) and (b) images, dark (bright) signifies higher (lower) elevations. In the (c) image, dark (bright) indicates a decrease (increase) in elevations between 1990 and 1993, if the ground control points were accurate and plentiful (more than 5). Since there were only four ground control points of dubious accuracy for this experiment, some of the differences seen in Figure 9.11c might be due to differences in the final plane to which the two DEM were resampled. In other words, there may be some inadvertent "roll" or other rotational difference between the two images. If this is not the case, Figure 9.11c indicates (by darker pixels) that the ground near the lake coast subsided in the 13-year interval between the two photo collection times. Note also that a topographic low in the lower right (north of road) in Figure 9.11b seems to have subsided or had its vegetation cover substantially shortened between the two dates. This low could be evidence for a sinkhole, since glacial till is underlain by carbonate rocks in this region.

This third of three chief causes for less-than-optimum results stems from errors in elevation extraction that occurred because of wave action along the beach, the wave front being at a different place in the left and right photos (taken a few seconds apart in time) for a given stereo pair. The errors were hand-edited out with an ATOM subroutine called IEDIT, and the water was made one polygon with one elevation that was representative of water level in the DEM. In retrospect, the editing should have been done after the image subtraction. The upper-left part of the coastline that is bright in Figure 9.11c is probably caused by this editing problem, rather than a rising coastline. The lake level in Figure 9.11a is higher than it was originally determined to be, due to a boundary condition problem with the subtraction image processing algorithm.

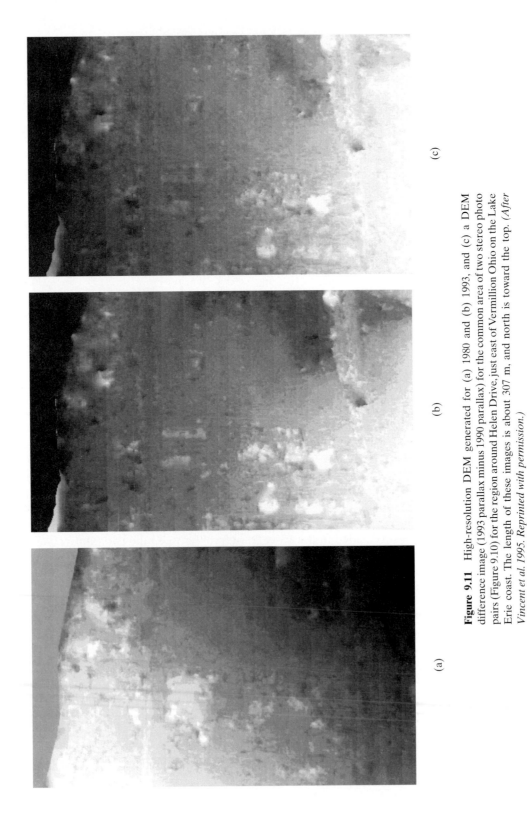

(a)

(b)

(c)

Figure 9.11 High-resolution DEM generated for (a) 1980 and (b) 1993, and (c) a DEM difference image (1993 parallax minus 1990 parallax) for the common area of two stereo photo pairs (Figure 9.10) for the region around Helen Drive, just east of Vermillion Ohio on the Lake Erie coast. The length of these images is about 307 m, and north is toward the top. *(After Vincent et al. 1995. Reprinted with permission.)*

Had photo transparencies been available for both stereo pairs, it would have been possible to maintain a pixel size of 15 cm or so, which would have yielded height differences between the two data sets of 25 cm or less. With higher spatial resolution DEM and GPS data on ground control points, it will be possible to map slumping of land by erosion of the coastal slope, permitting better short-term predictions of land loss than is possible with traditional procedures. Topographic elevation change detection will likely become a useful tool for coastal erosion research. It could also become useful for the detection of incipient sinkholes in karst terrain.

REMOTE SENSING APPLICATIONS FOR PIPELINES AND HIGHWAYS

The methane imaging method described in Chapter 5 and in the second section of this chapter has obvious advantages for the monitoring of natural gas leaks from gas pipelines, both onshore and offshore. From theoretical considerations, offshore methane plumes of 10 m thickness with concentration of 16,000 ppm or greater could be detected from space-borne hyperspectral infrared sensors (Vincent 1995). This would be equivalent to a 10,000-meter-thick plume with 16 ppm of methane in air. From airborne platforms at low altitudes, onshore plumes of 1 m thickness with concentrations of 2000 ppm or greater could be detected (Vincent et al. 1995). Since these concentrations represent fractional parts of the lower explosive limit of methane in air, which is 50,000 ppm (5% methane by volume), methane imaging theoretically shows potential as a safety tool for the monitoring of pipeline leaks.

The siting of pipeline and highway routes are similar problems. A recent example of pipeline siting that employed both remote sensing and GIS methods offers insight into siting procedures that could be applied for either application. A case study of the routing of the Caspian Pipeline, as reported by Feldman et al. (1994), follows.

The Caspian Pipeline Consortium (Sultanate of Oman, Republic of Kazakhstan, and the Russian Federation) sponsored a pilot project that employ remotely sensed data and GIS to assist the routing process for a pipeline that will carry oil from the Tengiz oil field in Kazakhstan, on the Caspian Sea, to Novorossiysk, Russia, on the Black Sea. Though the new pipeline will be 500 miles long and linked with 500 miles of existing pipeline, the pilot study area for the route siting involved only the last few percent of the total length of the new construction. The consortium had already employed general political and economic guidelines to constrain the area of study to an inverted L-shaped corridor, and had identified four points (A, B, C, and D) through which the pipeline was constrained to pass. The object of the study was to find a least-cost pathway and to compare its cost with the likely cost of three straight-line segments (totaling 26 miles in length) through these four points.

The idea employed by Feldman et al. (1994) was to extend the algorithms traditionally used in GIS for defining the paths of stream channels in a drainage basin, where a DEM is the input and water is assumed to always flow downhill. Feldman et al. extended the algorithms to define the least-cost pathway of a pipeline, with the path constrained to pass through specific nodes to specified endpoints. They generated a single weighted surface layer, based on input layers representing factors that contribute to the cost of traversing between two points with a pipeline. This single layer, called a combined weighted surface layer, is analogous to a topographic surface, in that it has peaks (areas of relatively high cost) and valleys (areas of relatively low cost). Therefore, a two-dimensional grid of this combined weighted surface layer replaced the DEM as input to the drainage algorithm, and the route of the pipeline "flowed" to the least expensive adjacent grid cell, with the constraint that the pipeline has a specified beginning point and a specified endpoint. A cumulative cost surface layer was generated from the combined weighted surface layer, consisting of the sum of all cells that had been selected as "channel" grid cells between the points of origin and ending. The algorithm was run for each of the three sets of beginning and end points: A-B, B-C, and C-D.

The weighting factors that made up each of the individual component layers were actually cost factors associated with the pipeline's crossing of slopes, streams, wetlands, roads, railroads, rocky land, agricultural land, and urban/industrial areas. Aerial photography and large-scale maps were not available for use in this project. Slope data was derived from a DEM produced from a digitized 1:500,000 scale Defense Mapping Agency topographic map. Roads, railroads, existing pipelines, and stream channels were also digitized from the topographic maps. A large-scale 1:200,000 scale geologic map was used to outline polygons of geological formations that were expected to contain large amounts of rocky terrain. LANDSAT TM (30-meter spatial resolution) and SPOT (10-meter resolution) digital data were used to produce supervised classification maps of land-use cover (agricultural and urban/industrialized). When these individual layers were weighted and added to obtain a combined weighted surface layer, and the algorithm was run between the three pairs of beginning and end points, a least-cost pathway was found for the total path from A to D that resulted in approximately 14% less cost, but 23% greater length (32 miles versus 26 miles) than the shortest path of three straight-line segments through points A, B, C, and D. The Least-Cost Pathway and Straight-Line Route are shown in Figure 9.12 (from Feldman et al. 1994).

Had digital photogrammetry been used on SPOT stereo images to produce 10-meter-posted DEM instead of digitizing 1:500,000 topographic maps and had the resulting DEM and combined SPOT and TM images been used to locate roads, railroads, streams, and rocky soil, it is likely that the results would have had significantly higher quality. Large-scale maps, even if accurate for the time that they were produced, do not reflect accurate positions of manmade features changed or added since the maps were produced. The accuracy of slope information from 10-meter-posted DEM made with digital photogrammetry from SPOT stereo images will also be much greater than if produced from digitized 1:500,000 scale topographic maps, even if the maps meet map standards for that scale. The rms error in elevation with every-pixel-posted DEM from digital photogrammetry of SPOT stereo images is typically about 12.5 meters (Vincent et al. 1988), which is close to map accuracy standards for 1:50,000 scale maps.

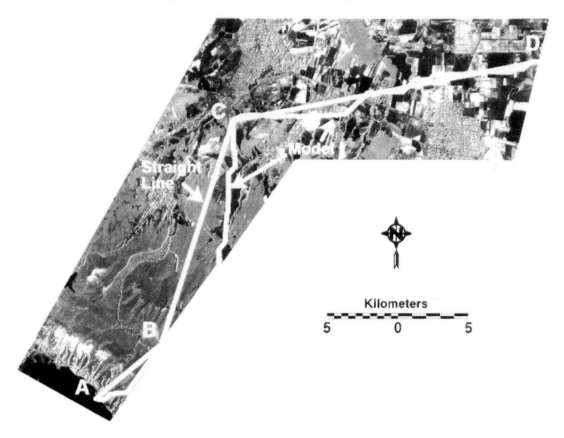

Figure 9.12 Caspian Pipeline route in Kazakhstan and Russia as derived from a least-cost pathway analysis model, compared with a straight-line route. A surface water run-off model was employed with estimated pipeline construction expense substituted for topographic elevation, such that the resulting "drainage" patterns represented the lower-cost pathways, and the largest "drainage" represented the least-cost pathway. *(After Feldman et al. 1994. Reprinted with permission of the Environmental Research Institute of Michigan.)*

The application of remote sensing to highway problems apparently still remains to be realized in the future. The most commonly read publication concerning highways is *Public Roads,* a quarterly publication of the U.S. Department of Transportation, Federal Highway Administration. A search of *Public Roads* for the years 1972 (when LANDSAT I was launched) through 1994 revealed not one article with a satellite image or aerial photo from a mapping camera and no articles discussing the application of remote sensing to highways. However, there are several applications of remote sensing to engineering geology related to highway monitoring that have excellent potential for high-cost effectiveness. Two of these involve high-resolution digital photogrammetry for the monitoring of roads and bridges, and a third involves image processing for automatic assessment of pavement condition.

It is possible to produce digital elevation models (DEM) with elevation errors on the order of 3 cm to 6 cm from aerial photography collected at altitudes of 300–600 meters with digital photography software like the ATOM program discussed in Chapter 6. There is, therefore, a distinct possibility that the amount and rate of ground subsidence between the dates of two aerial stereo photo collections over a bridge abutment or roadbed could be directly measured by digital photogrammetry. Bridges and highways over old underground mines, karst terrain, or growth fault areas are susceptible to ground subsidence over long periods of time. A DEM from time 1 subtracted from a DEM of time 2 would yield a measurement of the distance that the ground had subsided. When this distance is divided by the time between the two data collections, the rate of subsidence would be measured. The rate of erosion in gullies could be measured by this same method.

Whereas subsidence and gullying often cause localized pavement cracking in roadbeds, they are theoretically not supposed to cause much harm to bridge supports, which are assumedly extended downward to bedrock, beneath the weathering zone of soils. Bridge inspections appear to concentrate on searching for cracks in bridge supports, rather than searching for changes in their elevations. The comparison of high-resolution DEM of bridges and their support structures by means of low-altitude aerial photos collected one or more decades apart would be an interesting, possibly vital, transportation research investigation. Candidates for such research could come from the 33,400 bridges in the federal-aid system and 107,400 bridges in the nonfederal-aid system that were classified in 1985 as structurally deficient (Galambos 1987).

The third application of remote sensing for highways concerns pavement management. The automatic detection and counting of pavement cracks from digital camera images collected of the road surface from the back of moving vans has been performed, but not greatly published. In fact, an implementation of GeoSpectra's LIRA (Linear Recognition and Analysis) computer program on an Applied Intelligence System (AIS-5000) parallel processor was found to be capable of processing video tapes of a digital camera that collected 10% overlapped images as the van moved at 60 mph (97 km/hr) in twice the time it took to collect the images. The search for linear features was confined to two 45° segments, such that transverse linear features were counted in real time (at the data collection rate) on the first pass through the data, and longitudinal linear features were counted during the second pass through the data. That unpublished work, which was performed for a pavement management company named Pavedex, located in Spokane, Washington, occurred in the 1989–1990 time period. Given the great improvements in processing speed and cost of computers since that time, the automation of pavement-crack mapping should now be implementable as a cost-efficient tool for pavement management.

REVIEW

Remote sensing, which has been very useful for mineral and petroleum exploration for over two decades, should become equally useful during the next two decades for many different types of environmental and geological engineering applications.

Solid waste landfill monitoring and open pit mining, which are similar problems from a remote sensing standpoint, can be aided in two technically different ways. One is the use of high-resolution digital photogrammetry to perform detailed topographic mapping, and the other is the use of multispectral image data to map chemical compositional differences of the surface and of samples obtained by drilling from the subsurface.

The topography of active solid waste landfills and open pit mines is constantly changing. High-resolution digital photogrammetry offers the most accurate means for measurement of the volume change of a landfill, open pit mine, or tailings pile over a period of time, or for comparison of elevation contours before and after the period of landfill or open pit mine activity. High-resolution DEM and digital orthophotos can also be used with visible simulation software to produce simulated perspective views of a landfill or an open pit mine from any user-selected position of the observer, with the same resolution as the source stereo photos. This technology, called digital holography, can be useful for the production of technical presentations to management and to regulatory agencies.

Image processing of multispectral data can be used to map chemical composition differences in both solids and gases. Mappable solids of significance to landfills include clays in liners and caps, ferric and ferrous oxides that are by-products of low-temperature geochemical "cells" caused by contaminants in wastes, and possibly some unique compounds (such as cyanides and chromates) that indicate the presence of specific wastes. Mappable vegetation stress, such as chlorosis, can be indirect indicators of liquid and gaseous waste leaks. Multispectral imaging of gases is probably possible with properly filtered, multispectral imaging sensors, especially for methane, the most common gas escaping from landfills. For lead-zinc open pit mines, lead in tailings piles was found to be correlated with the occurrence of jarosite (an iron sulfate) and smectite (a disordered clay), both of which have spectral reflectances that are relatively easy to identify. The correlation of other toxic metals with spectrally unique minerals may likely benefit environmental monitoring of both open pit mines and solid waste landfill sites, whether active or abandoned.

The application of multispectral remote sensing to forest and wetland mapping will be improved by hyperspectral sensors. These can be used to determine folial biochemical content of a forest canopy. An application of this new technology to a slash pine forest has demonstrated that the chlorophyll, water, nitrogen, lignin, and (to a lesser extent) cellulose contents of needles in the upper canopy can be accurately determined from a one-time data collection pass of a sensor that includes 22 specially located, narrow spectral bands. Research is needed to link types of vegetation stress to those folial biochemical contents for various kinds of forests. If this effort is successful, applications would include the detection of toxic metal stress to trees (for mineral exploration, hydrocarbon exploration, and environmental monitoring) and the delineation of wetlands with a one-time data collection of hyperspectral data.

A commonly held misconception about color infrared (CIR) photography for the delineation of wetlands is that blue, dentritic patterns in a CIR photo collected during leaf-off conditions (outside the growing season) indicate the presence of wetlands. CIR laboratory experiments have shown that fallen leaves, whether wet or dry, can appear blue in a CIR photo, but this effect occurs primarily because the leaves lack chlorophyll and has nothing to do with their temperature. These experiments indicate that the blue dentritic patterns are likely drainage patterns, most or all of which may fail to fit the definition of a wetland given in government agency manuals. Therefore, CIR photos collected over deciduous forests outside the growing season yield insufficient data to delineate wetlands, which are defined according to the amount of time that the water table is near the surface during the growing season. Application of hyperspectral data for one-time data collection during the growing season to wetland delineation may be possible if the effects of near-surface ground water can be linked to folial biochemical content. The latter can now be accurately estimated from hyperspectral remote sensing data.

Farm land erosion and coastal erosion are both forms of mass wasting, though the latter usually has the faster erosion rate. Rates of erosion can be measured by digital photogrammetric production of high-resolution DEM from digitized stereo pairs collected on two or more dates that are widely separated in time. The volume of eroded material can be measured, and the rate can be derived from a division of the measured volume of material lost by the time between the two stereo photo collection dates. Surface water run-off models, which are important for farm land erosion studies, are heavily dependent on topographic elevation and slope information. This implies that high-resolution DEM will be found to be highly desirable for farm land erosion models because of the higher accuracy in elevation and slope information offered by this relatively new technology.

Remote sensing has considerable contributions to make to pipelines and highways, also. Pipeline and highway route selection schemes that employ GIS methods require accurate elevation and slope information, which high-resolution DEM can supply. When stereo photo pairs are available for two or more time periods, it is possible to search for quantitative evidence of subsidence and gullying, both which are important for highways and bridges. Subsidence can come from the collapse of old underground mine shafts or limestone caverns in karst terrain. Highway maintenance can also benefit from image processing software that can automatically recognize and count linear features (pavement cracks) seen in images of digital cameras mounted on vans that drive along the highway at the maximum legal rate of speed.

Exercises

1. What is the possible importance of being able to map the difference between disordered and ordered clays on Earth's surface for environmental remote sensing?

2. The company that you work for can win the job of monitoring all of the gas pipelines of Saudia Arabia if it can demonstrate its expertise on an offshore and an onshore pipeline on which there will be controlled releases of methane during your aircraft overflight. Assume that your company has an airborne imaging spectrometer that is capable of collecting data in three narrow spectral bands, all located in the 3–5 μm wavelength atmospheric window, and that you can produce an $R_{1+3,2}$ ratio image, where band 2 is in the methane absorption band and bands 1 and 3 are on either side of band 2, just outside the methane absorption band.

(a) What environmental conditions and times of day would you recommend for collecting data over the onshore and the offshore pipelines?

(b) Would the gas plume appear brighter or darker than the background in the $R_{1+3,2}$ ratio image for the environmental conditions that you selected for data collection?

Cited References

ABRAMS, M. J., L. ASHLEY, L. ROWAN, A. GOETZ, and A. KAHLE. 1977. Mapping of Hydrothermal Alteration in the Cuprite Mining District, Nevada, Using Aircraft Scanner Images for the Spectral Region 0.46 to 2.36 μms. *Geology* 5:713–718.

ABRAMS, M. J., J. E. CONEL, and H. R. LANG. 1984. *The Joint NASA/Geosat Test Case Project Final Report,* 1:8-70–8-82, 2:10-5–10-22. Tulsa, Okla.: American Association of Petroleum Geologists.

ARNETH, J. D., G. MILDE, H. KERNDORFF, and R. SCHLEYER. 1989. Waste Deposit Influences on Groundwater Quality as a Tool for Waste Type and Site Selection for Final Storage Quality. In *Lecture Notes in Earth Sciences,* ed. S. Bhattacharji, G. M. Friedman, H. J. Neugebauer, and A. Seilacher, 399–415. New York: Springer-Verlag.

AVERY, T. E. 1977. *Interpretation of Aerial Photographs.* Minneapolis: Burgess Publishing Co.

AYDAY, C., and GOKTAN, R. M. 1993. Evaluation of Geological and Geotechnical Data by Using GIS for the Diggability Classification of an Open-Pit Mine. In *Proceedings of the Ninth Thematic Conference on Geologic Remote Sensing,* vol. 1, 457–467. Ann Arbor: Environmental Research Institute of Michigan.

CARMAN, R. E. 1996. An Analysis of Methane Gas Concentrations for Three Sanitary Landfills in Northwest Ohio. Master's Thesis, Bowling Green State University, Bowling Green, Ohio.

CURRAN, P. J., and J. A. KUPIEC. 1995. Imaging Spectrometry: A New Tool for Ecology. In *Advances in Environmental Remote Sensing,* eds. F. M. Danson and S. E. Plummer, 71–88. New York: John Wiley & Sons.

ERB, T. L., W. R. PHILIPSON, W. T. TANG, and T. LIANG. 1981. Analysis of Landfills with Historic Airphotos. *Photogrammetric Engineering and Remote Sensing* 47, no. 9:1363–1369.

FELDMAN, S. C., R. E. PELLETIER, W. E. WALSER, J. C. SMOOT, and D. AHL. 1994. Integration of Remotely Sensed Data and Geographic Information System Analysis for Routing of the Caspian Pipeline. In *Proceedings of the Tenth Thematic Conference on Geologic Remote Sensing,* vol. 2, 206–213. Ann Arbor: Environmental Research Institute of Michigan.

GALAMBOS, C. F. 1987. Bridge Design, Maintenance, and Management. *Public Roads* 50, no. 4:109–115.

GHOLZ, H. L., S. A. VOGEL, W. P. CROPPER, JR., K. MCKELVEY, K. C. EWEL, R. O. TESKEY, and P. J. CURRAN. 1991. Dynamics of Canopy Structure and Light Interception in *Pinus elliottii* Stands of North Florida. *Ecological Monographs* 61:33–51.

GREEN, A. A., and M. D. CRAIG. 1985. Analysis of Aircraft Spectrometer Data with Logarithmic Residuals. In *Proceedings of the Airborne Imaging Spectrometer Data Analysis Workshop,* 227–231. Pasadena, Calif.: Jet Propulsion Laboratories.

HERMAN, J. D., J. E. WAITES, R. PONITZ, and P. ETZLER. 1994. A Temporal and Spatial Resolution Remote Sensing Study of a Michigan Superfund Site. *Photogrammetric Engineering and Remote Sensing* 60, no. 8:1007–1017.

JIAN, L. 1993. Establishing Analysis Model of Soil Erosion Using Remote Sensing Information and GIS in Gully-Hilly Area of Loess Plateau in China—Hou Gou Men Basin as Research Area. In *Proceedings of the Ninth Thematic Conference on Geologic Remote Sensing,* vol. 1, 419–430. Ann Arbor: Environmental Research Institute of Michigan.

KRUSE, F. A., W. M. BAUGH, and W. W. ATKINSON, JR. 1994. Mapping Alteration Minerals in Drill Core Using a Field Spectrometer and Hyperspectral Image Analysis Techniques. In *Proceedings of the Tenth Thematic Conference on Geologic Remote Sensing,* vol. 2, 37–43. Ann Arbor: Environmental Research Institute of Michigan.

KUBELKA, P. 1948. New Contributions to the Optics of Intensely Lightly-Scattering Materials, Part I. *Journal of the Optical Society of America* 38, no. 5:448–457.

LYON, J. G. 1982. Use of Aerial Photography and Remote Sensing in the Management of Hazardous Wastes. In *Hazardous Waste Management for the 80's,* eds. T. L. Sweeney, H. G. Bhatt, R. M. Sykes, and O. J. Sprout, 163–171. Ann Arbor, Mich,: Ann Arbor Science Publishers.

MACKEY, S. D. 1995. Private communication. Ohio Department of Natural Resources, Division of Geological Survey, Lake Erie Geology Section, Great Lakes Center, Sandusky, Ohio.

MUFTUOGLU, Y. V., and M. J. SCOBLE. 1985. Methods of Amenability to Excavation Determination in Open Cast Coal Mining. In *Proceedings of the Ninth Turkish Mining Congress, Ankara, Turkey,* 29–37.

MUNTS, S. R., P. L. HAUFF, A. SEELOS, and B. McDONALD. 1993. Reflectance Spectroscopy of Selected Base-Metal Bearing Tailings with Implications for Remote Sensing. In *Proceedings of the Ninth Thematic Conference on Geologic Remote Sensing,* vol. 1, 567–578. Ann Arbor: Environmental Research Institute of Michigan.

PLUMB, R. H., JR., and A. M. PITCHFORD. 1985. Volatile Organic Scans: Implications for Groundwater Monitoring. In *Proceedings of the National Water Well Association/American Petroleum Institute Conference on Petroleum Hydrocarbons and Organic Chemicals in Groundwater,* 13–15 Nov. Houston, Tex. p. 1–15.

ROWAN, L. C., A. F. H. GOETZ, and R. P. ASHLEY. 1983. Discrimination of Hydrothermally Altered and Unaltered Rocks in Visible and Near Infrared Multispectral Images. In *Remote Sensing, Geophysics Reprint Series No. 3,* eds. K. Watson and R. D. Regan, 288–301. Tulsa, OK: Society of Exploration Geophysicists.

SALOMONSON, V. V. 1983. Water Resources Assessment. In *Manual of Remote Sensing,* 2nd *Edition,* ed. R. N. Colwell. Falls Church, Va.: American Society of Photogrammetry.

SINGH, R. N., B. DENBY, I. EGRETLI, and A. G. PATHAN. 1986. Assessment of Ground Rippability in Open Cast Mining Operations. *Mining Dept. Magazine, University of Nottingham,* 38:21–34.

STOHR, C., W. J. SU, P. B. DuMONTELLE, and R. A. GRIFFIN. 1987. Remote Sensing Investigations at a Hazardous-Waste Landfill. *Photogrammetric Engineering and Remote Sensing* 53, no. 11:1555–1563.

STOHR, C., W. J. SU, L. FOLLMER, P. B. DuMONTELLE, and R. A. GRIFFIN. 1988. Engineering Geology Investigations of a Hazardous-Waste Landfill in West Central Illinois. *Bulletin of the International Association of Engineering Geology,* no. 37:77–88.

STOHR, C., R. G. DARMODY, T. D. FRANK, A. P. ELHANCE, R. LUNETTA, D. WORTHY, and K. O' CONNOR-SHORESMAN. 1994. Classification of Depressions in Landfill Covers Using Uncalibrated Thermal-Infrared Imagery. *Photogrammetric Engineering and Remote Sensing,* 60, no. 8:1019–1028.

U.S. ARMY CORPS OF ENGINEERS. 1987. *Wetlands Delineation Manual,* Technical Report Y-87-1. Environmental Laboratory, U.S. Army Engineer Waterways Experiment Station, Vicksburg, Miss.

U.S. ENVIRONMENTAL PROTECTION AGENCY. 1991. *Solid Waste Disposal Facility Criteria; Final Rule, Part II,* 40 CFR Parts 257 and 258. Federal Register, October 9, 1991, vol. 56, no. 196: 50978–51119.

U. S. FISH AND WILDLIFE SERVICE. 1981. National Wetland Inventory Map, La Plata, Md.

VINCENT, R. K. 1977. Chapter 10: Geochemical Mapping by Spectral Ratioing Methods. In *Remote Sensing Applications for Mineral Exploration,* ed. W. L. Smith, 251–278. Stroudsburg, Pa.: Dowden, Hutchison & Ross.

———. 1983. An ERTS Multispectral Scanner Experiment for Mapping Iron Compounds. In *Remote Sensing, Geophysics Reprint Series No. 3,* eds. K. Watson and R. D. Regan, 14–22. Tulsa, OK: Society of Exploration Geophysicists.

———. 1989. Digital Holography Provides the Ultimate in Three-Dimensional Image Processing. *Images,* The Image Society Newsletter, Fall Issue, 13 and 16.

———. 1994. Remote Sensing for Solid Waste Landfills and Hazardous Waste Sites. *Photogrammetric Engineering and Remote Sensing* 60, no. 8:979–982.

———. 1995. Flying Falcon: Multispectral Thermal IR Geological Imaging Experiment. *Remote Sensing for Oil Exploration and Environment, Proceedings of the Space Congress,* European Space Report, P. O. Box 140 280, 80452 Munich, Germany, 139–146.

VINCENT, R. K., P. K. PLEITNER, D. H. COUPLAND, H. SCHULTZ, and E. R. B. OSHEL. 1984a. New Digital Elevation Mapping Software Applied to SPOT Simulation Stereo Data. *1984 SPOT Symposium,* Scottsdale, Arizona, May 20–23:92–97.

VINCENT, R. K., P. K. PLEITNER, and M. L. WILSON. 1984b. Integration of Airborne Thematic Mapper and Thermal Infrared Multispectral Scanner Data for Lithologic and Hydrothermal Alteration Mapping. In *Proceedings of the International Symposium on Remote Sensing of Environment, Third Thematic Conference, Remote Sensing for Exploration Geology,* vol. 1, 219–226. Ann Arbor, Michigan: The Environmental Research Institute of Michigan.

VINCENT, R. K., M. A. TRUE, and D. V. ROBERTS. 1987. Automatic Extraction of High-Resolution Elevation Data Sets from Digitized Aerial Photos and Their Importance for Energy Mapping. *National Computer Graphics Association's Mapping and Geographic Information Systems '87 Proceedings,* San Diego, Ca. November 9–12, 203–210.

VINCENT, R. K., W. T. LEHMAN, R. L. HENRY, J. D. HERMAN, M. E. STIVERS, M. L. WILSON, and P. J. ETZLER. 1988. The Application of High-Resolution Digital Elevation Models to Petroleum and Mineral Exploration and Production. In *Proceedings of the International Symposium on Remote Sensing of the Environment, Sixth Thematic Conference, Remote Sensing for Exploration Geology,* 293–301. Ann Arbor, MI: The Environmental Research Institute of Michigan.

VINCENT, R. K., and E. B. SINGLETON. 1994. Methane Gas Concentrations Required for Infrared Imaging. Final Report for Hughes Santa Barbara Research Center Grant, Dept. of Geology, Bowling Green State University, Bowling Green, Ohio.

VINCENT, R. K., J. P. FRIZADO, and S. D. MACKEY. 1995. Comparison of Lake Erie Coastal Erosion Rates as Determined by Digital Photogrammetry and Aerial Photo Interpretation. Abstract and oral presentation, Geological Society of America, 1995 Annual Meeting (Nov. 5–9), New Orleans, La.

YAMAGUCHI, Y., T. OOKA, T. INOUE, and T. OHKURA. 1994. Automatic Field Identification of Alteration Minerals Using a New Portable Infrared Spectroradiometer. In *Proceedings of the Tenth Thematic Conference on Geologic Remote Sensing,* vol. 2, 73–77. Ann Arbor: Environmental Research Institute of Michigan.

10

Remote Sensing Applications to Global Monitoring

▼ ▼ ▼ ▼ ▼ ▼ ▼ ▼ ▼ ▼ ▼ ▼ ▼ ▼ ▼

INTRODUCTION

All of the geological remote sensing applications of the foregoing chapters involve relatively localized areas and sensors with spatial resolutions of 80 m or better. In this chapter, the subject will range from spot locations in regional searches, with high spatial resolution sensors, to the "big picture" applications of remote sensing for global monitoring, with spatial resolution as coarse as 150 km and areas of interest the size of continents. From a geological perspective, there are at least three good reasons why the globe should be monitored on an ongoing basis with multispectral sensors of both fine and coarse spatial resolution, and each is the subject of a separate section below.

First, there are geological hazards that have occurred in the past and will occur in the future, for which large-scale remote sensing monitoring may provide precursor warnings. The subject of the first subsection of this chapter is remote sensing designed to reduce the local effects of geological hazards that may be geographically scattered in a discontinuous fashion, such as flooding, mass wasting, volcanic eruptions, and earthquakes. There are some surprising recent studies that support the application of remote sensing toward the detection of geological hazard precursors. Reliable warnings that would reduce the local consequences of these hazards would alone be worth the cost of remote monitoring; considerable savings would also be achieved by reducing global consequences.

Second, human contributions to global changes can in some cases be measured by large-scale remote monitoring, but only after they are separated from changes caused by natural geological processes with which they sometimes interact.

An environmentalist is required to know what is natural before he or she can recognize what is unnatural. Figure 10.1 shows a graph of Earth's human population over the past two millennia. It took from the first human until the nineteenth century for one billion living humans to inhabit the globe at one time. Since then, the world's human population has grown to over 5.5 billion, which is geologically akin to an algal bloom. Only in this century has the possibility arisen that human activity could alter the planet, and human-caused global changes will in all probability continue to grow as the population increases. Two human-induced changes that interact and/or compete with natural geological processes are the addition of aerosols and greenhouse gases of anthropogenic origin to the atmosphere, that contribute to changes in global temperature, and the addition of anthropogenic chlorofluorocarbons to the upper atmosphere, which contribute to stratospheric ozone loss. Both types of changes and their relationship to remote monitoring are discussed in this chapter.

The final subsection of this chapter discusses the third application of global monitoring to geological processes: the search for those precursors of change in global weather (short term) and climate (long term) that are caused by natural contributions. Changes in ice and snow coverage (particularly glaciers and floating ice), which presage changes in sea level; changes in sea surface temperature, which appear to drive atmospheric changes; and changes in plate tectonic activity, which may be driving sea-surface temperature changes, are all considered with regard to their contributions to changes in global weather and climate. This discussion of global monitoring is necessarily more speculative than the other two, since natural

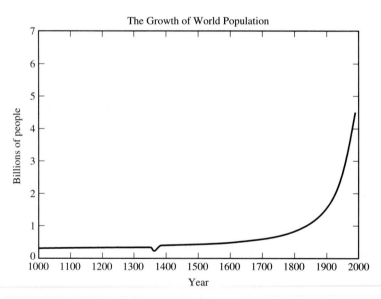

Figure 10.1 Growth of world population from A.D. 1000–2000. The drop in population about A.D. 1350 was caused by the Black Death (bubonic plague), which killed about a third of the population in Europe and Asia. *(After A. Getis and J. M. Getis, 1982, Geography, Houghton Mifflin Co., Boston, Mass. Reprinted with permission.)*

contributions to global weather and climate require inferences over far longer periods than the quarter-century for which remote sensing satellites have been available and because many of the hyperspectral imaging sensors that are well suited for these tasks have been flown on aircraft platforms, but have not yet been orbited.

GLOBAL MONITORING OF GEOLOGICAL HAZARDS

Although the damage caused by geological hazards is localized, geological factors of a broader scale can cause a widespread pattern of localized hazards. Thus, global monitoring for the collection of precursor warning information is required before the events, and damage assessment information is also required after the events. For instance, three of the Cascade Range volcanoes (Mounts Hood, Baker, and Rainier) in the U.S. Pacific Northwest (USA) erupted one at a time between 1865 and 1885, all caused by episodic motions of the North American Plate as it subducted the Juan de Fuca Plate, which forms part of the Pacific Ocean bottom. Had satellite data been available then, it may have been possible to predict those eruptions and to more quickly assess the damage caused by them. Mt. St. Helens erupted in the 1980s, when satellites were available but were not well utilized for monitoring such disasters. It is necessary therefore for the world's governmental and scientific communities to advance research on this general topic and to create a global monitoring capability for geological hazards as quickly as possible.

The stakes are high. The cost of geological hazards to the global economy exceeds $50 billion per year, two-thirds of which represents the direct cost of the damage and one-third of which represents costs for predicting, preventing, and mitigating disasters (Alexander 1993). Major disasters worldwide kill an average of 140,000 people a year, with about 95% of those deaths occurring in the Third World, where more than 4.2 billion people live (van Westen 1995).

One of the most important tools in hazard mitigation is a risk map, which outlines hazardous zones according to expected frequency, character, and magnitude of hazardous events in an area. This mapping requires information, called hazard information, on the probability of occurrence of a potentially damaging phenomenon within a specified period of time and within a given area (van Westen 1995). Remote sensing data permits the mapping of variations in both time and space of terrain properties, such as vegetative cover, location of surface water, topographic elevations, and surface composition of exposed rocks and soils. Geographic Information Systems (GIS) are a "powerful set of computer tools for collecting, storing, retrieving at will, transforming, and displaying spatial data from the real world for a particular set of purposes" (Burrough 1986). The combination of remote sensing and GIS technology into what could be called GIS/RS technology, plus field work, is required for the production of risk maps, as well as for the assessment of damage once the hazardous event has occurred. Risk maps are used as a planning tool by urban, state, federal, and international agencies for disaster preparedness and for routing infrastructure elements (roads, water and utility lines, and so on), as a premium estimation tool by insurance companies, and as a lending risk-assessment tool by financial institutions. The same agencies and companies employ damage assessment maps after a hazardous event for financial relief estimates and rebuilding activities.

Flooding and Mass Wasting

The number of people affected by floods are greater than any other type of hazardous event that occurred in the 1980–1992 time period (Reinhardt 1994). The most straightforward use of remote sensing for monitoring floods is for an estimation of the amount of area that is flooded. The Great Mississippi Basin Flood of 1993 prodded the U.S. Federal Emergency Management Administration (FEMA) into contracting private companies to provide maximum flood extent coverage of flooded areas in the Mississippi Basin to several federal agencies for the past Great Flood of 1993, as well as for future floods that may occur in that area. Those companies will use multispectral sensor data from satellites, like LANDSAT TM and SPOT, as well as synthetic aperture radar (SAR) images, such as are available from ERS-1. Radar, which can see through clouds, is particularly effective for early analysis of the extent of flooding, when multispectral sensors may be hampered by the same cloud cover that was responsible for the flood-causing precipitation.

Even when flood records are available for a long historical period, such as in China, satellite data can help visualize the locations and effects of manmade flood-control mechanisms. Such data collection is part of risk mapping for flooding. Woldai (1995) employed LANDSAT TM and SPOT data to create a map showing the flood diversion canals and dikes in the Wuhan area and its surroundings; the map is given in Figure 10.2. The pre-1949 map, given in Figure 10.3, shows historic flood-prone areas for the same region. The two maps are quite different as a result of significant manmade flood-control mechanisms and to changes in the course of the river over time. The amount of flood-prone land has been greatly reduced by the flood-control mechanisms, which have also reduced the amount of fresh water lake area. This remote sensing study made possible both the location and the evaluation of various flood-control mechanisms (the building of dikes, dams, canals) that were designed to mitigate flooding in the area.

The most significant development in flood prediction that will likely occur in the near future is the combination of high-resolution DEM generation by digital photogrammetric methods and the application of surface run-off models that have been altered to accept high-resolution DEM as input. Surface run-off models like that of Saghafian (1993) are currently available for handling raster format data, such as DEM, but the model input assumes that streams are all narrower than the DEM posting interval, and they are inefficient research computer programs, i.e., they have not been rewritten a number of times for the sole purpose of decreasing run times. Typical input for such programs are DEM with 100 m posting intervals.

Since the entire United States is covered by 1:80,000 scale NHAP (National High Altitude Photography) stereo photos, it would be possible to produce DEM of 2 m posting interval, just by scanning those photos at 1000 dpi (25 μm spot size) and running them through an every-pixel-posted digital holography database generator like ATOM (Automatic Topographic Mapper), as discussed in Chapter 6. Such DEM

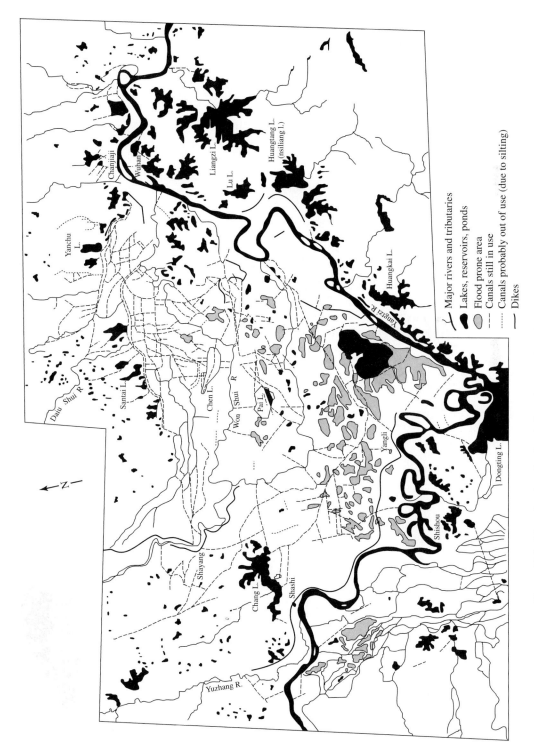

Figure 10.2 Map showing the flood diversion canals and dikes in the Wuhan area of the People's Republic of China and its surroundings as mapped from available remote sensing data sets. (*After Woldai 1995.*)

Map legend:
— Major rivers and tributaries
◼ Lakes, reservoirs, ponds
▨ Flood prone area
- - - Canals still in use
······· Canals probably out of use (due to silting)
— Dikes

Labels on map: Chanjiaji, Wuhan, Liangzi L., Lu L., Huangtang L. (nsiliang l.), Huangkai L., Yanchu L., Santai L., Chen L., Won Shui R., Pai L., Dauu Shui R., Yangli, Dongting L., Shishou, Shayang, Chang L., Shashi, Yuzhang R., Yangtzi R.

N

295

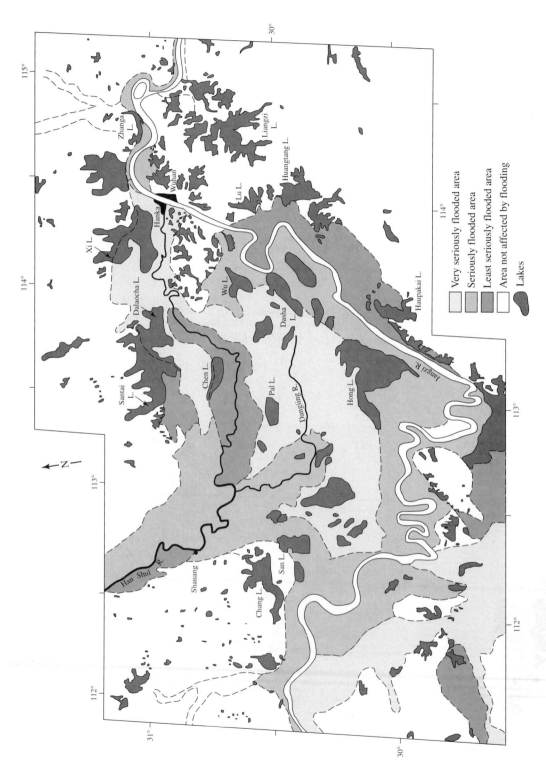

Figure 10.3 Pre-1949 flood-proned regions of the Wuhan area of the People's Republic of China and its surroundings (same area as Figure 10.2). *(After Woldai 1995.)*

Very seriously flooded area
Seriously flooded area
Least seriously flooded area
Area not affected by flooding
Lakes

would contain 2500 times as much data per square km of land area, which would require efficient surface run-off models for reasonable computer run times. Since most streams would be wider than 2 meters, this would also require a change toward greater detail in the manner that surface run-off models treat the stream channels.

Surface water run-off models depend primarily on slope information, which continues to improve with increasing spatial resolution of the DEM. For regions with little topographic relief, high-resolution DEM become all the more important for predicting surface water run-off, because it is imperative that the DEM resolution in elevation remain much smaller (at least 10 times smaller) than the maximum relief (highest minus the lowest elevation) in the watershed under investigation. However, a computer program like ATOM can calculate a DEM with a posting interval of 0.33 m (1 ft) and an rms error in elevation of approximately 0.42 m (1.25 ft) from 1:12,000 scale stereo photos that are scanned at a resolution of 1000 dpi (25 μm spot size). With high-resolution digital photogrammetry, it will be possible to calculate detailed surface water run-off in almost any watershed, even those with little topographic relief, where flooding affects the largest areas.

Flooding and mass wasting are often related, since flooding can increase soil moisture, which decreases friction, and can increase undercutting. Li and Li (1994) have created a Mud-Rock Flow Predicting Information System (MFPIS) model, still undergoing verification, for predicting mud flows caused by heavy rainfall along the Lower Jinshajiang River, which runs through the southeastern edge of the Tibet plateau in China. This is an area of frequent mud flows, especially during the rainy season from May to October, when 80% of the total annual rainfall occurs; rain can be as much as 13 cm per day in an area geomorphically characterized by steep valley walls in unconsolidated sediments. The MFPIS model results from the combination of a Rainstorm Runoff Model (RRM), which is a surface water run-off model that uses DEM and soil characteristic parameters as inputs and a Potential Mud-Rock Flow Damage Information System (PMDIS), which uses DEM and remote sensing information as input. The PMDIS includes a division of the entire study area into three classes of subregions, then calculates the degree of potential mud-rock flow danger for each of the smallest type of subregion (Class III). This degree of potential mud-rock flow danger is a calculated parameter that consists of the sum of 16 different indexes that are defined in Table 10.1. For each of the 16 indexes, higher values are assigned to the index as the contributing factor of that index to mud flow increases. Most of those could likely be predicted from a combination of DEM and multispectral scanner data from satellite stereo sensors. Although the testing of this engineering model has yet to be done, the information inputs are likely to be a general requirement for similar models. Where possible, however, such models would benefit by testing whether the exclusion of all indexes that do not lend themselves to measurement from remote sensing data would significantly affect the accuracy of the predictions. If not, a model that represents the best that one can do by remote sensing methods would emerge as the most easily automated and least expensive version for predictions of mud flows.

A highly desirable goal is the development of software technology that creates a landslide risk map from remote sensing and GIS information, but considerable research will be required before a successful version of it is widely accepted.

TABLE 10.1   THE ITEMS AND INDEXES OF THE FACTORS OF MUD-ROCK FLOW DANGER.

No.	The Factors of Mud-Rock Flow Danger	Classification	Index
1	Watershed areas of gully (A) (km)	0.2–2	4
		2–6	3
		6–100; < 0.2	2
		> 100	1
2	Maximum height difference (ΔH) (m)	> 1000	6
		600–100	4
		300–600	3
		100–300	2
		< 100	1
3	Average gradient of slope (β) (degree)	> 40°	8
		32°–40°	6
		25°–32°	4
		15°–25°	3
		< 15°	1
4	Gradient of gully (J) (%)	> 26.7 (> 16°)	11
		21.2–26.7 (12°–16°)	8
		10.6–21.3 (6°–12°)	6
		5.2–10.5 (3°–6°)	4
		< 5.2 (< 3°)	1
5	Gully shape (1 is the length of gully)	< 0.5	3
		0.6–0.5	2
		> 0.6	1
6	Feeder length rate (—length of feeder reach)	> 60	12
		30–60	9
		10–30	6
		< 10	1
7	Shape of gully section	V shaped	3
		U shaped	2
		open valley	1
8	Dynamic state of sediment-runoff	active	5
		obvious	3
		not obvious	1
9	Dynamic state of accumulation at stream outlet	active	12
		obvious	9
		not obvious	6
		no debris cone	1
10	Assemblage of lithostrata	loose Cenozoic group	5
		sandstone and shale with flysch bedding	3
		hard and massive	1
11	Attitude (strike-dip) (α − dip angle)	low dip angle $\alpha < \beta$	4
		$\alpha > \beta : \beta > 35°$	3
		others	0

TABLE 10.1   *(continued)*

No.	The Factors of Mud-Rock Flow Danger	Classification	Index
12	Faults	large active fault across	10
		many faults	5
		several faults	3
		few faults	1
13	Landslide	many big and deep-cut landslides	12
		many medium and small	9
		a few landslides	6
		not obvious	1
14	Store of loose material	> 10	4
		5–10	3
		1–5	2
		< 1	1
15	Plant cover rate	< 10	8
	(%)	10–30	6
		30–60	4
		> 60	1
16	Manmade excavations	rock broken in wide area	8
	and heaps	broken in local area	6
		a few broken rock masses	4
		not obviously broken	1

A training package entitled Geographic Information Systems in Slope Instability Zonation (GISSIZ) is available as an educational tool on the subject, but it is not capable of generating a new data set to create a real landslide-hazard analysis project (van Westen et al. 1994).

Huang and Chen (1991) used co-registered DEM and LANDSAT TM data to attempt a supervised classification of areas prone to landslides by using previously identified landslide areas as training sets. They included digitized map and DEM information, such as elevations, drainage systems, bedrock formations, and geological structures (like major faults), with the multispectral data, as if each piece of GIS information were a separate spectral band. They then used various supervised classification schemes to classify the data. The one that resulted in the highest classification accuracy was Bayesian classification with a threshold probability of 0.1, which resulted in a classification accuracy of 16.6%. The low value of their best effort provides a valuable indication that this type of classification approach will probably not suffice for assessing landslide risks.

Remote sensing can sometimes be used directly to detect earth flows in forested areas, where such flows typically are too slow and too well covered to be noticed on the ground. McKean et al. (1991) applied the Airborne Thematic Mapper Simulator scanner to the Jude Creek, Oregon earthflow area and found in this area forested by Douglas fir that the brightness of the thermal infrared band (the 10.9–12.2 μm wavelength band that corresponds to band 6 on LANDSAT TM) increased with the flow velocity within a slide. This brightness increase appeared to be related to the creation of openings in the old-growth coniferous forest canopy as the earth flow slowly moves. The tree tops, which are the coolest of anything in the scene, fill less of each pixel as earth flow causes the trees to lean over. Thus, the average of each pixel becomes warmer.

Multitemporal, topographic change detection also offers two other possibilities for the direct observation of mass wasting by remote sensing methods. One is the subtraction of high-resolution DEM from different data-collection times. These not only detect the flow, but also determine of the rate of mass wasting, which is equal to the volume change divided by the time interval between the two data collection dates. This approach would require the use of every-pixel-posted digital photogrammetric techniques and a high-resolution stereo sensor that has many repeat cycles. It is likely that pixel sizes on the order of 0.33 m (1 ft) will be required for this job; therefore, periodic aerial stereo photography would be the most likely source of data. However, once available, a 1 m stereo system in orbit would be attractive as a data source, because of its reliable repeat cycles, fewer frame boundaries, and the inexpensiveness of data per unit area covered, compared with aerial photography.

A second possibility for the direct measurement of mass wasting by topographic change detection is by interferometric methods applied to SAR (Synthetic Aperture Radar) images collected on different overpasses. This technique has been applied toward the measurement of motion along a fault from three satellite radar images collected before and after the quake that caused the motion (Peltzer and Rosen 1995). The technique will be discussed under the topic of earthquake hazards later in this chapter. A discussion of interferometric radar, in which both the radar image and elevation contours are obtained from the same sensor on a single overpass and presented on the same map, is given by Cumming and Gray (1989). However, the one-pass method for obtaining elevation contours from interferometric radar is heavily dependent on precise knowledge of the attitude of the sensor platform during the collection of a radar image. Traditionally, such knowledge has not been available to a level that would permit lower root-mean-square errors in elevation that can be achieved with low-altitude stereo photography.

Also, elevations extracted from almost any type of data are averaged over a one-pixel area, which is typically 25 m $\times$ 25 m for current SAR satellites and from 1–5 m square for aircraft radar. For digitized aerial stereo photos, pixel areas less than 1 m $\times$ 1 m are common, but the precision in elevation is inversely proportional to pixel size (linear dimension). Quantitative comparisons between elevations extracted by radar interferometry and elevations extracted from digitized stereo photography by every-pixel-posted DEM generation software are required for the same test areas (including flat and rugged terrain) before their relative accuracies can be determined. Meanwhile, high-resolution digital photography will likely remain the first choice in accessible, reasonable-weather areas because a photo is needed in addition to the elevations for most applications anyway, particularly for digital holography. For regions where cloud cover is ever-present and no visible image is necessary, however, interferrometric radar will be the clear choice for measuring topographic (elevation) change detection.

Volcanic Eruptions

Volcanoes tend to be clustered in certain areas, rather than randomly spread out. Since volcanoes and plate tectonics both depend on heat generated by Earth, it is not surprising that volcanic eruptions and plate motions are often correlated. For instance, composite volcanoes are located near edges of plates where subduction occurs, such as the Cascade Range volcanoes of the U.S. Pacific Northwest, which owe their existence to the subduction of the Juan de Fuca Plate by the North American Plate, as mentioned earlier in the introduction to this chapter. Shield volcanoes, such as the Hawaiian volcanoes, are caused by a mantle hot spot over which the Pacific Plate has moved, leaving an L-shaped string of extinct island volcanoes in its path. Fissure eruptions, like the Deccan Traps in India or the Columbia River Plateau in the northwestern United States, are regions where spreading centers once started that could have (but in these two cases did not) split a plate apart. Other areas of volcanic fields are not so easily explained.

There are two phases of volcanic monitoring to which remote sensing can be applied. First, remote sensing from satellites can be used to map the locations of existing volcanoes and their attendant cinder cones. Second, satellite remote sensing can be used to monitor known volcanoes for precursors to eruptions.

An example of the former was reported by Chagarlamudi et al. (1991) for volcanic features in Harrat Khaybar, on the Arabian Plate in the Kingdom of Saudi Arabia. The Arabic word *harrat* means volcanic field in English. Harrat Khaybar is also an example of volcanic fields that are not so easily explained because the earliest of its five lava flows, which occurred 9.1 million years ago (mya), is clearly alkaline, with relatively high concentrations of Na_2O and K_2O and low concentrations of SiO_2. This relatively rare basanite type of flow was followed by four later basaltic eruptions about 3–6 mya, 0.3–3 mya, 0.03–0.3 mya, and less than 0.03 mya (maybe as recent as A.D. 1254 in historical times). Chagarlamudi et al. (1991) found the cinder cone density in the central area of Harrat Khaybar to be 4.3 cones/100 km^2, with an average (geometric) mean spacing of 2.31 km. Another not so easily explained feature of these cinder cones was that they have a prevalent linear trend of N10°W, which parallels the trends of the neighboring Cenozoic volcanic fields, but differs from both the axis of the Red Sea (N30°W) and the trend (NW) of the Precambrian Najd fault system. The N10°W trend agrees with near-North-trending Cenozoic faults recognized in nearby Harrat Rahat by Camp and Roobol (1989), which are thought to be products of east-west extensional forces (a spreading center).

In a similar vein, Alvarez and Bonifaz (1994) used a DEM and LANDSAT TM data to map 319 cones in the Michoacan-Guanajuato Volcanic Field in central Mexico, where they found that the trend in volcanic cone distribution was NNE, the same as the tectonic trends in the area. The main point of this type of investigation is that the mapping of volcanic fields leads to a better understanding of plate motions; this understanding, in turn, leads to better predictive capability for volcanic eruptions in the future.

The second type of volcanic monitoring with remote sensing, also involving the search for eruption precursors of known volcanoes, has great potential for saving lives and minimizing property damage. At the present time, however, it is at an earlier stage of research than the mapping of volcanic fields. There are at least three precursors of volcanic eruptions for which remote sensing has some chance of detection. First, there is often a rise in temperature over parts of the surface of the volcano before it erupts. Second, gases (such as steam, sulfur dioxide, carbon dioxide, and methane) are often emitted before the eruption. Third, the topography often changes as the underground magma chamber swells.

An example (Bhattacharya et al, 1993) of the first of these precursors is given in Figure 10.4, which shows six LANDSAT TM images (black and white versions of bands 7, 5, and 4 color composite images) of Barren Island, situated in the Andaman Sea of the Bay of Bengal, India, for six overpasses between March 3 and May 22, 1991. The eruption started on April 1, 1991, ending a quiescent period of 188 years; lava reached the western coast of the island on May 6, 1991. A giant umbrella cloud over the caldera, probably caused by a vulcanian type of eruption, is visible in the image of May 22, 1991.

Figure 10.5 is a plot of thermal radiance of a black body versus wavelength, superimposed by the wavelength coverage of LANDSAT TM bands 4, 5, 6, and 7. Band 6 of TM is the long wavelength thermal IR band that would first come to mind for temperature mapping. However, Figure 10.5 shows that objects with a temperature of 1,000°C, such as volcanic lava, emit greater energy in TM bands 7 and 5 than in TM band 6. The spatial resolution of TM band 6 is 120 m, as compared to the much better resolution (30 m) of the other TM bands, and the dynamic range of band 6 is set to record temperatures below 200°C. For these

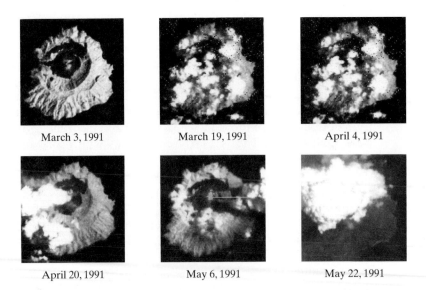

March 3, 1991   March 19, 1991   April 4, 1991

April 20, 1991   May 6, 1991   May 22, 1991

Figure 10.4   Co-registered multitemporal, LANDSAT TM imagery (originally color composite images of TM bands 7, 5, and 4) of Barren Island, India. Collection dates for images range from March 3, 1991 to May 22, 1991. *(After Bhattacharya et al, 1993. Reprinted with permission of the Environmental Research Institute of Michigan.)*

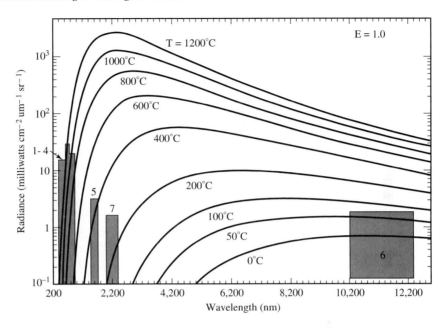

Figure 10.5 Wavelength dependence of thermal radiance of thermal infrared from a black body. Wavelength regions covered by LANDSAT TM bands 1–4, 5, 7, and 6 are darkened. *(After Bhattacharya et al. 1993. Reprinted with permission of the Environmental Research Institute of Michigan.)*

reasons, a color composite image of TM bands 7, 5, and 4 displayed as red, green, and blue is superior to band 6 for mapping hot lava, which is demonstrated by color plate CP30 for the September, 1986 eruption of Lascar volcano, Chile (Rothery and Francis, 1990). This color plate shows both a band 6 image and a color composite image of LANDSAT TM bands 7, 5, and 4 displayed in red, green, and blue, respectively, from a frame collected on July 8, 1991. Another example, this time with TM bands 7, 5, and 2 displayed as red, green, and blue, is given in color plate CP31 for Mt. Etna, Italy (Pieri et al, 1990). The color image, which shows hot lava as orange, has been draped over a DEM to yield a simulated perspective view of the volcano. CP31 is an example of how digital holography can be used for global monitoring.

Even AVHRR data, with 1.1 km spatial resolution, can be useful as a remote sensing tool for volcano temperature rises that are precursors to eruptions. Though of coarser resolution than LANDSAT, AVHRR has a data collection repeat cycle of twice daily, as opposed to once every 16 days for LANDSAT. Bhattacharya et al (1993) were able to extract reasonable temperatures of sub-pixel areas of hot lava from AVHRR data in the Barren Island experiment. After noting from the TM images in Figure 10.4 that an eruption had occurred between the TM overpass dates of March 19 and April 4, 1991, they used AVHRR data to pinpoint April 1, 1991 as the most likely date of initial recent eruption. A global watch for precursors to volcanic eruptions that employs AVHRR and LANDSAT TM data would appear to be supported by these findings. Indeed this is one of the functions planned for

NASA's ASTER (Appendix B), which is scheduled to be launched in 1998 on EOS AM-1 with several 90-meter-resolution thermal infrared bands.

The second type of precursor listed above could also be served by remote sensing methods. Imaging of gases from space has already been done by meteorological satellites for erupting volcanoes (Mouginis-Mark et al. 1993). For example, sulfur dioxide emitted by the El Chichon eruption on April 12, 1982, was mapped by Kreuger (1983) with the Nimbus-7 TOMS (Total Ozone Mapping Spectrometer) instrument. Kreuger employed two of the six TOM ultraviolet spectral bands near 0.3 μm, with a 3.1 km spatial resolution. Although TOMS is useful for mapping high-altitude, sulfur dioxide plumes of erupting volcanoes, it has too poor a spatial resolution to be of much use for detecting precursors of eruptions; these will likely require a spatial resolution of 100 m or better. Furthermore, the use of ultraviolet light, on which TOMS relies, limits monitoring operations to daytime and to higher altitude gas plumes than would be afforded by the use of gaseous absorption bands in the thermal infrared wavelength region. There are plans for a higher spatial resolution satellite experiment (Vincent 1995) for imaging several gases (including methane and sulfur dioxide) that exhibit strong absorption bands in the thermal infrared atmospheric windows of 3.0–5.0 μm and 8.0–14.0 μm.

The third type of precursor calls for topographic change detection in often-cloudy volcanic regions, for which interferometric SAR (Peltzer and Rosen 1995), discussed below in the next subsection, seems ideally suited. Digital photogrammetry applied to thermal infrared stereo images could be considered an alternate means of monitoring topographic change in volcanoes because of the possibility of day and night operations. However, the ability of radar to both work at night and penetrate clouds would assure that the required minimum number of overpasses for the sensor to obtain different look angles and to detect topographic changes would be achieved.

Earthquakes

Remote sensing is very useful for mapping suspected fault zones along which major earthquakes have occurred or will be likely to occur. Rokos et al. (1993) mapped such faults in an earthquake-prone region of Kalamata, Greece. Bayasgalan and Galsan (1993) mapped about 25 fault traces along which strong earthquakes have occurred in western Mongolia from satellite images and aerial photos. Fernandez-Alonso and Hanon (1993) created a GIS database of active faults for the Central and Southern Africa Sheet of the World Map of Active Faults, which was compiled from numerous sources.

One of the uses of remote sensing related to prediction of earthquakes is determination of the frequency interval, or return periods for large earthquakes. An example of this was given by Pinter (1993) for northern Owens Valley in the Eastern California–Central Nevada Seismic Belt, where several large earthquakes, but no ground-rupturing ones, have occurred over the last 120 years. Aerial photos and satellite images were photo-interpreted to map 226 fault scarps in the Volcanic Tableland. This area preserves a record of large, ground-rupturing earthquakes that have occurred since the volcanic lava was deposited over the northern Owens Valley about 738,000 years ago. By calculating the seismic moment, which is directly proportional to the displacement of a fault times the square of the length of the fault, Pinter (1993) was able to categorize 121 of those 226 faults as events of similar magnitude and extent as the 1872 quake that disrupted the ground in a part (southern) of Owens Valley not surrounded by lava. Therefore, over the 738,000 years since the lava flow first occurred, there have been 121 quakes of similar magnitude and extent as the 1872 quake. This information yields a repeat period of approximately 6100 years for quakes of that size.

Another example showing how remote sensing images have aided in estimating return periods for earthquakes of various magnitudes was given by Murphy and Bulmer (1994) for the Wairarapa Valley in New Zealand. They mapped landslides that were likely caused by earthquakes and modeled a subset of these landslides along the West Wairarapa Fault. With assumptions that the changes in original slope would have altered the total stress conditions, that the phreatic surface (ground water table) was coincident with the surface, and that the earthquakes that triggered the landslides occurred on the closest possible point on the West Wairarapa Fault, they calculated the return period in years versus the magnitude of the quake and combined the model data with recorded earthquakes, as shown by the dashed line in Figure 10.6. Their calculation resulted in significantly longer return periods for large earthquakes (every 200 years instead of every 40 years for magnitude 7 quakes) in this region than the estimates derived from recorded earthquakes only (shown in Figure 10.6 as a solid line). Thus, remote sensing assisted the investigation of seismic activity in the Wairarapa Valley prior to European colonization.

One of the precursors of earthquakes is a change in surface topography caused by increasing stress. There are two remote sensing methods that could prove beneficial for measuring this change, as alluded to previously in this chapter. One is

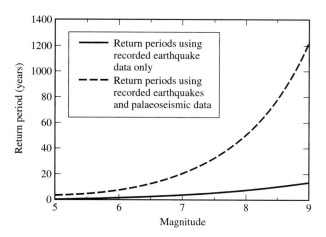

Figure 10.6 A plot of return periods versus magnitude for moderate to large earthquakes on the West Wairarapa Fault, New Zealand. (*After Murphy and Bulmer 1994. Reprinted with permission of the Environmental Research Institute of Michigan.*)

high-resolution digital photogrammetry, which would require two sets of digitized stereo images with pixel sizes on the order of 1 m or smaller and data collection times separated by a few years. Calculating high-resolution DEM for the two data collection times and subtracting the later from the earlier DEM would yield an elevation difference image that would permit the investigator to measure bulging or sinking of the terrain.

An even more dynamic method for accomplishing a similar objective involves interferometric synthetic aperture radar (SAR), which can be performed from satellite SAR images collected from two different orbital passes. A demonstration of this method, albeit applied to topographic changes along a fault caused by an earlier confirmed quake, is given by Peltzer and Rosen (1995). By employing a three-pass interferometric method (Gabriel et al. 1989; Zebker et al. 1994) on ERS-1 SAR images collected on September 14, 1992, November 23, 1992, and November 8, 1993, Peltzer and Rosen mapped displacement along a fault in Eureka Valley, California that was caused by a May 17, 1993 earthquake of magnitude 6.1 and 13 km depth. Color Plate CP32 shows three images, (a)–(c), which display SAR interferograms of the first two dates, of the second two dates, and the difference of the first two interferograms, respectively. Black areas are artifacts caused by areas of low coherence. The low coherence is caused by zones of major surface changes, such as cultivated fields, or zones of phase ambiguities, such as steep slopes facing toward the satellite. In image (a) of CP32, the elevation difference corresponding to the full range of colors displayed in the image is 50 m. In image (b), the full color range could be caused by an elevation difference of 78 m, a line-of-sight surface displacement of 28 mm (half the radar wavelength), or a combination of the two. In image (c), which is a three-pass interferogram, the topography has been removed, such that the full color range displayed in the image corresponds to a displacement of the ground of 28 mm (1.1 in.) in the direction of the satellite. The maximum vertical displacement in the center of the depression, which occupies the colored "bullseye" in the center of image (c), is 9.5 cm.

Figure 10.7 shows surface displacement profiles AA′, BB′, and CC′, the transects of which are shown in image (C) of CP32 as white lines. The radar results are given by solid lines and predicted displacements from an elastic dislocation model (Okada 1985) are given by dashed lines. The arrow A1 along the AA′ transect is where surface cracks were observed, and the arrow A2 along the CC′ transect is where the main rupture reached the surface. Peltzer and Rosen (1995) estimate that the radar-observed displacement map is accurate to within ±3 mm over short spatial scales, even though the spatial resolution of the ERS-1 radar is much coarser (25 m). They were able to calculate the fault plane strike and dip, as well as the trend of the slip during the earthquake from these results.

This three-pass interferometric SAR method is excellent for measuring displacement near the epicenter of a past known quake. However, the method may be too expensive and/or difficult for large-scale monitoring of topographic bulges or sags that are earthquake precursors, since it requires three satellite overpasses per site investigated and applies only to a limited subregion within an ERS-1 frame.

A late-breaking, yet incomplete story has emerged about a satellite remote sensing method that can reportedly be used to predict earthquakes several days in

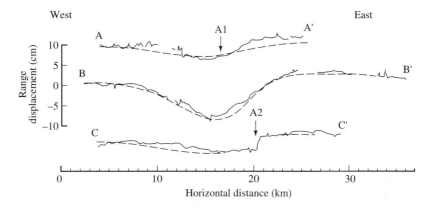

Figure 10.7 Profiles of surface displacement observed with the radar (solid lines), corrected for the geometric distortion induced by topography, and predicted with an elastic dislocation model (dashed lines). Arrow A1 in profile AA′ indicates location where surface cracks were observed and arrow A2 in profile CC′ is where the main rupture from the earthquake (occurring at 13 km depth) reached the surface. *(After Peltzer and Rosen 1995.)*

advance. The story is incomplete because the results of the initial investigators have not yet been confirmed (or denied) by other research teams, but it is compelling enough to be included in this chapter because of its potential for being one of the greatest contributions that remote sensing could make to mankind. As in the case of many remote sensing experiments, the story began with field measurements on the ground.

Two scientists (Qiang Zuji and Dian Changgong 1993) from the Institute of Geology and the Satellite Meteorological Center, both located in Beijing, People's Republic of China, reported that ground meteorological measurements had found temperature increases ranging from 2.6°C to 15.5°C from two to seven days before 11 of 15 earthquakes studied. All of the 15 quakes, which ranged in magnitude from 5.1 to 6.6, occurred on the mainland of China during the year 1989. The four exceptions, which showed little or no temperature rise prior to the quake, were all located in the high-elevation, Qinghai-Tibet Plateau.

Using that ground-based evidence, they conducted an experimental earth-quake-prediction study in which they searched for thermal infrared anomalies in METEOSAT image data that could conceivably be precursors to earthquakes or volcanic eruptions. METEOSAT is a European Space Agency weather satellite that has two thermal infrared spectral bands with 5 km spatial resolution in the 10.5–12.5 μm and 5.7–7.1 μm wavelength regions and a visible/reflective IR panchromatic band in the 0.4–1.1 μm wavelength region, with 2.3 km spatial resolution. It has a data collection rate of twice per hour. From color-composite images of one or more of these METEOSAT bands, Qiang Zuji and Dian Changgong (1993) were able to identify large-area thermal anomalies (regions of higher-than-average temperature) several days prior to three earthquakes and two volcanic eruptions, with epicenters that ranged from China to the Philippine island of Luzon. Earlier papers on this same subject were given in China (Qiang Zuji et al. 1991; Qiang Zuji et al. 1992).

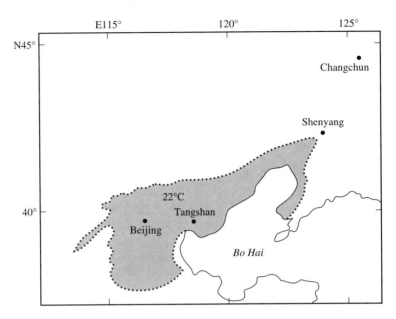

Figure 10.8 Temperature distribution of a thermal infrared anomaly (shaded area) near Tangshan and Beijing, People's Republic of China, with the boundary (dotted line) of the anomaly as it appeared in METEOSAT thermal infrared images on May 9, 1991. *(After Qiang and Dian 1993. Reprinted with permission of the Environmental Research Institute of Michigan.)*

Figure 10.8 shows a map of an area that surrounds Tangshan, which is near the epicenter (39.5°N, 118.2°E) of an earthquake of magnitude 5.1 that occurred on May 30, 1991. The warm temperature anomaly (the 22°C contour line) that started on May 9, 1991, in the southwest and spread northeastward through Tangshan is shown as the shaded region in Figure 10.8. Qiang Zuji and Dian Changgong (1993) do not state the mean background temperature on that date. They first predicted an earthquake in or near this large temperature anomaly on May 9. Figure 10.9 shows a second warm anomaly, displayed as a shaded region, that occurred in approximately the same area on May 17, 1991. The warmest part of this anomaly (outlined with a dotted line) was measured at 22°C, and the remainder of the anomaly was at 18.5°C. Once again, the background temperature was not stated. A second earthquake prediction was made on May 17, based on this information. On May 30, 1991, an earthquake occurred at the epicenter marked "x" in Figure 10.9, near Tangshan.

At almost the same time, there were precursors to a volcanic eruption that occurred about 1800 km ESE of Tangshan, at Unzedeka volcano on Kyushu, Japan, on May 24, 1991. Figure 10.10 (Qiang Zuji and Dian Changgong 1993) shows thermal infrared anomalies interpreted from METEOSAT images of May 16, 17–18, and 21, 1991. On the early mornings of those four dates, a NE-trending thermal anomaly spread from the northeast side of Taiwan to the southwest side of Kyushu island, becoming about 800 km long and 150 km wide. The average temperature increase

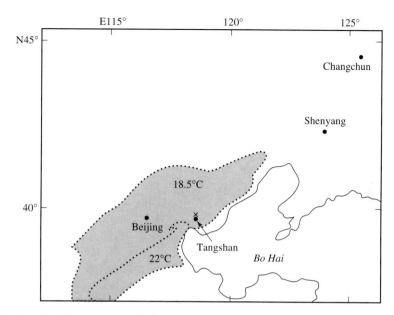

Figure 10.9   Temperature distribution of a thermal infrared anomaly (shaded area) near Tangshan and Beijing, People's Republic of China, with the boundary (dotted line) of the anomaly as it appeared in METEOSAT thermal infrared images on May 17, 1991. The X marks the epicenter of an earthquake that occurred on May 30, 1991. *(After Qiang and Dian 1993. Reprinted with permission of the Environmental Research Institute of Michigan.)*

above background was 3.5°C. On the afternoon of May 17 (the same day as the second thermal anomaly before the Tangshan quake), the southwest end of the thermal anomaly began to expand, and the temperature increased to 22.5°C. It expanded until a diameter of 300 km width was reached. By early morning of May 18, the anomaly had moved 200 km to the northeast, reaching Kyushu island. Unzedeka volcano erupted on May 24, after 200 years of dormancy.

A few days later, on June 1–2, 1991, a thermal anomaly of 25°C was observed in METEOSAT images of the South China Sea, about 4400 km SE of Kyushu, Japan. From June 2 to June 8, 1991, the warm anomaly moved east and southeast, until it reached the east side of Luzon island, in the Philippines, covering the eastern half of the island. On June 9, 1991, Pinatubo volcano began its eruption. Figure 10.11 (Qiang Zuji and Dian Changgong 1993) shows the spatial relationship between these thermal anomalies and Pinatubo.

Possible precursors were detected for two other events reported in that same paper (Qiang Zuji and Dian Changgong, 1993), a magnitude 6.0 earthquake in the Sea of Japan (41.8°N, 130.7°E) on May 11, 1990 (May 8–9 thermal anomalies) and a magnitude 4.0 earthquake at Shahe, near Beijing (40°N, 116.4°E) on September 22, 1990. Qiang Zuji and Dian Changgong predicted the Shahe quake on September 13, based on a thermal anomaly with 3°C–5°C higher temperature than its surroundings that appeared in the METEOSAT images on the mornings of September 10, 11, and 12, 1990.

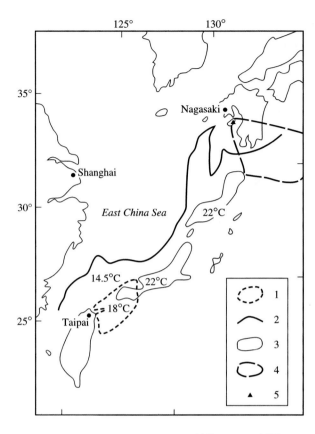

Figure 10.10 Temperature distribution of a series of thermal infrared anomalies in the East China Sea, People's Republic of China, with the anomaly boundaries as they appeared in METEOSAT thermal infrared images on May 16, 1991 (1), May 17 (2), May 18 (3), and May 21 (4). The triangle (5) marks the location of volcano Unzedeka, which erupted on May 24, 1991. *(After Qiang and Dian 1993. Reprinted with permission of the Environmental Research Institute of Michigan.)*

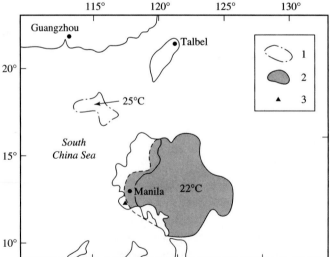

Figure 10.11 Temperature distribution of a series of thermal infrared anomalies in the South China Sea, Philippines, with the anomaly boundaries as they appeared in METEOSAT thermal infrared images on June 1–2, 1991 (dash-dot line) and June 8, 1991 (solid line with shaded interior). The triangle marks the location of volcano Pinatubo, which began erupting on June 9, 1991. *(After Qiang and Dian 1993. Reprinted with permission of the Environmental Research Institute of Michigan.)*

In a later paper, Qiang Zuji et al. (1995) reported on 38 earthquake predictions that they made from 1990–1994. They claimed that 10 were "comparatively accurate," 18 were "comparatively good," 7 were "not so good," 3 were false, and 10 earthquakes failed to be predicted in the area that they were investigating. They did not define the terms in quotations, nor did they specify exactly what kind of METEOSAT image they were using, though it appears from this later paper that it may have been a color level slice of the 10.5–12.5 μm spectral band, with red showing highest temperatures. In this paper, they gave four features that characterize the abrupt temperature increases that they believe were precursors to impending earthquakes:

1. The thermal anomaly is 3°C to 5°C warmer than its surroundings and is an isolated body with various shapes (circles, ellipses, strips, stripe-blocks, belts, and so forth).
2. The dimension of the thermal anomaly varies with time.
3. The most common time interval between the appearance of a thermal anomaly precursor and its impending earthquake is 5–7 days, although some begin as early as 15–20 days before the quake.
4. There are two types of thermal anomalies, stable (temperature increases steadily until the earthquake occurs) and pulsed (the temperature fluctuates between increases and decreases).

What would cause such a thermal anomaly prior to an earthquake? Qiang Zuji and Dian Changgong (1993) cited laboratory work on the electrodynamics of rock fractures by Brady and Rowell (1986) and Martelli et al. (1989). They concluded that light emission prior to rock failure is caused mainly by escaped gases, which are bombarded by ions, particles, and high temperature powder flow that also come from the rock prior to fracture. Qiang Zuji and Dian Changgong (1993) also reported that prior to the May 30, 1991 Tangshan earthquake, they measured a sharp rise in the low-altitude atmospheric methane content in Beijing (180 km SW of the epicenter) from a normal reading of 2 ppm (this is higher than the global atmospheric average of 1.7 ppm) to values of 2.92 ppm on May 21 and 2.86 ppm on May 29, 1991. These corresponded with an early morning (6 A.M.) temperature increase of 2.5°C and 2.0°C, respectively, measured on the ground. They also measured (in Beijing) an abrupt change in the electric field in the air near the ground surface of −422 volts/meter on May 10 and May 20, 1991.

Qiang Zuji et al. (1995) cite evidence that the temperature of rocks increases as the rocks are placed under pressure, but those temperature increases are usually less than 1°C (Geng Naiguang et al. 1993). However, the amount of temperature increase will vary with the experimental conditions, including the amount and type of rock involved.

Qiang Zuji et al. (1995) believe that the mechanism involves the release of methane and carbon dioxide (both are greenhouse gases) and charged particles while the rocks are under stress but before they rupture. They believe that the temperature rise (the thermal anomaly) is caused by the dual action. First is the action of the greenhouse effect. (Solar-absorbed heat radiated from Earth's surface is absorbed by methane and carbon dioxide in the troposphere, primarily in the 3.0–5.0 μm wavelength region, and the extra atmospheric heat is then reradiated back to the surface.) The second action is the excitation of methane and carbon dioxide by the electric field caused by charged particles. As noted earlier, ground measurements in Beijing have shown increases in methane and strong electric field potential changes in the lower atmosphere a few days prior to earthquakes with nearby epicenters ($\leq$ 180 km away).

It would greatly strengthen their case for earthquake precursors if methane increases could be detected directly from space, because it would help separate an earthquake precursor from other weather phenomena. In fact, Zhao Gaoxing and Wang Hongqi (1995) of the Institute of Atmospheric Physics, Chinese Academy of Sciences in Beijing, presented a paper recently about just that subject. Table 10.2 shows the increase in methane (CH_4) concentration of the near surface air in parts per million (ppm) by volume for dates prior to four earthquakes, the epicenters of which ranged from 158 km to 400 km from Beijing. Considering that the background methane content of the global troposphere is 1.7 ppm, these increases ranged from 1.4 to 5.5 times the background, with a grand average of 4.0 times the background for the 22 dates of anomalous methane content. They did some calculations with assumed atmospheric models and tropospheric methane concentrations that ranged from normal to twice as much methane as normal in the lower 5 km of the atmosphere. These

TABLE 10.2 VARIATIONS OF CH_4 CONCENTRATION IN THE NEAR SURFACE ATMOSPHERE IN BEIJING BEFORE FOUR EARTHQUAKES IN TANGSHAN, LUANXIAN, HEJIAN, AND BOHAI. CONCENTRATION IS IN PARTS PER MILLION BY VOLUME (PPMV), AND MAGNITUDE (M) IS ACCORDING TO THE RICHTER SCALE.

Earthquake	CH₄ Concentration (ppmv) in Beijing								
Tangshan May 30, 1991 $M = 5.1$ Δ 180 km	May 21 2.92	29 2.36							
Luanxian July 27, 1991 $M = 4.6$ Δ200 km	July 9 5.4	10 6.2	11 3.1	16 9.3	19 4.5	23 5.6	24 3.0	25 3.16	26 2.6
Hejian Aug. 10, 1991 $M = 3.8$ Δ 158 km	July 28 4.5	29 3.59	Aug. 2 4.43	3 3.28	4 3.2	5 2.7	6 3.7	8 3.9	
Bohai Sept. 2, 1991 $M = 4.6$ Δ 400 km	Aug. 13 3.6	21 3.2	31 3.65						

Source: After Zhao and Wang 1995.

assumptions are conservative, based on the Beijing atmospheric measurements of Table 10.2. They then called for remote sensing of two spectral bands with spectral resolution of 0.1 cm^{-1}, one centered at a methane absorption band at 2895.8 cm^{-1} (3.453 μm) and the other centered outside a methane absorption band at 2898.3 cm^{-1} (3.450 μm), which could be used together to detect methane increases of a factor of 2 greater than the background for the lowest 5 km of the atmosphere. However, they did not perform signal-to-noise calculations for the sensor, which would likely have shown that such a narrow-band sensor would be too insensitive to detect that amount of methane increase.

Vincent (1995) presented a paper on methane imaging from an upcoming Space Shuttle Get-Away-Special experiment called the Flying Falcon (after the Bowling Green State University's mascot) in a different section of the same Space Congress to which Zhao Gaoxing and Wang Hongqi (1995) had submitted their paper. Although the reasons given for the experiment involved the mapping of methane leaks from pipelines, gas wells, LNG carriers, and possibly from natural gas seeps in offshore areas, that experiment could be used for the purposes that Zhao Gaoxing and Wang Hongqi (1995) proposed, i.e., the mapping of methane and thermal anomalies prior to earthquakes. He proposed the use of three spectral bands, rather than two, and chose to employ the strongest absorption band of methane centered at a wavelength of 3.314 μm, which Zhao Gaoxing and Wang Hongqi (1995) rejected because water absorption is stronger there than at their chosen methane absorption band at 3.453 μm. He also chose wider band-width, such that the spectral limits of bands 1, 2, and 3, were 3.100–3.132 μm, 3.298–3.330 μm, and 3.500–3.532 μm, respectively, and relied on the use of a three-band ratio given by the following equation:

$$R_{1+3,2} = \frac{M_1^T + M_3^T}{2M_2^T}$$ (Eqn. 10.1)

where

M_1^T = the total exitance in spectral band 1
M_2^T = the total exitance in spectral band 2
M_3^T = the total exitance in spectral band 3

Since band 2 is the methane absorption band, this ratio increases as the concentration of the methane "plume" or cloud increases, if Earth's surface is much warmer than the plume, and the ratio decreases with increasing methane concentration if the cloud is much warmer than the surface. If both are at the same temperature, the methane will be undetectable, except for sunlight reflected off Earth's surface, which would be zero at night or almost zero during the day over water. Figure 10.12 shows the calculated relationship between the three-band ratio of Equation 10.1 and methane concentration given as ppm in a 10-meter-thick plume. The concentration in ppm times the thickness of the plume is the factor that controls the exponential term of the plume transmission or emission. Therefore, a 10-meter-thick plume with 10,000 ppm would have the same transmission or emission as a 10,000-meter-thick plume with 10 ppm methane. The Flying Falcon experiment would produce a three-band spectral ratio image that would make methane

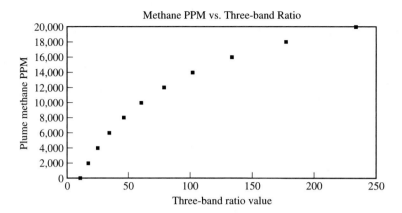

Figure 10.12 Concentration of methane (in ppm by volume) in a methane/air plume of 10 meter thickness versus a three-band ratio ($R_{1+3,2}$) as observed over the ocean by a three-band (3.100–3.132 μm, 3.298–3.330 μm, and 3.500–3.532 μm) thermal infrared sensor from space, under conditions of the ocean surface temperature at 25°C and temperature of the plume (and sea level atmosphere) at 5°C or less. *(After Vincent 1995. Reprinted with permission.)*

brighter than the background, if the background is warmer than the methane (darker if the methane is warmer than the background). An imaging device is preferred, such that the spatial configuration of the methane can be discerned, just as for the thermal anomalies. Any single band of the Flying Falcon experiment would also be able to detect thermal anomalies.

When Vincent (1995) calculated the lowest methane concentration limit (above the background level of 1.7 ppm for a 10,000-m-thick troposphere) detectable for most infrared sensors with a spectral band-width that was 0.032 μm wide, the answer was 16,000 ppm for a 10-meter-thick plume, which is the same as 16 ppm for a 10,000-m-thick plume. However, the infrared sensor that will be aboard the Flying Falcon experiment is a new type: an uncooled, thermal infrared return beam tube, with a selectable filter in front of it. Return beam tubes, heretofore available only in the visible wavelength region, can produce noiseless gain with gain factors up to 50,000 times. This capability means that the signal-to-noise ratio is likely to be substantially higher for this sensor than standard infrared sensors. Therefore, it is possible that an amount of methane as low as 1 to 8 ppm for a 10-kilometer-thick methane plume may be detectable from space. Of course, much lower methane concentrations could be detected from low-altitude aircraft, but a space platform is much preferred for the monitoring of large areas.

The Flying Falcon is scheduled for launch on the Space Shuttle in 1997, but it will be in orbit two weeks or less. If it works well, the payload conceivably could be placed aboard a geostationary satellite for earthquake-precursor detection. On the Space Shuttle, at an altitude of approximately 160 km, it will have a spatial resolution of 20 m, but will cover only about a 10 km swath width (512 pixels × 512 pixels). The same sensor package aboard a geostationary satellite would have a pixel size of about 4.5 km and cover an area of approximately 2300 km $\times$ 2300 km. The package would be similar in spatial resolution and area coverage to METEOSAT and GOES, but with superior spectral resolution and multispectral coverage than existing sensor packages aboard those weather satellites. Since the filter is selectable, the same sensor package could also image sulfur dioxide, carbon monoxide, ozone, and many other gases from a geostationary vantage point. It would also be capable of thermal anomaly detection.

MONITORING OF HUMAN CONTRIBUTIONS TO GLOBAL CHANGES

Humans may be only the second single species in geological history capable of producing global changes, the first such species being oceanic blue-green algae, which changed Earth's atmosphere from anaerobic to oxygen-rich more than 3 billion years ago. It is clear that before the end of the nineteenth century, humans were too few and too technologically backward to create planetary changes. It is also clear from the population growth curve in Figure 10.1 and from the accelerated growth of technology in the second half of this century that in the next century humans will be capable of making planetary changes, and will also have to be careful that they do not inadvertently create undesirable changes in the global atmosphere and oceans. The burning question in this transitional period at the end of the twentieth century is, Have humans unintentionally already changed the globe?

The purpose of this subsection is to discuss possible man-induced changes to Earth that have already occurred, changes that compete with natural geological events. Thus far humans have caused great changes in other living species. For example, tropical forests have been affected by clear cutting that has greatly reduced the amount of land area covered by tropical forests. Fish have also been affected; annual tonnage of the commercial fishing harvest worldwide started declining in the early 1990s, despite ever-increasing numbers and sophistication of fishing vessels. But the emphasis here is direct change to Earth itself, and its atmosphere and oceans. It is in this latter capability that our species will differ most from other temporarily dominant terrestrial life forms, such as the dinosaurs.

The next two subsections are about human-induced effects of greenhouse gases and aerosols on global temperature and of ozone-destroying gases on the stratosphere.

Greenhouse Gases and Aerosols Added to the Atmosphere by Human Activities

Gases that absorb thermal infrared radiation are called greenhouse gases because they tend to warm Earth's surface by absorbing thermal infrared radiation emitted by Earth's surface and reradiating it back to the surface. This process is similar to the way that glass on a greenhouse roof transmits visible sunlight, but absorbs and reradiates heat emitted by objects in the greenhouse, thus warming the greenhouse. The greenhouse gases in the atmosphere that absorb the most thermal infrared radiation are water vapor and carbon dioxide.

Human activities have contributed significantly to increases in three key greenhouse gases (carbon dioxide, methane, and nitrous oxide) since the Industrial Age of the nineteenth century. Table 10.3 (Harriss 1993) compares the concentrations of these three gases in the total atmospheric column between pre-industrial times (A.D. 1750–1800) and current times (A.D. 1990). Carbon dioxide is the most common gas of the three in the atmosphere, and it is increasing at a rate that will double its atmospheric content in 200 years. Methane is a much lesser atmospheric constituent, but it is increasing at almost twice the rate of carbon dioxide and could be the greatest contributor to greenhouse warming of the three gases. Nitrous oxide is an even lesser constituent and is increasing at a lesser rate than either of the other two gases.

Another potential greenhouse gas that is added to the atmosphere by human activity is carbon monoxide, the atmospheric content of which was not measured until recently. Although it was expected that carbon monoxide would be mainly found downwind of major industrial areas, a MAPS (Measurement of Air Pollution from Satellites) gas filter correlation radiometer carried on Space Shuttle flights in 1981 and 1984 surprisingly showed that the highest atmospheric concentrations of carbon monoxide during the 1984 data-collection flight were located in South America and Africa over agricultural areas that yearly employ extensive biomass burning (Harriss 1993). The highest levels of carbon monoxide were in the 100 ppb

TABLE 10.3 SUMMARY OF KEY GREENHOUSE GASES INFLUENCED BY HUMAN ACTIVITIES.[a]

Parameter	CO_2	CH_4	N_2O
Pre-industrial atmospheric concentration (1750–1800)	280 ppmv[b]	0.8 ppmv	288 ppbv[b]
Current atmospheric concentration (1990)[c]	353 ppmv	1.72 ppmv	310 ppbv
Current rate of annual atmospheric accumulation	1.8 ppmv (0.5%)	0.015 ppmv (0.8%)	0.8 ppbv (0.25%)
Atmospheric lifetime[d] (years)	(50–200)	10	150

[a]Ozone has not been included in the table because of lack of precise data.

[b]ppmv = parts per million by volume; ppbv = parts per billion by volume; pptv = parts per trillion by volume.

[c]The current (1990) concentrations have been estimated based upon an extrapolation of measurements reported for earlier years, assuming that recent trends remained approximately constant.

[d]For each gas in the table, except CO_2, the "lifetime" is defined here as the ratio of the atmospheric content to the total rate of removal. This time scale also characterizes the rate of adjustment of the atmospheric concentrations if the mission rates are changed abruptly. CO_2 is a special case since it has no real sinks, but is merely circulated between various reservoirs (atmosphere, ocean, biota). The "lifetime" of CO_2 given in the table is a rough indication of the time it would take for the CO_2 concentration to adjust to changes in the emissions.

Source: IPCC 1990. After Harris 1993.

range for altitudes of 3 km to 18 km, which is much lower than the three gases in Table 10.3.

While it is true that the addition of these gases to the atmosphere since the Industrial Revolution must have contributed toward heating of the planet, it also true that a competing type of pollutant, namely aerosols, would have had a cooling effect on Earth. A new book (Karl et al. 1995), recently reviewed by Kerr (1995), has presented data from the last few decades showing that although the globe has warmed since mid-century, some areas of the globe have been cooled, such as eastern North America, central Europe, and eastern Asia. Their hypothesis was that this cooling is caused by aerosols, primarily of human origin.

Karl et al. divided the northern hemisphere of the globe into nine latitudinal bands and for each band correlated the change in sulfur emissions between 1966 and 1980 with the change in temperature over the same period. In every band, sulfur emission increased and summer temperature decreased, a relationship they say supports a link between aerosols and cooling, though it says nothing about the relative amounts of anthropogenic versus natural aerosols. They then looked at daily maximum and minimum temperatures recorded on the ground for the continental United States before and after the start of the U.S. Clean Air Act enforcement in 1970. Aerosols would be expected to lower the maximum temperature, which usually occurs during the day, by reflecting sunlight away from Earth's surface. The minimum temperature, which usually occurs at night, would not be expected to be affected by increased aerosols, since the absence of sunlight would leave nothing for the aerosols to reflect.

Thus, their hypothesis would predict that daily maximum temperatures would have increased after 1970, when fewer manmade aerosols were in the air, and that the daily minimum temperatures would remain about the same before and after 1970, since they were not affected much by aerosols. This is exactly what the data showed, with the added evidence that the maximum daily temperature decreased during the unregulated years from 1950–1970, as aerosol content due to human activity was increasing. They estimate that roughly 21 million tons of sulfur are now released every year in the United States, which is keeping the mean temperature about 1°C cooler than it would be otherwise. The estimate of aerosol cooling for the entire northern hemisphere is 0.5°C cooler than the temperature would otherwise be without aerosols. Thus, one of the keys to understanding greenhouse warming is to understand aerosol cooling (Kerr 1995). Natural aerosols are discussed in the next section of this chapter.

Anthropogenic carbon dioxide comes primarily from the burning of fossil fuels. Even short of switching to nonfossil fuels, there is a lot that can be done to decrease carbon dioxide emissions, the most important being switching from coal to natural gas, which would result in only half as much carbon dioxide released per Btu of energy created. This reduction is particularly important in countries such as China that have large quantities of both coal and natural gas. A switch to natural gas would likely reduce aerosol content concomitantly, however, and might not have a great net effect on global warming.

Anthropogenic methane comes from oil and gas field development, pipeline leaks, waste decomposition (sewers and landfills), and livestock animals. The amount of methane escaping from producing oil and gas fields was greatly reduced when the practice of flaring was begun a few decades ago, but it is still a source of methane release. The amounts of methane released by sewers and landfills could be largely curtailed by the use of the methane generated by the decomposition of organic wastes for fuel purposes. Most methane entering the atmosphere is now thought to come from natural sources, a topic that will be discussed in the next section. However, one of the questions that remote sensing technology will be expected to answer in the next decade is, What are the relative amounts of anthropogenic methane and natural methane being added to the atmosphere annually? If the results are imaged, it should be possible to determine the difference between anthropogenic and natural methane by the location of their sources and, perhaps, by the detection of other gases present with the methane.

Ozone-Destroying Gases Added to the Atmosphere by Human Activities

The region of Earth's atmosphere called the stratosphere, which ranges from altitudes of 12 km to 50 km, is important to life on the surface of Earth because the stratosphere is where the bulk of solar ultraviolet radiation between the wavelengths of 0.29 μm and 0.32 μm is absorbed (Kaye 1993). Most of this ultraviolet light absorption is done by ozone (O_3). If ozone in the stratosphere were to be severely depleted, there would be a great increase in skin cancers among humans, and many other serious biological effects could occur (Bower and Ward 1982).

Farman et al. (1985) made ground-based measurements at Halley Bay Station, Antarctica that gave evidence for the first recorded Antarctic ozone hole, a large region centered near the South Pole where ozone was unexpectedly depleted. Those measurements were subsequently confirmed by satellite measurements. The TOMS sensor (Total Ozone Mapping Spectrometer) measured the reflection of ultraviolet sunlight by the entire atmosphere and found the ozone hole depletion to be seasonal, with the greatest time of depletion occurring in the Antarctic spring (September and October). Even though scientists had expected chlorofluorocarbons from anthropogenic sources to have destroyed some ozone at the highest levels of the stratosphere, they were shocked to see the total atmospheric ozone column almost halved in the Antarctic ozone hole. Figure 10.13 (Schoeberl 1993) shows the Antarctic minimum total ozone during the month of October in Dobson units, which is a measure of ozone in the total atmospheric column, for the years 1979–1990. Antarctic ozone depletion increased significantly over that time period.

What causes the depletion of ozone? The immediate cause (Kaye 1993) is the interaction of ozone with free radicals in the stratosphere, the most important of which are hydroxyl (OH), nitric oxide (NO), chlorine (Cl), and bromine (Br). Hydroxyl comes from the decomposition of water (H_2O) or methane (CH_4) in the atmosphere. Nitric oxide comes from nitrous oxide (N_2O). Chlorine, the principal ozone destroyer, comes from methyl chloride (CH_3Cl), which occurs naturally, or from manmade chlorofluorocarbons, such as dichlorodifluoromethane (CF_2Cl_2). About 80% of chlorine in the stratosphere has originated from manmade chemicals

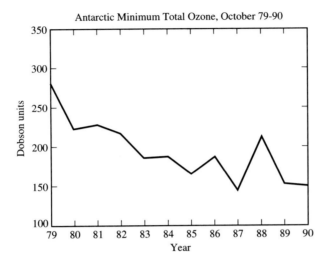

Figure 10.13   October mean minimum values of total ozone over Antarctica for each year since 1979. The decreasing overall trend is clearly shown, although it is superimposed on significant interannual variability. *(After Schoeberl 1993.)*

(Schoeberl 1993). Most of the chlorine that enters the atmosphere from the ocean is in the form of water-soluble compounds, such as NaCl, that get carried back down by rain and never make it to the stratosphere. The manmade chlorofluorocarbons, however, are water insoluble, and tend to accumulate in the stratosphere. Bromine comes from methyl bromide (CH_3Br), which can be natural or man-caused from forest fires and brush fires, or from manmade halons, such as the fire-extinguishing gas, trifluoromethylbromide (CF_3Br). Even if anthropogenic bromine is discounted, the ozone hole can be called a manmade phenomenon because anthropogenic chlorine has increased from nothing to four times the natural chlorine levels in the stratosphere over the last few decades (Schoeberl 1993).

However, volcanoes have also contributed to the ozone hole, as was dramatized by the eruption of Mount Pinatubo (Luzon Island, Philippines) in June, 1991. As early as February, 1991, Solomon et al. (1993) had begun ultraviolet spectroscopic measurements from the ground at McMurdo Station, Antarctica. They measured chlorine dioxide (OClO), which breaks down in a two-step process to the ozone-destroying chlorine radical ClO. The first detectable chlorine dioxide occurred on April 13, 1992, and it was observed every day until May 8, 1992, when the sun began its stay below the horizon for the rest of the Antarctic winter. The aerosols from the Pinatubo eruption were comprised mostly of frozen crystals of sulfuric acid that were formed by water mixing with volcanic sulfur and sulfur dioxide spewed to high altitudes. Rather than adding more chlorine, the aerosols acted as catalysts by providing a conducive surface area on which chlorine dioxide could form from methyl chloride and chlorofluorocarbons (Tolbert 1994). The frozen sulfuric acid aerosol particles also appear to serve as the substrates on which polar stratospheric clouds form, and such clouds are known to provide surface areas for the same chlorine-dioxide-forming reactions. Thus, our current state of knowledge about ozone depletion is that volcanic eruptions are amplifying what humans have created. Fortunately, chlorofluorocarbon compounds have been made illegal in most countries of the world by recent international treaty agreements. It will still take decades, perhaps a century, for the current amount of anthropogenic chlorine in the stratosphere to dissipate, even if the international ban on chlorofluorocarbons is perfectly effective.

MONITORING OF NATURAL CONTRIBUTIONS TO GLOBAL CHANGES

Natural contributions to global changes require our understanding so that humans will have time to adapt to them. At some point in the future, however, this adaptation may require such great effort that the previously unthinkable task of planetary engineering may become the path of least resistance. In either case, the monitoring of natural global changes is an absolute necessity, because a great lead time will be required for the massive engineering efforts that will either adapt human environments to global changes or prevent global changes from occurring.

Following are three subsections on changes in ice and snow coverage (the cryosphere) of Earth, changes in sea surface temperature, and changes in the tectonic plates of Earth's crust; the latter may be the principal change agent for the other two. Currently, the possibility that plate tectonic activity could be the primary causal factor of natural global changes in weather and climate is a matter of great debate. However, it is a debate that can never be settled without crucial data. The geological record and global monitoring by multispectral remote sensing from satellites together can provide that data.

Changes in Ice and Snow Coverage

One of the most sensitive gauges of global temperature increases is the amount of Earth's surface (called the cryosphere) covered year-round by ice and snow. Because of the hostile environment and the remoteness of the north and south polar regions, measurements from the surface of Earth are inadequate for global monitoring of the cryosphere. Thus, satellite remote sensing is proving to be the only viable way of monitoring the year-round ice and snow cover of the world.

However, not all satellites are equally well suited to the task. As has been alluded to in earlier chapters, the primary technology bottleneck for any remote sensing sensor package is the number of pixels that can be sent back to the user. Regardless of the wavelength region or the field of view of the remote sensing package, the number of pixels that can be communicated to the user is fixed for a given state of the art in communication and data processing technology. The practical meaning of this limitation is that for global monitoring, the pixel sizes must be on the order of 1 km to 150 km square, such that the area covered by those pixels is thousands of km square in one image. There is an optimum pixel size for any remote sensing task. The corollary of this statement is that for a sensor of any given pixel size, there is a remote sensing problem that is ideally served by that sensor. The global mapping of ice and snow cover is a classic example of a remote sensing problem that has little need for high spatial-resolution sensors, such as the LANDSAT and

SPOT satellite series, except as a complement to lower spatial-resolution sensors. However, the weather satellites, with pixel sizes up to 150 km wide, have ideal spatial resolution for most ice and snow cover tasks.

Even though the visible and reflective wavelength regions can be used to map snow and ice cover in the daytime, they cannot be used at night, and the polar regions, where most of the world's snow and ice reside, are in darkness approximately half of each year. Thermal infrared measurements can be used night or day to map snow and ice cover, but cannot see through clouds. The most useful sensors for monitoring the cryosphere have been passive microwave sensors, which can discern temperature and emissivity differences between snow or ice and water, land, or vegetation cover, even through clouds. The first passive microwave imager from space was the Electrically Scanning Microwave Radiometer (ESMR) aboard the Nimbus-5 weather satellite, which operated from 1973–1976. It was a single-channel instrument which looked only at horizontally polarized radiation and operated at a wavelength of 1.55 cm, where open water has an emissivity of 0.44 and sea ice has an emissivity of 0.80–0.98 (Parkinson and Gloersen 1993). Since it made cross-track images, the spatial resolution varied from pixel sizes of 30 km (at nadir) to 100 km. ESMR was used to create two atlases of sea ice cover for the years 1973–1976 of the south polar region (Zwally et al. 1983) and the north polar region (Parkinson et al. 1987).

A more sophisticated microwave sensor, called the Scanning Multichannel Microwave Radiometer (SMMR) was orbited on the Nimbus-7 platform in late 1978 (Parkinson and Gloersen 1993). It collected dual polarization at five different wavelengths of 0.81, 1.4, 1.7, 2.8, and 4.6 cm. Spatial resolution varied between 30 km for the 0.81 cm channel to 150 km for the 4.6 cm channel. It was able to tell the difference between first-year ice and multiyear ice that had undergone a summer melt period (Cavalieri, Gloersen, and Campbell 1984). An atlas of monthly averages of sea ice concentrations from both poles for the 1978–1987 time period, plus multiyear ice concentrations for the winter months were produced by Gloersen et al. (1992). Color Plate CP33 (Parkinson and Gloersen 1993) shows the monthly average sea ice concentrations for the south polar region for the four seasons represented by February, May, August, and November, 1985, as recorded by the Nimbus-7 SMMR.

The Defense Meteorological Satellite Program (DMSP) picked up the challenge in 1987 by including a seven-channel Special Sensor Microwave Imager (SSMI) on its platform. Dual-polarization was measured in bands at wavelengths of 0.35, 0.81, and 1.55 cm, and a band at 1.35 cm wavelength was recorded for only one polarization. Cavalieri et al. (1991) discussed the validation of sea ice concentration measurements from SSMI. Hall and Martinec (1985) discussed the use of higher-resolution sensors, such as LANDSAT, SPOT, and SAR (synthetic aperture radar) sensors for complementing these coarse-resolution, passive microwave data sets.

One of the most interesting findings of the passive microwave monitoring efforts is that a *polynya* (a Russian word for a large open water region surrounded by sea ice) occurred in Antarctica's Weddell Sea over the southern winters of 1974–1976, although it moved slightly westward and changed shape over that time period. This polynya has failed to occur since 1976 at anything close to the same size and duration of that event.

All of the above discussion has been about sea ice. However, ice sheets, which occur over land masses, are more important drivers of climate change, because ice over land is not now supported by water. This means that if ice sheets were to melt, sea levels would rise far more than if an equivalent amount of sea ice were to melt. The Antarctica ice sheet contains about 70% of Earth's fresh water locked up in the form of snow and ice. Color Plate CP34 shows a satellite mosaic image of the Antarctic continent from 1 km resolution AVHRR data obtained from U.S. NOAA weather satellites (Merson 1989). If this entire ice sheet were to melt, Earth's sea level would rise about 70 meters (Thomas 1993). If the Greenland Ice Sheet were to melt along with it, only about 5 meters would be added to the sea level rise.

However, the Antarctic Ice Sheet doesn't have to melt completely to cause lesser, yet significant, rises in sea level. In early 1995, a piece of 300-meter-thick ice with an area the size of Rhode Island broke free and became the world's largest known iceberg. This ice mass had previously extended from an island to the mainland of Antarctica and was only partially supported by land. Had it been 100% supported by land masses, it would have caused a worldwide rise in sea level of about 2.5 mm within days of its breaking free. Since most of Antarctica is above sea level and some of the continent's edges have steeply graded terrain, small temperature rises capable of reducing friction to the point of making portions of the ice sheet slide into the sea would become sufficient (with an assist from gravity) to cause similarly sudden rises in sea level. Along a rugged coast, this means that only enough thermal energy to melt the bottom few centimeters of the ice sheet, not the entire thickness of it, could be sufficient to cause relatively fast rises in sea level. More on this subject will be discussed in the final subsection of this chapter.

Despite the availability of AVHRR, LANDSAT, and SPOT visible and infrared data, GEOS-3 and SEASAT radar altimeter data, and ERS-1 SAR data, very little is known about what is happening to the ice sheets of Greenland and Antarctica (Thomas 1993). Radar and laser altimeters, and possibly stereo thermal infrared images employed with digital photogrammetry software, are required to measure the thickness of ice over both of those land masses. How that thickness changes with time is a very important indicator of future sea level and climate changes. The higher-resolution imaging devices, particularly the stereo imagers like SPOT, are needed to measure changes in ice movement along rugged shorelines. All of these types of measurements need to be made on a continuing basis for decades, if the effects of ice sheets on world climate and sea level changes are to be properly monitored.

Changes in Sea Surface Temperature

For more than two decades, thermal infrared imaging sensors have been available from weather satellites for mapping sea surface temperature (SST) with a precision of 0.5°C or better. Before that time, measurements by ships at sea and coastal stations were used to determine temperature fluctuations of the sea surface by contact measurements, but large gaps between stations were common. It has long been understood that the seas are warmer near the equator than near the poles because of solar insolation. More heat is stored in equatorial waters by the absorption of sunlight because the sun's rays strike much more vertically there. The atmosphere, which is mostly transparent in visible wavelength regions and opaque in certain thermal infrared wavelength regions, gets a large portion of its heat from thermal infrared radiation emitted by Earth's surface, which is mostly water. Therefore, the latitudinal asymmetry in SST due to differences in solar insolation is the principal cause of what we consider to be normal atmospheric weather patterns.

There are short-term (several months to a few years) and long-term (centuries) perturbations to those atmospheric weather patterns, referred to as climatary changes, that upset these "normal" atmospheric weather patterns. The most famous of the short-term climatary perturbations that could be called a long-term weather pattern is the El Nino-Southern Oscillation (ENSO), which was first noticed in the mid-1970s. Color Plate CP35 (Njoku and Brown 1993) shows SST of the Pacific and Atlantic Oceans during an ENSO event and during normal conditions. This shows that the finger of cooler water that extends westward from S. America into the equatorial Pacific Ocean in normal times is nearly absent during the ENSO event. During ENSO, warm water piles up against the western S. American coast much farther south than it normally does, and weather patterns in N. America change drastically from their normal patterns. Torrential rains in normally dry areas and droughts in normally wet areas have been associated in N. America with ENSO events in 1957–1958, 1965, 1972–1973, 1982–1983, 1987–1988, and 1991–1993.

For most of the time since ENSO was first recorded, orthodox thinking among meteorologists held that atmospheric patterns controlled SST. More recently (Graham 1995), atmospheric models forced only with observed SST have been shown to closely simulate global temperature records collected since 1970, i.e., SST seems to be controlling atmospheric patterns. Graham's (1995) results imply that at time scales of years to decades, the changes in global average surface air temperature observed during the 1970s and 1980s were driven almost entirely by changes in tropical SST. The changes in SST were communicated to the atmosphere via changes in tropical precipitation and evaporation (Graham 1995).

Color Plate CP36 (from Graham 1995) shows the observed difference in 6-year-mean SST (given in °C) between the 1977–1982 time period and the 1971–1976 time period for the entire world. Redder areas were warmer in the 1977–1982 time period, which includes part of the 1982–1983 ENSO, than they were in the 1971–1976 time period, which includes all of the 1972–1973 ENSO. In essence, this map of SST and land temperature changes reflects a longer-term trend than ENSO occurrences. It should be noted that the temperature increases in the tropical Pacific waters were on the order of 1°C or less. The central and eastern parts of N. America were cooler during the later time period, in agreement with the results (Karl et al. 1995) given in the previous section about the cooling effect of anthropogenic aerosols in industrialized areas of the world. However, the temperature increases in Russia and China from CP36 do not agree with those longer-term findings.

Certainly, SST measurements by remote sensing (thermal infrared and microwave) methods, and studies of those data, are important enough to continue, and even to expand.

Changes in Plate Tectonic Activity

When a continental plate starts one of its episodic movements, several volcanoes along the active margin of the overriding plate become active. It was mentioned earlier that the Cascade Mountain range composite volcanoes of Mounts Hood, Rainier, and Baker all erupted in the 1865–1882 time period, preceded a few years earlier by Mt. St. Helens in 1857. This was caused by a subduction episode of the Juan de Fuca Plate by the N. American Plate. Much more recently, Mt. St. Helens erupted in 1980, El Chichon (Mexico) erupted in 1982, and Mt. Pinatubo (Philippines) erupted in 1991, all because of subduction events beneath the plate margins on which they are located. The name *composite volcano,* the class of which two of the above volcanoes are a member, is derived from the fact that such volcanoes, which occur above subduction zones, issue forth lavas of different composition and gaseous contents during different eruptive events.

Because of their potential for high gaseous content, composite volcanoes often erupt so explosively that they can create short-term climate changes from the gases and aerosols that they spew into high-altitude portions of the atmosphere. The primary aerosol that composite volcanoes emit is volcanic ash, which can cool the earth by reflecting back sunlight before it hits the ground, particularly if the ash is carried by the eruption to high altitudes in the atmosphere. However, there are also secondary aerosols that are produced from emitted volcanic gases. One of the gases that some volcanoes emit is sulfur dioxide, which combines with atmospheric water to create fine-grained frozen droplets of sulfuric acid in the stratosphere. These, in turn, cause more polar stratospheric clouds to form and provide catalytic surfaces on which hydroxyl, nitric oxide, chlorine, and bromine radicals can destroy ozone (Tolbert 1994). These secondary volcanic aerosols also tend to cool Earth by backscattering sunlight.

The amount of sulfur dioxide released by volcanoes has been underestimated in the past because common assumptions were that the amount of sulfur dioxide released should be equal to the amount dissolved in the volcanic magma. However, Wallace and Gerlach (1994) recently found that the total sulfur dioxide released by Mt. Pinatubo was much greater than the sulfur dioxide dissolved in its magma. This implies that pre-eruptive magmatic vapors are a major source of the sulfur dioxide released during many volcanic eruptions.

The effect of volcanic effluents on greenhouse gases appears to be the opposite of what one might expect. Instead of adding methane and carbon dioxide to the atmosphere, the Mt. Pinatubo eruption in 1991 was followed by a sharp decrease in the amount of atmospheric methane and carbon dioxide (Kerr 1994). It was hypothesized in that brief report that the aerosols from Pinatubo could have cooled Earth, causing less methane and carbon dioxide generation by wetlands. One possible conclusion to this finding, if substantiated during future volcanic eruptions, is that the principal effect on global weather or on climate (if the effect is long-term) of aerosols and gases issued from volcanoes during eruptions is to cool the planet, not to heat it up.

A recent study of seismic events listed by the International Seismological Center (ISC) of Berkshire, England, and the U.S. National Earthquake Information

Service (NEIS) has indicated that subsea eruptions, particularly along a spreading ridge, may affect sea surface temperature (SST). D. A. Walker (1988) reported on observations of long-term seismic activity on a study area that covered from lat. 20°–40°S and long. 100°W–120°W, which includes Easter Island and the triple junction among the northern East Pacific Rise, the Chile Rise, and the southern East Pacific Rise. He used ISC data from 1964–1984 and NEIS data from April, 1984 through September, 1987. In the 23.75-year period covered by the study there were 512 events recorded in the study area, for an average of 1.8 seismic events per month. Walker (1988) calculated the energy of each quake from the magnitude, according to Richter (1958), and plotted the accumulated sum of the square roots of the energies with the trend of the plot removed, which Benioff (1951) claims is proportional to the elastic strain preceding the earthquake. Such a plot is called a strain release curve.

Walker (1988) then plotted the inverse of the strain release curve versus time and plotted the Southern Oscillation Index versus time, as shown in Figure 10.14. The Southern Oscillation Index (SOI) is defined as the Easter Island atmospheric

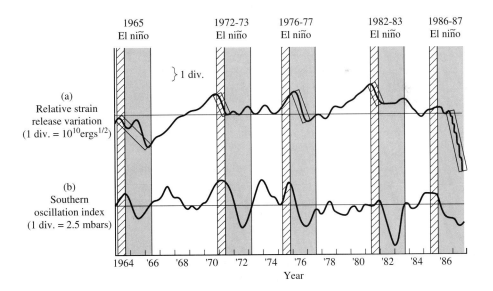

Figure 10.14 (a) The inverse of the smoothed strain release curve as calculated from historical earthquake data and (b) the Southern Oscillation Index curve for the years from 1964–1988 in a study area bounded by lat. 20°–40°S and long. 100°–120°W, which includes Easter Island and a triple junction formed by the northern and southern arms of the East Pacific Rise and the Chile Rise. Shading indicates the times including and immediately preceding the El Ninos of 1965, 1972–1973, 1976–1977, 1982–1983, and 1986–1987. Bars on the strain release curve show the five portions of that curve having the greatest rates of strain release (indicated by rapidly declining values). *(D. A. Walker, EOS, vol. 69, 1988, copyright by the American Geophysical Union. Reprinted with permission.)*

pressure minus the Darwin, Australia atmospheric pressure on the same day, and it originated as a means of describing fluctuations in the atmospheric and hydrospheric circulation over the Indo-Pacific region (G. T. Walker 1924; Berlage 1961). Since the atmospheric pressure deviations are much greater on Easter Island than at Darwin, Australia, the SOI is essentially just the atmospheric pressure on Easter Island. The two plots in Figure 10.14 were smoothed by the same filter and procedure, but the inverse plot of the seismic strain release curve makes the sharply declining values of this plot correspond to increasing tectonic activity. The five El Nino Southern Oscillation (ENSO) events between 1965 and 1987 are marked in Figure 10.14 as shaded areas on both plots. The association with ENSO events and SOI lows has been known for many years, but the association of increased tectonic activity (lows in the inverse seismic strain release curve) was first noted by D. A. Walker (1988).

Walker (1988) had no direct evidence for what the physical mechanism could be between an ENSO occurrence and increased tectonic activity around this triple junction of a spreading center, but he suggested a possible mechanism. He suggested that the ridge system, which is a reservoir of heat, gases, and minerals, could heat the surface water through hydrothermal venting during episodic periods of increased tectonic activity. In turn, this could reduce atmospheric pressure, destroying the normal Easter Island high pressure formation that accounts for the normally strong East–West trade winds, which normally keep warm waters of the western Pacific from approaching the coast of S. America. During an ENSO occurrence, warm (about 1°C warmer than its surroundings) surface waters of the Pacific extend all the way to the western coast of S. America, as shown in CP35.

A more recent paper (D. A. Walker 1995) has extended his study to include the 1991–1993 ENSO event, and tied in SST data from the Climate Monitoring Bulletin of the Australian Bureau of Meteorology, which show a maximum SST anomaly of +3.13°C in July, 1991 (centered at about 17°S, 116°W). He also showed that there was a swarm of 12 quakes recorded in the study area in February, 1991 and a swarm of 6 reported earthquakes at 18°S, 116°W in August, 1991, just one month after the SST anomaly. Both of the months of February and August, 1991 had many more seismic events than the 23.75-year average of 1.8 events per month. The 1991–1993 ENSO event began in very late 1990, or in January, 1991.

The conclusion that this evidence nudges us toward is pretty surprising: plate tectonics may be the principal controller of changes in climate, some of which (the ENSO events) have periodicities of only 5 to 7 years. There are some small pieces of evidence in previous parts of this chapter that might nudge us a bit further. First, Graham (1995) has shown by simulations that changes in sea surface temperature (SST) control changes in atmospheric weather patterns, and not vice versa. Thus, all plate tectonics has to do is to change SST to change weather or climate. Certainly, there is a great deal of energy expended in plate tectonics by the cooling interior of Earth, and the spreading ridges and subduction zones of the Pacific Plate (the "ring of fire") could be losing enough of that energy to ocean water in those localities to cause perturbations in the SST.

Second, Qiang Zuji and Dian Changgong's (1993) data indicate that SST thermal anomalies occur in vast areas prior to earthquakes and volcanic eruptions in short-termed warmings of 3°–5°C. Even if these areas cool to 1°C, that could be enough to start an ENSO event, and as D.A. Walker (1988) says, "the triggering phenomenon would not necessarily have to sustain the forces necessary for (ENSO events) because of the unstable nature of the ocean and atmospheric circulation (of the South Pacific)." Also, the methane precursor that has been measured in Beijing prior to nearby earthquakes (Qiang Zuji and Dian Changgong 1993) offers another possible mechanism for the transfer of heat to sea surface waters. The hydrothermal vents could pass methane, which rises to the top of the water and into the air, warming the water by the greenhouse effect and/or by interaction with charged particles that accompanied them from the vents.

Third, the spatial pattern of the largest area of the increased temperature in the Pacific Ocean shown in CP36 (Graham 1955) appears to be a wedge bounded on the southern side by a very long linear feature that in the western Pacific passes through several aligned trenches where the Pacific Plate is being subducted. Color Plate CP37 shows that same plot, overlaid with over 30,000 seismic epicenters that occurred between the years of 1961 and 1967 (Barazangi and Dorman 1969).

Fourth, there are some seismic event timings outside of D. A. Walker's (1988) study area that appear suspicious. For instance, the two volcanic events that affected long-term weather the most in recent times were El Chichon, which started its eruption on April 12, 1982, and Mt. Pinatubo, which began erupting on June 9, 1991. These both occurred during the early stages of an ENSO event. Furthermore, the Pinatubo eruption followed the Tangshan, China earthquake (May 30, 1991) and the Unzedeka volcano eruption on Kyushu, Japan (May 24, 1991) by only two to three weeks. Such geographically widespread events would support the case for an episodic motion that involved a large portion of the Pacific Plate.

Plate tectonic activity at the poles, where ice and snow are resident, could have a leveraged effect on climatary changes. For instance, consider the steep-banked shores of Antarctica. If the Antarctic Plate were to start an episodic movement, volcanoes underneath the thick ice sheets would have to supply only enough energy to melt the bottom meter or so of the ice sheet to enable gravity to move the ice into the sea. This would cause a sudden (from a geological perspective) rise in sea level, which would only be reversed after the Antarctic Plate motion slowed or stopped long enough for reaccumulation of snow and ice to lower the sea level again to normal levels.

There is evidence in the geological record for relatively sudden, repeated episodes of sea level changes. The cyclothems (Montgomery and Dathe 1994) of Pennsylvanian age are repeated series of nonmarine sediments overlain by a series of marine sediments, which indicate repeated regressions and transgressions of the sea. The nonmarine sediment sequence typically begins with fluvial (river) deposits, often with cross-bedded sands and gravels, that progress to sandy shales. These shales are overlain by fresh-water limestones that contain both terrestrial and fresh-water fossils. Following next is usually a clay-rich layer, overlain by a coal horizon, both of which are deltaic-type deposits. The overlying marine sediments typically display a marine-fossil-rich sequence of shale/limestone/shale/limestone/shale, and they are represented by sediments deposited in a rapidly changing marine environment. The most plausible causes of the cyclothems (Montgomery and Dathe 1994) were two-fold: (1) The rapid transgressions and regressions appear to be the result of repeated glaciation in the southern continent of Gondwana (which included Antarctica); and (2) since the craton was close to sea level, small rises and falls of sea level caused very large areas to become either flooded or exposed. Montgomery and Dathe (1994) state that the timing of glacial cycles were right for the transgressions and regressions, and there are abundant glacial sediments, such as tillites, of this age on the southern continents, including Antarctica.

Ice that is already on the sea can be melted by subsea volcanism, especially in shallow seas. The polynya that occurred in Antarctica's Weddell Sea over the southern winters of 1974–1976 (see the first subsection of this section) could have been caused by undersea volcanic events. Both Arctic and Antarctic sea ice could conceivably be thinned or melted entirely by subsea volcanic events. Part of the Mid-Atlantic Ridge is located in the Arctic Ocean, and some of its shallower parts are located over the northern extension of that spreading ridge.

One thing is clear: global monitoring by multispectral remote sensing methods at several different spatial resolutions over long durations of time is needed to build or refute the case for the control of short-term and long-term climate changes by plate tectonics. It is easy to see that high spatial-resolution sensors would be effective for monitoring geologic hazards such as composite volcanoes along active plate margins. However, high spatial-resolution data would also be effective over the rugged coasts of Antarctica and over shallower portions of spreading ocean ridges (particularly the East Pacific Rise triple junction west of Easter Island). LANDSAT TM data would be very useful for these purposes because of the 120 m thermal infrared band, which is co-registered with six visible and reflective IR bands. Such data should be collected regularly and searched for thermal or opaqueness anomalies at the sea surface in the East Pacific Rise area studied by D. A. Walker (1988), discussed above. Recent reports (Kaiser 1995) that thrice-repeated applications, days apart, of iron fertilization to sea water made the sea water turn green with phytoplankton point to the possibility that mineralized fluids from hydrothermal vents could cause phytoplankton blooms that warm the sea surface by absorbing more sunlight at shallower depths than normal. Both thermal and visible opacity anomalies should be compared with seismic activity for possible correlations.

LANDSAT TM images from two different overpasses have already been used to determine the rate of movement of glaciers in Antarctica (Luchitta and Ferguson 1986), and this type of work should continue. SPOT data and newer stereo satellite data would be useful for monitoring elevation changes with the help of digital photogrammetry. ERS-1 and RADARSAT provide very useful high-resolution images of the Antarctic coasts, greatly aid the monitoring of sea ice, and can be used in interferometric form to measure small changes in elevation for a number of applications.

Coarse-resolution monitoring of the planet should be continued with the geostationary orbiting weather satellites, but hyperspectral imaging devices should be included on them such that methane, sulfur dioxide, and other gases could be imaged, in addition to the thermal imaging that is currently performed. Thermal infrared return beam cameras with selectible filters, discussed in a previous section

(Vincent 1995) of this chapter, would be especially useful for monitoring gases from geostationary orbit because, unlike the TOMS sensor (which looks at ultraviolet wavelengths), the thermal infrared sensor would be useful in daytime or nighttime. If detectable methane anomalies are found to be reliable precursors of earthquakes, the contribution of remote sensing to mankind will be substantial because of that one application. However, there are much broader applications of remote sensing to be tested that could result in even greater contributions. A combination of geostationary hyperspectral imaging sensors and global seismic data could become part of long-term weather forecasts of the future.

REVIEW

There are at least three categories of global monitoring applications requiring a geological perspective that are important for the long-term habitation of Earth. First, the remote monitoring of geological hazards can reduce the damage caused by such hazards on both a local and a global basis. Second, the remote monitoring of anthropogenic contributions to global changes, and their relationship with naturally caused changes, can be used to change human behavior, so as to curtail those contributions, when necessary. Third, the remote monitoring of natural changes to global climate can yield early warnings about serious short-term and long-term changes in Earth's atmospheric, cryospheric, and hydrospheric environments that may require long-term technological efforts to overcome.

Flooding, mass wasting, volcanic eruptions, and earthquakes are types of geological hazards that remote sensing methods can help in the monitoring of precursors before an event and damage assessment after an event. The use of digital elevation models (DEMs) as inputs to surface water run-off models is being used to predict flooding, but will improve with the use of high-resolution DEMs produced by digital photogrammetry as inputs to newer models that can handle such large data sets. The application of interferometric synthetic aperture radar (SAR) and high-resolution DEMs for topographic change detection is helpful for mass wasting and erosion studies. These techniques permit calculation of the volume of material that moved.

The mapping of existing volcanic cones and faults by remote sensing methods is important for relating them to past plate movements and for determining the frequency of eruptions and earthquakes in the past. The search for precursors of both volcanic eruptions and earthquakes can pay for itself by providing warnings that permit the mitigation of damages by either type of event. Volcanoes often display temperature increases, gaseous emissions, and topographic changes (swells and collapses) at various times before eruption. LANDSAT TM bands 4, 5, 6, and 7 have been used to image hot spots prior to eruption. AVHRR weather satellite images have been used to map a plume of sulfur dioxide from the erupting volcano El Chichon, Mexico, as well as to map temperature increases on Barren Island, India. ERS-1 SAR images from three overpasses have been used to map slight changes in topography along a fault where an earthquake occurred. Large-area thermal anomalies, typically 3°–5°C, have been detected in METEOSAT thermal infrared images of regions in China, Japan, and the South China Sea; these anomalies could be precursors to earthquakes, typically occurring 7–9 days before the quake. Methane, warm temperature anomalies, and electric field changes have been measured in the near-surface atmosphere near Beijing prior to earthquakes with epicenters that are less than 180 km away. The cause of the precursor thermal anomalies has been hypothesized to be the release of methane and charged particles by Earth as pressure builds before an earthquake, with the methane heating Earth's surface by the Greenhouse effect and by interaction with electric fields produced by the charged particles. If the two anomalies were to be considered singly, methane would likely be a more unique precursor than a thermal anomaly. Considered together, the two could considerably reduce ambiguity in the mapping of earthquake precursors.

Human beings contribute to global changes primarily by the addition of greenhouse gases and aerosols to the atmosphere and by the destruction of natural sinks for these gases. Anthropogenic gases are added to natural greenhouse gases, thus enhancing global warming. Anthropogenic aerosols have been demonstrated to have caused cooling of the eastern part of N. America prior to passage of the U.S. Clean Air Act of 1970, and cooling has been noted over the second half of this century in all of the parts of the world where such manmade aerosols have been produced in greatest quantities. Therefore, some of the human contribution to global warming has been positive and some has been negative. On the other hand, chlorofluorocarbon and methyl bromide gaseous contributions by human activities have had a purely negative effect on stratospheric ozone. Volcanos have also contributed to ozone depletion. High-powered volcanic emissions of sulfur dioxide, such as provided by the El Chichon (Mexico) and Pinatubo (Philippines) volcanoes, have yielded frozen sulfuric acid aerosol particles in the stratosphere, which provide catalytic surfaces on which the radicals of chlorine, bromine, hydroxyl, and nitric oxide from both natural and unnatural sources can more efficiently destroy stratospheric ozone. It is important, therefore, to monitor ozone and the atmospheric aerosols and radicals that tend to destroy ozone. Past remote sensing imaging of ozone has depended on reflected sunlight in the ultraviolet wavelength region, which cannot be performed under nighttime conditions. Thermal infrared imaging of ozone, aerosols, and radicals in the stratosphere are recommended in the future as complements to the ultraviolet-imaging TOMS sensor.

Three types of surface-level monitoring subjects are most important for the detection of natural changes to the planet: the cryosphere, sea surface temperature, and plate tectonic activity. The cryosphere (snow and ice cover of Earth) has been monitored successfully from passive microwave radiometers aboard weather satellites at very coarse spatial resolution for the past two decades. Greater numbers of higher-resolution images from both radar and thermal infrared wavelengths, as well as more altimeter data, are needed in the future. The ice-covered, rugged coastlines of Antarctica, which contain ice sheets that could slide into the sea, require monitoring because of the possible effects on geologically sudden rises in sea level that such off-loading would create. The total area covered by both sea ice and ice sheets on land are sensitive indicators of global warming or cooling. The extent of coverage therefore makes ice monitoring an important remote sensing objective.

The importance of monitoring sea surface temperature (SST) from space has been heightened by a recent weather simulation study that has provided evidence that SST drives the weather (but not climate), rather than vice versa. The El Nino-Southern Oscillation (ENSO), which is a periodic occurrence of unusual weather patterns in the Western Hemisphere, is now thought to be caused by warm water patches or cells that move eastward from the western Pacific to the S. American coast near the Equator. Imaging of these warm water patches helps not only the prediction of ENSO events, but also the understanding of what causes the warm water patches in the first place.

The remote monitoring of plate tectonic activity from space is important for all of these reasons and more. Since the Pacific Plate episodically moves as one unit or as a few large units, several "ring of fire" volcanic eruptions and earthquakes can be expected to occur within a short time of one another at the edges of the plate, and other events (earthquakes and hydrothermal venting) can be expected near the spreading ridges. First, remote monitoring is important for geological hazard warnings in the vicinities of these events. Second, it is important to learning about long-term global climate changes. For instance, the sum total of all the volcanic aerosols that reach high altitudes, which are mostly from composite volcanoes, is an important determining factor in both the amount of ozone depletion that occurs in the stratosphere and in the amount of cooling that is caused by aerosol reflection of sunlight. Mount Pinatubo's eruption alone was found to simultaneously increase ozone depletion and decrease the effect of global warming for a period of several years. If plate movements caused several of such eruptive events over a sustained time interval, long-term effects to global climate would then occur. Third, remote monitoring is important for short-term climate changes. A recent historical seismic study of a region in the Pacific Ocean near a triple junction in the East Pacific Rise, within 300 km of Easter Island, has found initial evidence for a temporal correlation between periods of maximum release of seismic stress in that area and the occurrence of ENSO events. With the recent discovery that SST drives the weather, we need only to find direct evidence for a connection between eruptive events on the East Pacific Rise and warming of surface ocean water by something on the order of 1°C to make this correlation understandable. Also, episodic motions by the Antarctic Plate could cause under-ice volcanic activity to melt the bottoms of ice sheets; this melting would assist gravity in moving ice from rugged Antarctic coastlines into the sea, giving rise to short-term sea level rises. This process is a possible cause of polynyas (open water surrounded by sea ice), such as occurred in 1974–1976 in the Weddell Sea, off the coast of Antarctica.

Low-spatial-resolution, hyperspectral, thermal infrared sensors in geostationary orbit and additional high-spatial-resolution, multispectral and radar sensors with stereo imaging capabilities in low Earth orbit are needed for global monitoring of geologic hazards and global climate changes. The monitoring of plate tectonic activities is an intermediate step toward performance of those practical, highly valuable tasks.

Exercises

1. **(a)** When and where does the "ozone hole" occur?

   (b) Describe the contributions of man and nature to the ozone hole.

2. The center wavelengths of AVHRR spectral bands 1–5 are as follows: 0.63 μm, 0.86 μm, 3.74 μm, 11.0 μm, and 12.0 μm, respectively. For a given wavelength λ, the ratio of radiances given off by two targets, one at temperature T_1 covering a fraction f of one pixel area (f is a number between 0 and 1.0) and the other at temperature T_2 covering a whole pixel is given by

$$R(T_1, T_2) = f \frac{e^{\frac{14{,}388}{\lambda T_2}} - 1}{e^{\frac{14{,}388}{\lambda T_1}} - 1}$$

   where λ is in μm and T is measured in K.

   (a) If a forest fire with temperature $T_1 = 800$ K covers an entire pixel ($f = 1.0$), and if the normal forest background has a temperature of $T_2 = 300$ K, calculate the above ratio for band 5 of AVHRR, using the center wavelength (12.0 μm) as λ in the above equation. Calculate the same ratio for band 4 and band 3 of AVHRR. (*Note:* This is not a spectral ratio, but a ratio of radiances of two targets of different temperatures in the same spectral band).

   (b) Assuming that a forest fire covering a fraction $f_{minimum}$ of a pixel would be detectable in a given spectral band as long as that pixel were at least twice as bright as the normal forest background pixels, meaning that $R(T_1, T_2) = 2$, what minimum fraction of pixel would be required for the forest fire to be detected in AVHRR bands 3, 4, and 5?

   (c) Which of these spectral bands can detect the smallest forest fires (i.e., has the smallest $f_{minimum}$), and how big must the forest fire (at $T_2 = 800$ K) be in square meters for it to be detectable, given that an AVHRR pixel is 1.1 km $\times$ 1.1 km in area?

3. Fill a cylindrical glass with tap water to a depth of 10 cm, then add enough ice shavings until the top of the ice is at a height of 15 cm above the bottom of the glass. (*Note:* Ice has a specific gravity of approximately 0.9).

   (a) Where is the water line height after the ice is added, but before the ice melts?

   (b) After all of the ice is melted, what is the level of the water in the glass?

   (c) What does this say about changes in sea level caused by the melting of sea ice that forms from the freezing of water already in the sea, as opposed to the addition of ice to the sea by the calving of glaciers and ice sheets in coastal regions?

Cited References

ALEXANDER, D. 1993. *Natural Disasters*. London: University College London Press, Ltd.

ALVAREZ, R., and R. BONIFAZ. 1994. Defining Volcanic Field Distributions using LANDSAT TM and DTM Data. In *Proceedings of the Tenth Thematic Conference on Geologic Remote Sensing*, vol. 1, 359–374. Ann Arbor: Environmental Research Institute of Michigan.

BARAZANGI, M., and J. DORMAN. 1969. World Seismicity Maps Compiled from ESSA, Coast and Geodetic Survey, Epicenter Data, 1961–1967. *Bulletin of the Seismological Society of America* 59:369–380.

BAYASGALAN, A., and P. GALSAN. 1993. The Investigation of the Seismogenic Structures with the Help of Remote Sensing Data. In *Proceedings of the Ninth Thematic Conference on Geologic Remote Sensing,* vol. 2, 983–989. Ann Arbor: Environmental Research Institute of Michigan.

BENIOFF, H. 1951. Earthquakes and Rock Creep. *Bulletin of the Seismological Society of America* 41:31.

BERLAGE, H. P. 1961. Variations in the General Atmospheric and Hydrospheric Circulation of Periods of a Few Years Duration Affected by Variations of Solar Activity. *Annals of the New York Academy of Science* 95:354.

BHATTACHARYA, A., C. S. S. REDDY, and S. K. SRIVASTAV. 1993. Remote Sensing for Active Volcano Monitoring in Barren Island, India. In *Proceedings of the Ninth Thematic Conference on Geologic Remote Sensing,* vol. 2, 993–1003. Ann Arbor: Environmental Research Institute of Michigan.

BOWER, F. A., and R. B. WARD, eds. 1982. *Stratospheric Ozone and Man.* Boca Raton, Fl: CRC Press.

BRADY, B. T., and B. A. ROWELL. 1986. Laboratory Investigation of the Electrodynamics of Rock Failure. *Nature* 321, 488–492.

BURROUGH, P. A. 1986. *Principles of Geographical Information Systems for Land Resource Assessment.* Oxford, U.K.: Clarendon Press.

CAMP, and ROOBOL. 1989. In *Proceedings of the Eighth Thematic Conference on Geologic Remote Sensing,* vol. 1. Ann Arbor: Environmental Research Institute of Michigan.

CAVALIERI, D. J., J. P. CRAWFORD, M. R. DRINKWATER, D. T. EPPLER, L. D. FARMER, R. R. JENTZ, and C. C. WACKERMAN. 1991. Aircraft Active and Passive Microwave Validation of Sea Ice Concentration from the Defense Meterological Satellite Program Special Sensor Microwave Imager. *Journal of Geopyhysical Research* 96, no. 2:1989–2008.

CALVALIERI, D. J., P. GLOERSEN, and W. J. CAMPBELL. 1984. Determination of Sea Ice Parameters with the Nimbus-7 SMMR. *Journal of Geophysical Research* 89:5355–5369.

CHAGARLAMUDI, P., F. A. ZAKIR, and M. R. MOUFTI. 1991. Application of Aerial Photograph, LANDSAT TM and Radar Images in Delineating Volcanic Features in Harrat Khaybar, Kingdom of Saudi Arabia. In *Proceedings of the Eighth Thematic Conference on Geologic Remote Sensing,* vol. 1, 613–626. Ann Arbor: Environmental Research Institute of Michigan.

CUMMING, I., and L. GRAY. 1989. Interferometric Radar: A Better Tool for Exploration Geology? In *Proceedings of the Seventh Thematic Conference on Geologic Remote Sensing,* vol. 1, 561–566. Ann Arbor: Environmental Research Institute of Michigan.

FARMAN, J. C., B. G. GARDINER, and J. D. SHANKLIN. 1985. Large Losses of Total Ozone in Antarctica Reveal Seasonal ClO_x/NO_x Interaction. *Nature* 315:207–210.

FERNANDEZ-ALONSO, M., and M. HANON. 1993. The Benefits of Compiling the ILP-World Map of Active Faults (Sheet Central and Southern Africa) in a G.I.S. Environment. In *Proceedings of the Ninth Thematic Conference on Geologic Remote Sensing,* vol. 2, 963–974. Ann Arbor: Environmental Research Institute of Michigan.

GABRIEL, A. G., R. M. GOLDSTEIN, H. A. ZEBKER. 1989. Mapping Small Elevation Changes Over Large Areas: Differential Radar Interferometry. *Journal of Geophysical Research,* 94, no. B-7:9183–9191.

GENG NAIGUANG, CUI CHENYU, and DENG MINGDE. 1993. The Remote Sensing Observation in Experiments of Rock Failure and the Beginning of Remote Sensing Rock Mechanisms. *Acta Seismologica Sinica* 6, no. 4:971–980.

GETIS, A., and J. M. GETIS. 1982. *Geography.* Boston: Houghton Mifflin Co.

GLOERSEN, P., W. J. CAMPBELL, D. J. CAVALIERI, J. C. COMISO, C. L. PARKINSON, and H. J. ZWALLY. 1992. *Arctic and Antarctic Sea Ice, 1978–1987: Satellite Passive-Microwave Observations,* NASA SP-511. National Aeronautics and Space Administration, Washington, D.C.

GRAHAM, N. E. 1995. Simulation of Recent Global Temperature Trends. *Science* 267:665–671.

HALL, D. K., and J. MARTINEC. 1985. *Remote Sensing of Ice and Snow*. London: Chapman and Hall, Publishers.

HARRISS, R. C. 1993. Tropospheric Chemistry. In *Atlas of Satellite Observations Related to Global Change,* eds. R. J. Gurney, J. L. Foster, and C. L. Parkinson, 181–189. Cambridge, U.K.: Cambridge University Press.

HUANG, S. L., and B. K. CHEN. 1991. Integration of LANDSAT and Terrain Information for Landslide Study. In *Proceedings of the Eighth Thematic Conference on Geologic Remote Sensing,* vol. 2, 743–754. Ann Arbor: Environmental Research Institute of Michigan.

KAISER, J. 1995. Random Samples: Oceanographers' Green Thumb. *Science* 269:759.

KARL, T. R., R. KNIGHT, G. KUBLA, and J. GAVIN. 1995. Evidence for Radiative Effects of Anthropogenic Sulfate Aerosols in the Observed Climate Record. In *Aerosol Forcing of Climate,* eds. J. Charlson and J. Heintzenberg, 363–382. Chichester, U.K.: John Wiley & Sons.

KAYE, J. A. 1993. Stratospheric Chemistry, Temperature, and Dynamics. In *Atlas of Satellite Observations Related to Global Change,* eds. R. J. Gurney, J. L. Foster, and C. L. Parkinson, 41–57. Cambridge, U.K.: Cambridge University Press.

KERR, R. A. 1994. Did Pinatubo Send Climate-Warming Gases Into a Dither? *Science* 263:1562.

KERR, R. A. 1995. Study Unveils Climate Cooling Caused by Pollutant Haze. *Science* 268:802.

KRUEGER, A. J. 1983. Sighting of El Chichon Sulfur Dioxide Clouds with the Nimbus-7 Total Ozone Mapping Spectrometer. *Science* 220:1377–1379.

LI, D., and J. LI. 1994. A Mud-Rock Flow Predicting Information System. In *Proceedings of the Tenth Thematic Conference on Geologic Remote Sensing,* vol. 2, 558–567. Ann Arbor: Environmental Research Institute of Michigan.

LUCHITTA, B. K., and H. M. FERGUSON. 1986. Antarctica: Measuring Glacier Velocity from Satellite Images. *Science* 234:1105–1108.

MARTELLI, G., P. N. SMITH, and A. J. WOODWARD. 1989. Light, Radiofrequency Emission and Ionization Effects Associated With Rock Fracture. *Geophysical Journal International* 98: 397–401.

MCKEAN, J., S. BUECHEL, and L. GAYDOS. 1991. Remote Sensing and Landslide Hazard Assessment. In *Proceedings of the Eighth Thematic Conference on Geologic Remote Sensing,* vol. 2, 729–742. Ann Arbor: Environmental Research Institute of Michigan.

MERSON, R. H. 1989. An AVHRR Mosaic Image of Antarctica. *International Journal of Remote Sensing,* 10:669–674.

MONTGOMERY, C. W., and D. DATHE. 1994. *Earth Then and Now*. Dubuque, Iowa: Wm. C. Brown Publishers.

MOUGINIS-MARK, P. J., D. C. PIERI, and P. W. FRANCIS. 1993. Volcanoes. *Atlas of Satellite Observations Related to Global Change,* eds. R. J. Gurney, J. L. Foster, and C. L. Parkinson, 341–357. Cambridge, U.K.: Cambridge University Press.

MURPHY, W., and M. H. K. BULMER. 1994. Evidence of Pre-Historic Seismicity in the Wairarapa Valley, New Zealand As Indicated By Remote Sensing. In *Proceedings of the Tenth Thematic Conference on Geologic Remote Sensing,* vol. 1, 341–351. Ann Arbor: Environmental Research Institute of Michigan.

NJOKU, E. G., and O. B. BROWN. 1993. Sea Surface Temperature. *Atlas of Satellite Observations Related to Global Change,* eds. R. J. Gurney, J. L. Foster, and C. L. Parkinson, 237–249. Cambridge, U.K.: Cambridge University Press.

OKADA, Y. 1985. Surface Deformation Due to Shear and Tensile Faults In a Half-Space. *Bulletin of the Seismological Society of America* 75, no. 4:1135–1154.

PARKINSON, C. L., J. C. COMISO, H. J. ZWALLY, D. J. CAVALIERI, P. GLOERSEN, and W. J. CAMPBELL. 1987. *Arctic Sea Ice, 1973–1976: Satellite Passive-Microwave Observations.* NASA SP-489. Washington, D.C.: National Aeronautics and Space Administration.

PARKINSON, C. L., and P. GLOERSEN. 1993. Global Sea Ice Coverage. *Atlas of Satellite Observations Related to Global Change,* ed. R. J. Gurney, J. L. Foster, and C. L. Parkinson, 371–383. Cambridge, U.K.: Cambridge University Press.

PELTZER, G. and P. ROSEN. 1995. Surface Displacement of the 17 May 1993 Eureka Valley, California, Earthquake Observed by SAR Interferometry. *Science* 268, no. 5215:1333–1336.

PIERI, D. C., L. S. GLAZE, and M. J. ABRAMS. 1990. Thermal Radiance Observations of an Active Lava Flow During the June 1984 Eruption of Mount Etna, *Geology,* vol. 18, no. 10:1018-1022. (front cover)

PINTER, N. 1993. Estimating Earthquake Hazard from Remotely Sensed Images, Eastern California–Central Nevada Seismic Belt. In *Proceedings of the Ninth Thematic Conference on Geologic Remote Sensing,* vol. 1, 251–256. Ann Arbor: Environmental Research Institute of Michigan.

QIANG ZUJI, XU XIUDENG, and DIAN CHANGGONG. 1991. Thermal Infrared Anomaly-Precursor of Impending Earthquakes. *Chinese Science Bulletin (Kexuetongbao)* 36, no. 4:319–323.

QIANG ZUJI, DIAN CHANGGONG, WANG XUANJI, and HU SIYI. 1992. Thermal Infrared Anomalous Temperature Increase and Impending Earthquake Precursor. *Chinese Science Bulletin* 37, no. 19:1643–1646.

QIANG ZUJI and DIAN CHANGGONG. 1993. The Thermal Infrared Anomaly of METEOSAT-Precursor of Impending Earthquakes. In *Proceedings of the Ninth Thematic Conference on Geologic Remote Sensing,* vol. 2, 1005–1013. Ann Arbor: Environmental Research Institute of Michigan.

QIANG ZUJI, DIAN CHANGGONG, ZHAO YONG, and GUO MANHONG. 1995. Satellite Thermal Infrared Temperature Increase Precursor—Short Term and Impending Earthquake Prediction. In *Environmental Assessment of Geological Hazards, Proceedings of the Space Congress,* 53–57. Munich, Germany: European Space Report.

REINHARDT, D. 1994. Satellite Data in Change Analysis for Disaster Monitoring. In *Proceedings of the Tenth Thematic Conference on Geologic Remote Sensing,* vol. 2, 11–604. Ann Arbor: Environmental Research Institute of Michigan.

RICHTER, C.F. 1958. *Elementary Seismology.* San Francisco: W. H. Freeman.

ROKOS, D., J. SPYRAKOS, D. ARGIALAS, and N. FYTROLAKIS. 1993. Evaluation of Analog and Digital Image Analysis Techniques for Mapping Suspected Fault Zones in the Earthquake-Prone Region of Kalamata, Greece. In *Proceedings of the Ninth Thematic Conference on Geologic Remote Sensing,* vol. 2, 951–962. Ann Arbor: Environmental Research Institute of Michigan.

ROTHERY, D. A. and P. W. FRANCIS. 1990. Short Wavelength Infrared Images for Volcano Monitoring, *International Journal of Remote Sensing,* vol. 11, no. 10:1665-1667. London: Taylor & Francis. (front cover)

SAGHAFIAN, B. 1993. Implementation of a Distributed Hydrologic Model within Geographic Resources Analysis Support System (GRASS). In *Second International Conference/Workshop on Integrating GIS with Watershed Modeling Proceedings,* NCGIA, Breckenridge, Colo.

SCHOEBERL, M. R. 1993. Stratospheric Ozone Depletion. In *Atlas of Satellite Observations Related to Global Change,* eds. R. J. Gurney, J. L. Foster, and C. L. Parkinson, 59–65. Cambridge, U.K.: Cambridge University Press.

SOLOMON, S., R. W. SANDERS, R. R. GARCIA, and J. G. KEYS. 1993. Increased Chlorine Dioxide Over Antarctica Caused by Volcanic Aerosols from Mount Pinatubo. *Nature* 363:245–248.

THOMAS, R. H. 1993. Ice Sheets. In *Atlas of Satellite Observations Related to Global Change,* eds. R. J. Gurney, J. L. Foster, and C. L. Parkinson, 385–400. Cambridge, U.K.: Cambridge University Press.

TOLBERT, M. A. 1994. Sulfate Aerosols and Polar Stratospheric Cloud Formation. *Science* 264: 527–528.

VAN WESTEN, C. J. 1995. Remote Sensing and Geographic Information Systems for Geological Hazard Mitigation. In *Environmental Assessment of Geological Hazards, Proceedings of the Space Congress,* 63–71. Munich, Germany: European Space Report.

VAN WESTEN, C. J., R. SOETERS, and N. RENGERS. 1994. GISSIZ: Training Package for the Use of Geographical Information Systems in Slope Instability Zonation. In *Proceedings of the Tenth Thematic Conference on Geologic Remote Sensing,* vol. 1, 386–397. Ann Arbor: Environmental Research Institute of Michigan.

VINCENT, R.K. 1995. Flying Falcon: Multispectral Thermal IR Geological Imaging Experiment. In *Remote Sensing for Oil Exploration and Environment, Proceedings of the Space Congress,* 139–146. Munich, Germany: European Space Report.

WALKER, D. A. 1988. Seismicity of the East Pacific Rise: Correlations With the Southern Oscillation Index? *EOS* 69, no. 38:857, 865–867.

———. 1995. More Evidence Indicates Link Between El Niños and Seismicity. *EOS Transactions,* American Geophysical Union 76:33–36.

WALKER, G. T. 1924. World Weather II. *Mem. India Meteorol. Dep.* 24:275.

WALLACE, P. J., and T. M. GERLACH. 1994. Magmatic Vapor Source for Sulfur Dioxide Released During Volcanic Eruptions: Evidence from Mt. Pinatubo. *Science* 265:497–499.

WOLDAI, T. 1995. Satellite Remote Sensing: Flood Hazard and Management in the Area Around Wuhan, Hubei Province, China. *Environmental Assessment of Geological Hazards, Proceedings of the Space Congress,* 35–51. Munich, Germany: European Space Report.

ZEBKER, H. A., P. A. ROSEN, R. M. GOLDSTEIN, A. G. GABRIEL, and C. L. WERNER. 1994. *Journal of Geophysical Research,* vol. 99. 19617.

ZHAO GAOXIANG and WANG HONGQI. 1995. The Possibility of Monitoring Increased CH_4 Concentration in the Atmosphere and Its Potential Use in Earthquake Prediction. In *Environmental Assessment of Geological Hazards, Proceedings of the Space Congress,* 58–62. European Space Report. Munich, Germany.

ZWALLY, H. J., J. C. COMISO, C. L. PARKINSON, W. J. CAMPBELL, F. D. CARSEY, and P. GLOERSEN. 1983. *Antarctic Sea Ice, 1973–1976: Satellite Passive-Microwave Observations.* NASA SP-459, Washington, D.C.: National Aeronautics and Space Administration.

Epilogue

▼ ▼ ▼ ▼ ▼ ▼ ▼ ▼ ▼ ▼ ▼ ▼ ▼ ▼ ▼

The human species has not only survived, it has thrived in the first four million years or so of its existence. One of the reasons that we have done so well is that humans have learned to cooperate in groups, first as pack hunters, then as more complex societies, where different members performed services or produced products needed by the group.

Throughout much of this twentieth century, groups operated as exclusive teams most of the time and made wars on other groups. As the century comes to a close, in the early years of the Space Age, it is time for us to consider our species as one supergroup.

We are now populous enough and sufficiently technologically advanced to be able to change the globe. In the past, this ability has been primarily viewed in a negative light. However, the geological record provides evidence for two important threats to the survival of our species, threats that will require all of the cooperative skills and technological prowess that humans can muster. First, the third planet from the Sun has released enormous amounts of internal heat in the past and will continue to dissipate it in the foreseeable future, in the form of plate tectonics. Second, our solar system has had countless numbers of orbiting projectiles that have kept the planet disturbed, and many asteroids and comets are still roiling up Earth after all these eras. In other words, the human species is riding a bucking bronco and dodging boulders at the same time.

Remote sensing is one of the tools that will be required to avoid extinction, for both technical and social reasons. The technical reasons are given in this book. The social reasons are reflected in the wonderful variety of names of the contributors, cited in the references, investigators who come from all over the globe. To be sure,

U.S. Presidents Eisenhower and Carter are owed a thank-you for the Open Skies policy that began with the launch of LANDSAT I in 1972 and has been maintained ever since. However, it has been the scientists of many nations who have picked up the baton and run with it to whom primary thanks are due. We must now use remote sensing for the supergroup, rather than for subgroup competition. However, remote sensing can only warn and monitor; it cannot construct. It is equally important that the world's engineering and scientific skills advance, in anticipation of those days that will surely come when we will have to race against time to save our species.

It is essential for us to cooperate for the good of the whole. Long-term projects of any sort require excellent rules of behavior, or ethics, that govern behavior all the way up and down the chain of command. Poor ethics is a formula for long-term chaos. Of course, cooperation is a group term; on an individual basis, it is equivalent to care and concern for fellow members. From the geological record, this strategy of mutual concern appears to be a successful one for species survival.

Integral Equations

▼ ▼ ▼ ▼ ▼ ▼ ▼ ▼ ▼ ▼ ▼ ▼ ▼ ▼

CHAPTER 4

A Lambertian surface, one that is perfectly diffuse, reflects light such that the reflected radiance L_λ^r (with units of watts/m²-steradian-μm) from that surface is constant from any angle of reflectance with respect to the surface normal and for any azimuth angle. The reflected radiance is equivalent to what the human eye perceives as brightness in the visible wavelength regions. In other words, a Lambertian surface has equal brightness in all directions, i.e., it reflects light equally into all solid angles in the hemisphere centered about the mean normal to the sample surface.

The total radiant flux (in watts) reflected into the hemisphere from a Lambertian surface of area dA, is given by:

$$d\Phi_\lambda^h = \int_0^{2\pi}\int_0^{\pi/2} L_\lambda^r dA \, \cos\theta \sin\theta \, d\theta \, d\phi$$

$$= 2\pi\int_0^{\pi/2} L_\lambda^r dA \, \cos\theta \sin\theta \, d\theta \qquad \text{(Eqn. 4.2a)}$$

$$= \pi L_\lambda^r dA$$

The ratio of total reflected radiant flux to the radiant flux incident upon the target, $d\Phi_\lambda^i = sE_\lambda dA$, defines the *diffuse reflectance* or *hemispherical reflectance* of the target surface. The ratio assumes that both the incident electromagnetic radiation and the reflected radiation are constant for any solid angle over the hemisphere above the sample plane, and it is given by:

$$\rho^h(\lambda) = \frac{d\Phi_\lambda^h}{d\Phi_\lambda^i} = \frac{\pi L_\lambda^r}{sE_\lambda} \tag{Eqn. 4.2b}$$

where

> s = a unitless *shadow-slope factor* that varies between 0 and 1.0 according to the percentage of each picture element (pixel) on the target that is in shadow ($s = 0$ for 100% shadow, $s = 0.6$ for 40% shadow, and so on)
>
> E_λ = direct solar spectral irradiance impinging on the target at wavelength λ with units of watts/m²-μm

This Equation 4.2b is included in the first terms of Equations 4.2 and 4.3 in Chapter 4. Equation 4.5 reduces to the following for the thermal infrared wavelength region:

$$L_\lambda = \epsilon(\lambda)\tau(\lambda)L_\lambda^e(bb, T)$$

$$= \frac{2hc^2\epsilon(\lambda)\tau(\lambda)}{\lambda^5\left[e^{\frac{hc}{\lambda kT}} - 1\right]} \tag{Eqn. 4.5a}$$

where Planck's black-body equation has been substituted for $L_\lambda^e(bb, T)$, in which c is the velocity of light, h is Planck's constant, and k is Boltzmann's constant.

After substitution of Equation 2.7 into Equation 4.5a, the rigorous form of Equation 4.6 results as follows:

$$L_\lambda = \frac{2hc^2[1 - b'\rho^c(\lambda)]\tau(\lambda)}{\lambda^5\left[e^{\frac{hc}{\lambda kT}} - 1\right]} \tag{Eqn. 4.6a}$$

For Equations 4.7 and 4.8 in the text, the integral equations are given as follows:

$$L(i) = \frac{g(i)}{\pi}\left[bs\int_{\lambda_l^i}^{\lambda_u^i} f^i(\lambda)\rho(\lambda)\tau(\lambda)E_\lambda d\lambda \right. \tag{Eqn. 4.7a}$$

$$\left. + \pi\int_{\lambda_l^i}^{\lambda_u^i} f^i(\lambda)L_\lambda^r(Path)d\lambda\right] + a(i)$$

for the 0.4–4.0 μm wavelength region, and

$$L(i) = 2hc^2g(i)\int_{\lambda_l^i}^{\lambda_u^i} \frac{f^i(\lambda)\tau(\lambda)[1 - b'\rho^c(\lambda)]}{\lambda^5\left[e^{\frac{hc}{\lambda kT}} - 1\right]} d\lambda + a(i) \tag{Eqn. 4.8a}$$

for the greater than 4.0 μm wavelength region, where in both equations,

> $g(i)$ = electronic gain for the ith spectral band
> $a(i)$ = electronic offset for the ith spectral band
> λ_l^i = lower wavelength limit of the ith spectral band
> λ_u^i = upper wavelength limit of the ith spectral band
> $f^i(\lambda)$ = filter function for the ith spectral band (filter spectral transmittance)

Substitution of Equation 4.8a into Equation 4.9 and slight re-arrangement yields

$$DN(i) = q(i)L(i)$$

$$= q(i)g(i)2hc^2\tau(i)f(i)[1 - b'\rho^c(i)]\int_{\lambda_i^i}^{\lambda_u^i} \frac{f^i(\lambda)\tau(\lambda)[1 - b'\rho^c(\lambda)]}{\lambda^5\left[e^{\frac{hc}{\lambda kT}} - 1\right]} d\lambda \quad \text{(Eqn. 4.9a)}$$

$$+ q(i)a(i)$$

which holds in the thermal infrared for wavelengths longer than 4.0 μm, assuming a ground temperature of approximately 300 K. The exponential term in the denominator of this equation can be written more conveniently as follows:

$$x = \frac{hc}{\lambda kT} = \frac{14{,}388}{\lambda T} \quad (x \text{ is unitless if } \lambda \text{ is expressed in } \mu\text{m and } T \text{ in K}).$$

CHAPTER 5

The three-component integral equation that is represented by Equation 5.13 for the exitance coming from a plume of gas is given by the following:

$$M^T(i) = M^{be}(i) + M^e(i) + M^{br}(i)$$

$$= \epsilon(i)\int_{\lambda_{il}}^{\lambda_{iu}} e^{-\alpha(\lambda)w}M_\lambda^{bb}(T_b)\, d\lambda + \int_{\lambda_{il}}^{\lambda_{iu}}[1 - e^{-\alpha(\lambda)w}]M_\lambda^{bb}(T_c)\, d\lambda$$

$$+ [1 - \epsilon(i)]\int_{\lambda_{il}}^{\lambda_{iu}} e^{-2\alpha(\lambda)w}E_\lambda^{Sun} d\lambda \quad \text{(Eqn. 5.13a)}$$

where

$M^T(i)$ = total exitance in the ith spectral band

$M^{be}(i)$ = exitance emitted by the background of spectral emittance $\epsilon(i)$ (taken from Equation 2.7 and assumed to be constant over the wavelength interval λ_u to λ_l) and temperature T_b that has been transmitted once through the plume

$M^e(i)$ = exitance emitted by the plume of temperature T_p (assuming that plume emission absorbed by the ground is re-emitted and absorbed by the plume)

$M^{br}(i)$ = exitance from reflection of sunlight and atmospheric emission off the background with attenuation of both incoming and outgoing radiation by the plume

Spectral exitance from a "black body" is given in Equation 5.14 as π times the Planck radiation function $L_\lambda^{bb}(T)$, taken from Equation 4.5a, as follows:

$$M_\lambda^{bb}(T) = \pi L_\lambda^{bb}(T) = \frac{2\pi c^2 h}{\lambda^5}\frac{1}{(e^{\frac{hc}{kT\lambda}} - 1)} \quad \text{(Eqn. 5.14a)}$$

where $M_\lambda^{bb}(T)$ is radiation power per unit area of temperature T (in K) radiated into a hemisphere per unit wavelength interval, with dimensions of watts/(cm^2 − μm).

Multispectral Sensor Systems

▼ ▼ ▼ ▼ ▼ ▼ ▼ ▼ ▼ ▼ ▼ ▼ ▼ ▼ ▼

The purpose of this appendix is to list the characteristics and sales agents for digital satellite sensors that are producing or will be producing commercially available data in the near future. It is not intended to be a comprehensive list of all civilian satellites. Those satellites about which there is some doubt of commercial accessibility or that lack specific commercial information are not included here. The exception is ASTER, an important U.S. NASA remote sensing satellite that is scheduled to be orbited in the near future with the NASA EOS mission. Even though ASTER data are not currently slated for commercial accessibility, the LANDSAT precedent will likely enable NASA to make arrangements for commercial dissemination of the data.

Table B.1 lists the satellite sensors for which data are already commercially available as of the printing of this book. Table B.2 is a partial list of satellite sensors that are likely to produce commercially available digital multispectral data in the near future.

TABLE B.1 SATELLITE MULTISPECTRAL SENSORS WITH COMMERCIALLY AVAILABLE DATA.

Sensor Name	Bandpasses Band: (μm)	Earliest Date	Sales Agent	Pixel Size (m)	Swath Stereo
LANDSAT MSS	4: 0.5–0.6 5: 0.6–0.7 6: 0.7–0.8 7: 0.8–1.1	1972	EOSAT 4300 Forbes Blvd. Lanham, MD 20706-9954 (800) 344-9933	80	185 km No stereo
LANDSAT TM	1: 0.45–0.52 2: 0.53–0.60 3: 0.63–0.69 4: 0.76–0.90 5: 1.55–1.75 6: 10.4–12.5 7: 2.08–2.35	1984	EOSAT 4300 Forbes Blvd. Lanham, MD 20706-9954 (800) 344-9933	30 B6: 120	185 km No stereo
SPOT HRV	1: 0.54–0.59 2: 0.61–0.68 3: 0.79–0.89 P: 0.51–0.73	1986	SPOT Image Corp. 1897 Preston White Dr. Reston, VA 22091-4368 (800) 275-7768	20 P: 10	50 km Stereo
LISS I, II (India)	1: 0.46–0.52 2: 0.52–0.59 3: 0.62–0.68 4: 0.77–0.86	1991	EOSAT 4300 Forbes Blvd. Lanham, MD 20706-9954 (800) 344-9933	LISS I 37 LISS II 73	67 km No stereo
AVHRR	1: 0.58–0.68 2: 0.72–1.10 3: 3.55–3.93 4: 10.5–11.5 5: 11.5–12.5[*]	1984	EROS Data Center DAAC Program Sioux Falls, SD 57198 (605) 594-6151	1100 (Nadir)	2500 km Some stereo
JERS-1 (Japan)	1: 0.52–0.60 2: 0.63–0.69 3: 0.76–0.86 4: 0.76–0.86 5: 1.60–1.71 6: 2.01–2.12 7: 2.13–2.25 8: 2.27–2.40 L-Band Radar (15 Mhz)	1992	EOSAT 4300 Forbes Blvd. Lanham, MD 20706-9954 (800) 344-9933	18.3–24.2	75 km Stereo
ERS-1,2 (Europe)	C-Band Radar (5.3 Ghz)	1991	Radarsat International 3851 Shell Road, Suite 200 Richmond, British Columbia Canada V6X 2W2 (604) 231-4902	30	102 km Stereo VV Polar.
RADARSAT (Canada)	C-Band Radar (5.6 Ghz)	1995	Radarsat International 3851 Shell Road, Suite 200 Richmond, British Columbia Canada V6X 2W2 (604) 231-4902	Fine 10	50 km Stereo HH Polar.

*Only on NOAA 7, 9, and 11.

TABLE B.2 NEAR FUTURE SATELLITE MULTISPECTRAL SENSORS WITH COMMERCIALLY AVAILABLE DATA.

Sensor Name	Bandpasses Band: (μm)	Earliest Date	Sales Agent	Pixel Size (m)	Swath Stereo
SPOT HRVIR	1: 0.54–0.59 2: 0.61–0.68 3: 0.79–0.89 4: 0.51–0.73 5: 1.58–1.75	1997	SPOT Image Corp. 1897 Preston White Dr. Reston, VA 22091-4368 (800) 275-7768	20 P: 10	60 km Stereo
LISS III (India)	1: 0.52–0.59 2: 0.62–0.68 3: 0.77–0.86 4: 1.55–1.70 P: 0.50–0.75	1997	EOSAT 4300 Forbes Blvd. Lanham, MD 20706-9954 (800) 344-9933	24 P: 10	141 km Stereo
SEASTAR	1: 0.40–0.42 2: 0.43–0.45 3: 0.48–0.50 4: 0.50–0.52 5: 0.55–0.57 6: 0.66–0.68 7: 0.75–0.79 8: 0.85–0.89	1997	Orbital Sciences Corp. 21700 Atlantic Blvd. Dulles, VA 20166 (703) 406-5000	1000 (Nadir)	2500 km Some stereo
ASTER	Bands 1–3: 0.53–0.86 Bands 4–9: 1.60–2.43 Bands 10–14: 8.00–11.65	1998	Yet undetermined	1–3: 15 4–9: 30 10–14:90	60 km Stereo

LANDSAT TM and MSS Brightness and Ratio Codes of Minerals, Vegetation, and Snow

▼ ▼ ▼ ▼ ▼ ▼ ▼ ▼ ▼ ▼ ▼ ▼ ▼ ▼ ▼ ▼

There is so much information in reflectance spectra of natural materials that a form of shorthand data compression is needed to accentuate the information that is most important to a user. The purpose of this appendix is to reduce the LANDSAT TM and MSS spectral band reflectances to a few brightness and ratio codes that reveal at a glance whether a particular target would be bright or dark in a particular band or spectral ratio image. The following paragraphs describe how these codes were produced.

The LANDSAT TM and MSS spectral bands are given as follows:

LANDSAT TM Band No.	Wavelength (μm)	LANDSAT MSS Band No.	Wavelength (μm)
1	0.45–0.52	4	0.5–0.6
2	0.52–0.60	5	0.6–0.7
3	0.63–0.69	6	0.7–0.8
4	0.76–0.90	7	0.8–1.1
5	1.55–1.75		
6	10.4–12.5		
7	2.08–2.35		

The first step in creation of the brightness and ratio codes was a calculation of the average reflectance in each of the LANDSAT TM and MSS spectral bands for 146 mineral, vegetation, and snow/ice spectra that were included as members in this brightness/ratio code data set. The 0.4–2.5 μm wavelength spectra of 139 minerals included in the data set all came from a spectral library of minerals (Grove et al. 1992) produced by the Jet Propulsion Laboratory, California Institute of Technology,

Pasadena, California. The three vegetation spectra and the four snow/ice spectra (Salisbury et at. 1994) included in the brightness/ratio code data set all came from the Johns Hopkins FTP site: rocky.eps.jhu.edu, which is accessible via the Internet. (*Note:* There are many more spectra at that data site than are shown here.) The reflectance spectrum of each data set member was averaged over the wavelength limits of the TM and MSS bands given above. This assumed a square filter function for each band. The spectral ratios for all nonreciprocal ratios of reflectances (15 for TM and 6 for MSS) were then calculated for each member of the data set.

The second step was to sort all of the 146 members of the data set according to brightness, one spectral band at a time. The third step was to segregate the 146 members into 10 deciles (approximately 15 members per decile), assigning all members of the darkest decile a value of 0, the second darkest decile a value of 1, and so forth to the brightest decile, which was assigned a 9. This operation was repeated for each of the single bands of TM and MSS, as well as for the 15 spectral ratios of TM and the 6 spectral ratios of MSS. It should be noted that band 6 of TM (the thermal infrared band) was omitted from the brightness and ratio codes because it represents how warm objects are, not how much light is reflected off of them, as do all of the other TM bands. Also, it should be noted that the MSS bands start their numbering with band 4, not band 1, which represents NASA's conventional designation throughout most of MSS operating history (1972–present).

Table C.1 gives the LANDSAT TM and MSS brightness codes for the 146-member data set, 139 of which are minerals, 3 of which are vegetation, and 4 of which are snow and ice. The particle size of the spectrum chosen for each member is shown in the second column, after the name of the material. The next six columns give the brightness codes of each member for TM bands 1, 2, 3, 4, 5, and 7. The final four columns give the MSS brightness codes for MSS bands 4, 5, 6, and 7.

An example of how this table can be used will be given for the CONIFER (AVE.) member (fortieth listing Table C.1). With a brightness code of 1, conifers are brighter than only 10% (darker than 80%) of all the 146 members of the data set in TM bands 1, 2, 3, and 7. However, conifers are brighter than 40% of the data set in TM band 4, and brighter than 20% of the data set in TM band 5. If you were to make a color-composite image of TM bands 2, 3, and 4 displayed as blue, green, and red, respectively, you would expect conifers to be very dark in blue and green, but medium dark in red, giving it a dark red appearance. In fact, these bands are used in a false color-composite image that shows vegetation as red. As another example, note that the mineral cassiterite (see Table C.1) is darker at all wavelengths than almost all other members of the data set, with a zero brightness code in all TM and MSS bands. Anatase is brighter at all wavelengths than almost all other members of the data set, with a brightness code of 9 in all TM and MSS bands.

Table C.2 gives the LANDSAT TM ratio codes for the same data set. In columns 3–17 of that table, the first row identifies the numerator and denominator, respectively, of the spectral ratio, with the numerator always a longer wavelength band than the denominator. All 15 nonreciprocal spectral ratios of TM bands 1, 2, 3, 4, 5, and 7 are represented, such that the ratio code for each member is a 15-digit number. The reciprocal ratio code can be found by subtracting a ratio code from 9. For instance, the $R_{2,1}$ ratio code of actinolite is 6, which implies that the $R_{1,2}$ ratio

TABLE C.1 LANDSAT TM AND MSS BRIGHTNESS CODES (MINERALS, VEGETATION, AND SNOW).

Material	Particle Size	TM Bands						MSS Bands			
		B1	B2	B3	B4	B5	B7	B4	B5	B6	B7
ACTINOLITE IN–4A	125–500μm	3	2	2	2	4	4	2	2	2	2
ALBITE TS–6A	125–500μm	6	6	5	5	6	7	6	5	5	5
ALBITE TS–6A	125–500μm	6	6	5	5	6	7	6	5	5	5
ALMANDINE GARNET NS–04A	125–500μm	2	2	3	3	1	2	2	3	3	2
ALUNITE SO–4A	125–500μm	5	5	7	7	4	3	5	6	7	7
AMBLYGONITE P–3A	125–500μm	8	7	7	7	5	5	7	7	7	8
ANALCIME TS–18A	< 45μm	8	8	7	8	8	8	8	7	8	8
ANATASE, SYNTHETIC O–12A	< 45μm	9	9	9	9	9	9	9	9	9	9
ANDESINE TS–4A	125–500μm	7	6	6	6	7	8	6	6	6	5
ANGLESITE SO–10A	125–500μm	2	2	2	2	3	4	2	2	2	2
ANHYDRITE SO–1A	125–500μm	5	4	4	5	6	6	4	4	5	5
ANORTHITE TS–5A	125–500μm	4	4	4	3	4	5	4	4	3	3
ANTHOPHYLLITE IN–8A	125–500μm	5	4	4	3	4	4	4	4	3	3
ANTLERITE SO–11A	125–500μm	3	2	1	1	2	2	2	1	1	1
APATITE P–1A	125–500μm	5	4	4	4	8	8	5	4	4	4
APHTHITALITE SO–9A	125–500μm	8	8	8	8	9	9	8	8	8	8
ARSENOPYRITE S–5A	125–500μm	1	1	1	1	1	1	1	1	1	1
ATACAMITE H–4A	125–500μm	2	1	0	0	2	2	1	0	0	0
AUGITE IN–15A	< 45μm	3	2	2	1	2	5	2	2	1	1
AZURITE C–12A	125–500μm	3	2	0	0	2	2	2	0	0	0
BARITE SO–3A	125–500μm	9	9	8	9	9	9	9	8	9	9
BERYL CS–2A	125–500μm	6	5	4	3	5	6	5	4	3	3
BIOTITE PS–23A	125–500μm	2	2	2	2	3	5	2	2	2	2
BORAX B–6A	125–500μm	9	9	9	9	3	1	9	9	9	8
BORNITE S–9A	125–500μm	1	0	0	1	1	2	0	0	1	1
BRUCITE OH–1A	125–500μm	6	7	7	6	6	2	7	7	7	6
BUDDINGTONITE, FELDS TS–11A	125–500μm	3	4	3	4	4	4	4	3	4	4
BYTOWNITE TS–13A	< 45μm	6	6	6	6	7	8	6	6	6	6
CALCITE C–3E	125–500μm	8	8	8	8	9	7	8	8	8	7
CASSITERITE O–3A	125–500μm	0	0	0	0	0	0	0	0	0	0
CELESTITE SO–5A	125–500μm	8	7	7	7	9	9	7	7	7	8
CERUSSITE C–10A	125–500μm	7	7	7	8	9	8	7	7	7	8
CHABAZITE TS–15A	< 45μm	8	8	8	8	7	6	8	8	8	8
CHALCOCITE S–8A	125–500μm	1	0	1	1	0	0	1	1	1	1
CHALCOPYRITE S–4A	125–500μm	1	1	1	1	1	2	1	1	1	1
CHLORITE, RIPIDOLITE PS–12A	125–500μm	1	1	1	1	2	3	1	1	1	1
CLINOZOISITE SS–4A	125–500μm	2	3	3	3	5	4	3	3	3	3
COLEMANITE B–1A	125–500μm	8	8	8	8	2	1	8	8	8	7
COLUMBITE O–7A	125–500μm	0	0	0	1	0	1	0	0	0	0
CONIFER (AVE.)		1	1	1	4	2	1	1	1	2	4
COOKEITE PS–9A	125–500μm	4	4	5	4	4	4	4	5	4	4
CORDIERITE CS–3A	125–500μm	4	3	3	2	5	6	3	3	3	2
CORRENSITE PS–10A	125–500μm	1	1	2	2	3	3	1	2	2	2
CORUNDUM, SYNTHETIC O–15A	125–500μm	9	9	9	9	9	8	9	9	9	9
CRISTOBALITE TS–7A	< 45μm	5	5	5	5	4	5	5	5	5	4
CRYOLITE H–1A	125–500μm	8	8	8	8	9	9	8	8	8	8
CUMMINGTONITE IN–6A	125–500μm	0	0	0	0	0	0	0	0	0	0
DICKITE PS–3A	< 45μm	7	7	8	8	9	8	7	8	8	8
DIOPSIDE IN–9B	125–500μm	2	2	2	2	3	3	2	2	2	2

347

TABLE C.1 *(continued)*

Material	Particle Size	TM Bands						MSS Bands			
		B1	B2	B3	B4	B5	B7	B4	B5	B6	B7
DOLOMITE C–5C	125–500μm	7	7	7	7	8	6	7	7	7	7
ENSTATITE IN–10B	125–500μm	3	3	3	1	1	2	3	3	2	2
EPIDOTE SS–1C	125–500μm	2	2	2	2	4	5	2	2	2	2
FAYALITE NS–1A	125–500μm	0	0	1	0	0	1	0	1	0	0
FERROAXINITE CS–4A	125–500μm	2	2	2	2	1	0	2	2	3	2
FLUORITE, PURPLE H–2A	125–500μm	6	5	6	7	8	9	6	6	6	7
FORSTERITE, SYNTHETI NS–2A	< 45μm	5	5	5	4	4	6	5	5	4	4
GAHNITE O–11A	125–500μm	4	3	3	3	0	0	3	3	3	3
GALENA S–7A	125–500μm	3	2	2	1	1	1	2	2	1	1
GIBBSITE, SYNTHETIC OH–3A	< 45μm	9	9	9	9	6	4	9	9	9	9
GLAUBERITE SO–8A	125–500μm	6	6	5	6	7	9	6	5	5	6
GLAUCONITE PS–19A	125–500μm	1	1	0	0	0	1	1	0	0	0
GLAUCOPHANE IN–3A	125–500μm	1	1	1	1	3	3	1	1	1	1
GOETHITE OH–02A	< 45μm	1	2	2	2	5	6	2	2	2	3
GRAPHITE E–1A	125–500μm	0	0	0	0	0	0	0	0	0	0
GRASS (SENESCENT)		3	3	3	4	4	5	3	3	4	4
GRASS (VIGOROUS)		0	1	1	3	2	2	0	1	2	3
GROSSULAR GARNET NS–03B	125–500μm	6	6	7	7	8	8	6	7	7	7
GYPSUM SO–2B	125–500μm	9	9	9	9	3	3	9	9	9	9
HALITE HO–3A	125–500μm	8	8	8	8	9	9	8	8	8	8
HEMATITE O–1A	< 45μm	0	1	2	2	2	3	1	2	2	2
HEMIMORPHITE SS–2A	125–500μm	8	8	8	8	5	4	8	8	8	8
HOWLITE B–5A	125–500μm	9	9	9	9	5	3	9	9	9	9
HYDROXYAPOPHYLLITE PS–22A	125–500μm	8	8	8	8	3	2	8	8	8	8
HYPERSTHENE IN–14A	< 45μm	4	4	4	3	3	5	4	4	3	3
ICE		0	0	0	0	0	0	0	0	0	0
ILLITE PS–11A	< 45μm	3	4	3	4	7	8	4	3	4	4
JAROSITE SO–7A	125–500μm	2	3	3	3	6	6	3	3	3	3
JOHANNSENITE IN–12A	125–500μm	0	0	0	0	1	2	0	0	0	0
KAOLINITE, WELL ORDERED PS–01A	< 45μm	8	8	9	9	9	8	9	9	9	9
KERNITE B–2A	125–500μm	9	9	9	9	4	4	9	9	9	9
LABRADORITE TS–2A	125–500μm	7	7	7	6	5	8	7	7	7	5
LEPIDOLITE, YELLOW PS–13A	125–500μm	5	6	6	5	8	5	5	6	6	5
MAGNESIOCHROMITE O–8A	125–500μm	0	0	0	0	0	0	0	0	1	0
MAGNESITE C–6A	125–500μm	7	8	8	8	6	5	8	8	8	8
MAGNETITE O–4A	< 45μm	0	0	0	0	0	0	0	0	0	0
MALACHITE C–7A	125–500μm	3	3	1	0	2	3	3	1	0	1
MICROCLINE TS–17A	125–500μm	4	5	5	5	8	8	5	5	5	6
MIMETITE A–1A	125–500μm	3	5	6	6	7	7	4	5	6	7
MONTEBRASITE P–2A	125–500μm	6	6	6	5	4	4	6	6	5	5
MONTMORILLONITE PS–02B	< 45μm	8	8	9	9	8	6	8	9	9	9
MUSCOVITE PS–16A	125–500μm	4	4	4	4	6	5	4	4	4	4
NATROJAROSITE SO–7C	125–500μm	2	3	3	3	6	6	3	3	3	3
NATROLITE TS–8A	125–500μm	7	7	7	7	4	3	7	7	7	8
NEPHELINE TS–16A	125–500μm	6	6	6	5	7	7	6	6	5	6
NONTRONITE PS–6A	< 45μm	4	4	5	5	7	7	4	5	5	5
OLIGOCLASE TS–3A	125–500μm	6	6	5	5	6	7	6	5	5	5
ORTHOCLASE TS–12A	125–500μm	7	7	7	7	8	9	7	7	6	6

TABLE C.1 *(continued)*

Material	Particle Size	TM Bands						MSS Bands			
		B1	B2	B3	B4	B5	B7	B4	B5	B6	B7
PALYGORSKITE PS–4A	< 45μm	6	6	6	6	7	6	6	6	6	6
PERICLASE O–14A	< 45μm	9	9	9	9	9	9	9	9	9	9
PLUMBOJAROSITE SO–7B	125–500μm	2	3	3	3	5	5	3	3	3	3
PREHNITE PS–21A	125–500μm	7	7	6	6	7	4	7	6	6	6
PYRITE S–02A	< 45μm	1	1	1	1	1	1	1	1	1	1
PYROLUSITE O–6A	125–500μm	1	1	1	1	1	1	1	1	1	1
PYROPHYLLITE PS–7A	125–500μm	5	5	6	6	9	5	5	6	6	7
PYRRHOTITE	< 45μm	0	0	1	1	0	1	0	1	1	1
QUARTZ, SMOKY TS–1B	125–500μm	5	4	4	5	8	9	4	4	4	5
REALGAR S–3A	125–500μm	1	3	4	5	8	9	3	4	5	5
RHODOCHROSITE C–8A	125–500μm	5	5	5	5	7	5	5	5	5	5
RHODONITE IN–1A	125–500μm	3	3	4	3	3	4	3	4	4	2
RIEBECKITE IN–7A	125–500μm	0	0	0	0	1	3	0	0	0	0
RUTILE O–2A	125–500μm	1	1	1	2	1	2	1	1	1	1
SANIDINE TS–14A	< 45μm	5	5	5	6	7	9	5	6	6	6
SAPONITE PS–24A	< 45μm	8	8	8	8	9	7	8	8	8	9
SCHEELITE T–1A	125–500μm	4	4	4	4	6	7	4	4	4	4
SCORODITE A–2A	125–500μm	2	2	2	2	2	1	2	2	2	3
SEPIOLITE PS–5A	125–500μm	7	7	7	7	5	4	7	7	7	7
SERPENTINE PS–20A	125–500μm	4	5	4	4	5	3	5	4	4	4
SIDERITE C–9A	125–500μm	3	3	3	2	3	5	3	3	3	1
SILLIMANITE NS–08A	125–500μm	6	6	6	7	8	7	6	7	7	7
SMITHSONITE C–11A	125–500μm	8	8	7	5	7	5	7	7	5	4
SNOW (COARSE)	Coarse	9	9	9	8	0	0	9	9	8	6
SNOW (FINE)	Fine	9	9	9	9	2	1	9	9	9	9
SNOW (FROST)	Frost (finest)	9	9	9	9	3	3	9	9	9	9
SODALITE TS–10A	125–500μm	4	3	3	4	6	7	3	3	4	4
SPHALERITE S–1A	125–500μm	2	2	2	2	3	1	2	2	2	2
SPODUMENE IN–13A	125–500μm	6	6	6	6	7	9	6	6	6	6
STIBNITE S–6A	125–500μm	2	1	2	3	7	8	1	2	2	3
STILBITE TS–9A	125–500μm	7	7	7	7	3	2	7	7	7	7
STRONTIANITE C–1A	125–500μm	7	7	7	7	8	7	7	7	7	7
SULFUR E–2A	125–500μm	5	9	9	9	9	9	8	9	9	9
TALC PS–14A	125–500μm	9	8	8	7	9	8	8	8	7	7
TINCALCONITE B–4A	125–500μm	9	9	9	9	3	2	9	9	9	9
TITANITE NS–07A	125–500μm	4	5	5	6	5	6	5	5	6	6
TOPAZ NS–06A	125–500μm	5	4	4	4	4	3	4	4	4	4
TOURMALINE, DRAVITE-S CS–1A	125–500μm	0	0	0	0	0	0	0	0	0	0
TREMOLITE IN–5A	125–500μm	7	7	6	6	8	6	7	6	6	6
TRIPHYLITE P–4A	125–500μm	4	3	3	2	1	7	3	3	3	2
TRONA C–4A	125–500μm	5	5	5	6	2	0	5	5	5	6
TSCHERMIGITE SO–6A	125–500μm	9	9	8	8	1	0	9	8	8	7
ULEXITE B–3A	125–500μm	8	8	8	7	2	0	8	8	8	7
VERMICULITE PS–18B	< 45μm	3	4	4	4	5	6	4	4	4	4
VESUVIANITE SS–3A	125–500μm	4	5	5	4	6	4	5	5	5	5
WITHERITE C–2A	125–500μm	7	7	7	7	8	7	6	7	7	7
WOLLASTONITE IN–2A	125–500μm	7	6	6	5	6	7	7	6	6	5
ZINCITE O–13A	< 45μm	9	9	9	9	9	9	9	9	9	9
ZIRCON NS–09A	125–500μm	3	3	3	3	5	6	3	3	3	3

349

TABLE C.2 LANDSAT TM RATIO CODES (MINERALS, VEGETATION, AND SNOW).

Material Name	Particle Size (μm)	2,1	3,1	3,2	4,1	4,2	4,3	5,1	5,2	5,3	5,4	7,1	7,2	7,3	7,4	7,5
ACTINOLITE IN-4A	125–500μm	6	0	0	1	0	6	9	9	9	9	8	8	9	9	1
ALBITE TS-6A	125–500μm	5	4	3	4	4	4	3	3	3	3	5	4	5	5	7
ALBITE TS-6A	125–500μm	5	4	4	4	4	4	3	3	3	3	5	4	5	5	7
ALMANDINE GARNET NS-04A	125–500μm	9	9	9	9	9	9	2	1	0	0	5	2	1	1	9
ALUNITE SO-4A	125–500μm	8	8	8	8	7	7	2	1	1	1	1	1	1	1	1
AMBLYGONITE P-3A	125–500μm	3	4	4	4	5	5	1	1	1	1	1	1	1	2	2
ANALCIME TS-18A	<45μm	4	5	6	5	5	5	3	3	3	3	3	3	3	3	4
ANATASE, SYNTHETIC O-12A	<45μm	2	3	2	2	2	3	2	3	3	2	3	4	4	4	6
ANDESINE TS-4A	125–500μm	5	5	4	5	4	4	3	3	3	3	5	5	5	5	8
ANGLESITE SO-10A	125–500μm	6	7	8	7	8	8	8	8	8	8	8	8	8	8	8
ANHYDRITE SO-1A	125–500μm	3	5	6	6	7	8	6	6	6	5	6	6	6	5	5
ANORTHITE TS-5A	125–500μm	6	6	4	4	1	1	4	4	4	6	4	3	4	4	4
ANTHOPHYLLITE IN-8A	125–500μm	5	3	1	0	0	0	4	5	5	8	3	3	3	6	3
ANTLERITE SO-11A	125–500μm	0	0	0	0	0	1	6	7	9	9	3	5	8	9	2
APATITE P-1A	125–500μm	3	1	0	3	3	7	6	7	7	8	6	7	7	7	6
APHTHITALITE SO-9A	125–500μm	4	4	4	4	4	5	4	4	3	3	4	4	4	4	5
ARSENOPYRITE S-5A	125–500μm	4	5	5	5	5	6	5	5	5	5	6	6	6	6	8
ATACAMITE H-4A	125–500μm	0	0	0	0	0	0	8	9	9	9	7	8	9	9	2
AUGITE IN-15A	<45μm	7	1	0	3	0	0	7	7	8	9	9	9	9	9	9
AZURITE C-12A	125–500μm	0	0	0	0	0	0	6	7	9	9	4	7	9	9	2
BARITE SO-3A	125–500μm	3	3	3	3	4	5	3	3	4	4	4	5	5	5	6
BERYL CS-2A	125–500μm	1	0	0	0	0	0	3	4	6	8	3	4	5	7	5
BIOTITE PS-23A	125–500μm	7	7	6	7	7	8	9	9	9	0	9	9	9	9	9
BORAX B-6A	125–500μm	2	2	2	1	1	2	0	0	0	0	0	0	0	0	0
BORNITE S-9A	125–500μm	0	0	5	7	9	9	8	8	9	8	8	9	9	8	8
BRUCITE OH-1A	125–500μm	7	7	7	6	3	0	4	2	2	2	8	9	9	1	0
BUDDINGTONITE, FELDS TS-11A	125–500μm	8	8	8	8	8	8	7	7	7	6	6	6	4	2	2
BYTOWNITE TS-13A	<45μm	5	4	4	4	4	4	4	4	4	4	5	5	5	5	7
CALCITE C-3E	125–500μm	4	4	4	4	4	5	3	3	3	3	2	3	3	3	3
CASSITERITE O-3A	125–500μm	0	1	3	2	5	7	1	2	2	2	2	3	4	3	6
CELESTITE SO-5A	125–500μm	1	2	3	4	6	7	4	5	6	5	5	6	6	6	7
CERUSSITE C-10A	125–500μm	4	4	5	4	5	5	4	4	4	4	3	3	3	3	4
CHABAZITE TS-15A	<45μm	4	4	4	4	4	5	2	1	2	2	2	2	2	2	3
CHALCOCITE S-8A	125–500μm	0	1	0	1	1	4	2	2	3	3	3	5	6	6	8
CHALCOPYRITE S-4A	125–500μm	9	8	7	7	6	4	7	7	6	7	8	7	7	7	8
CHLORITE, RIPIDOLITE PS-12A	125–500μm	2	0	0	0	0	1	7	8	9	9	8	9	9	9	9

TABLE C.2 *(continued)*

Material Name	Particle Size (μm)	TM Spectral Ratios														
		2,1	3,1	3,2	4,1	4,2	4,3	5,1	5,2	5,3	5,4	7,1	7,2	7,3	7,4	7,5
CLINOZOISITE SS-4A	125–500μm	9	9	9	9	8	2	9	9	8	8	9	8	6	6	2
COLEMANITE B-1A	125–500μm	4	4	4	3	3	3	0	0	0	0	0	0	0	0	0
COLUMBITE O-7A	125–500μm	1	1	2	3	5	7	3	5	5	4	7	7	7	7	9
CONIFER (AVE.)		7	0	0	9	9	9	8	9	9	0	6	6	8	0	0
COOKEITE PS-9A	125–500μm	7	7	6	6	6	6	5	4	3	2	2	2	2	2	2
CORDIERITE CS-3A	125–500μm	0	4	7	0	1	0	8	8	8	9	7	8	8	9	5
CORRENSITE PS-10A	125–500μm	8	8	8	8	8	8	9	9	9	9	9	9	9	8	5
CORUNDUM, SYNTHETIC O-15A	125–500μm	2	2	2	1	2	3	2	2	2	2	2	2	2	3	4
CRISTOBALITE TS-7A	< 45μm	4	4	5	3	4	3	3	3	2	3	4	5	4	4	6
CRYOLITE H-1A	125–500μm	2	2	2	2	2	3	3	3	4	4	4	5	5	5	6
CUMMINGTONITE IN-6A	125–500μm	4	3	1	1	1	1	5	5	6	7	5	5	5	6	5
DICKITE PS-3A	< 45μm	6	6	7	6	6	6	6	6	5	5	4	3	3	3	3
DIOPSIDE IN-9B	125–500μm	7	6	5	7	7	8	8	8	9	9	8	8	8	8	3
DOLOMITE C-5C	125–500μm	5	6	6	6	6	6	4	5	4	3	2	2	2	2	2
ENSTATITE IN-10B	125–500μm	7	7	7	0	0	0	1	0	0	3	2	2	2	2	9
EPIDOTE SS-1C	125–500μm	9	9	8	9	9	0	9	9	8	8	9	9	8	8	4
FAYALITE NS-1A	125–500μm	8	7	6	5	1	0	7	6	7	8	7	7	7	8	8
FERROAXINITE CS-4A	125–500μm	7	9	9	8	9	8	4	2	1	1	1	0	0	0	0
FLUORITE, PURPLE H-2A	125–500μm	3	6	7	6	7	7	5	6	5	5	6	6	6	6	7
FORSTERITE, SYNTHETI NS-2A	< 45μm	2	2	2	2	3	4	3	3	3	4	4	5	5	5	7
GAHNITE O-11A	125–500μm	0	0	2	6	8	9	0	0	0	0	2	0	0	0	1
GALENA S-7A	125–500μm	0	1	1	0	1	1	1	2	2	2	2	3	4	4	8
GIBBSITE, SYNTHETIC OH-3A	< 45μm	2	2	3	2	2	2	1	1	1	1	1	1	1	1	1
GLAUBERITE SO-8A	125–500μm	5	5	5	5	6	6	5	5	6	6	6	6	6	6	7
GLAUCONITE PS-19A	125–500μm	1	0	0	0	0	0	0	1	1	6	6	7	8	9	9
GLAUCOPHANE IN-3A	125–500μm	1	1	1	6	7	9	9	9	9	9	9	9	9	9	5
GOETHITE OH-02A	< 45μm	9	9	9	9	9	9	9	9	8	8	9	7	9	8	7
GRAPHITE E-1A	125–500μm	1	1	2	3	6	7	6	7	7	7	7	7	7	8	9
GRASS (SENESCENT)		8	9	9	9	9	9	8	8	7	5	7	7	6	3	3
GRASS (VIGOROUS)		9	5	0	9	9	9	9	9	9	9	8	8	8	1	1
GROSSULAR GARNET NS-03B	125–500μm	6	6	7	6	6	4	5	6	5	5	5	5	4	4	5
GYPSUM SO-2B	125–500μm	2	2	3	2	2	2	0	1	0	1	0	1	1	1	2
HALITE HO-3A	125–500μm	3	3	3	3	3	3	3	3	4	4	5	5	5	5	7
HEMATITE O-1A	< 45μm	9	9	9	9	8	3	9	9	7	6	9	9	7	6	8
HEMIMORPHITE SS-2A	125–500μm	3	3	3	3	3	3	1	1	1	1	1	1	1	1	1
HOWLITE B-5A	125–500μm	4	4	4	3	3	3	1	1	1	1	1	0	1	1	1

TABLE C.2 *(continued)*

Material Name	Particle Size (μm)	2,1	3,1	3,2	4,1	4,2	4,3	5,1	5,2	5,3	5,4	7,1	7,2	7,3	7,4	7,5
HYDROXYAPOPHYLLITE PS–22A	125–500μm	2	2	2	1	2	2	0	0	0	0	0	0	0	0	0
HYPERSTHENE IN–14A	< 45μm	6	6	7	4	3	1	5	5	3	6	6	6	6	7	9
ICE		1	2	1	1	1	2	1	2	2	2	2	2	2	2	4
ILLITE PS–11A	< 45μm	7	8	8	8	8	9	8	8	8	8	8	8	8	7	6
JAROSITE SO–7A	125–500μm	9	9	9	9	9	6	9	9	8	9	9	9	8	8	3
JOHANNSENITE IN–12A	125–500μm	6	7	7	7	8	8	9	9	9	9	9	9	9	9	9
KAOLINITE, WELL ORDERED PS–01A	< 45μm	4	5	5	5	5	5	4	4	4	3	2	3	3	3	3
KERNITE B–2A	125–500μm	2	3	3	2	2	2	1	1	1	1	1	1	1	1	2
LABRADORITE TS–2A	125–500μm	6	6	6	5	4	2	2	1	1	2	4	4	3	4	8
LEPIDOLITE, YELLOW PS–13A	125–500μm	8	7	7	6	3	1	6	6	6	7	3	2	2	2	1
MAGNESIOCHROMITE O–8A	125–500μm	1	1	1	1	2	6	6	6	3	2	2	1	2	3	4
MAGNESITE C–6A	125–500μm	6	6	5	5	5	5	2	2	2	2	2	1	2	2	3
MAGNETITE O–4A	< 45μm	1	1	2	1	1	1	4	6	6	7	6	7	7	8	9
MALACHITE C–7A	125–500μm	0	0	0	0	0	0	1	2	9	9	2	4	9	9	8
MICROCLINE TS–17A	125–500μm	8	8	8	8	7	6	7	6	6	6	7	6	6	5	5
MIMETITE A–1A	125–500μm	9	9	9	9	8	7	8	7	5	4	8	6	5	4	5
MONTEBRASITE P–2A	125–500μm	3	3	3	3	3	4	1	1	2	2	1	1	1	2	2
MONTMORILLONITE PS–02B	< 45μm	4	5	5	5	5	4	2	2	2	2	2	2	2	2	3
MUSCOVITE PS–16A	125–500μm	7	7	8	7	7	5	7	7	6	6	6	5	4	4	2
NATROJAROSITE SO–7C	125–500μm	9	9	9	9	8	1	9	8	8	7	9	8	7	8	4
NATROLITE TS–8A	125–500μm	5	4	4	4	4	4	1	1	1	1	1	1	1	1	1
NEPHELINE TS–16A	125–500μm	4	3	3	3	4	5	4	4	5	5	4	4	4	4	5
NONTRONITE PS–6A	< 45μm	9	9	8	8	7	7	8	7	6	6	7	6	5	5	4
OLIGOCLASE TS–3A	125–500μm	3	4	4	3	4	4	3	4	4	4	4	4	5	5	6
ORTHOCLASE TS–12A	125–500μm	5	4	4	4	4	4	4	4	4	4	5	5	6	6	8
PALYGORSKITE PS–4A	< 45μm	6	6	5	6	5	6	5	5	4	4	3	3	3	3	4
PERICLASE O–14A	< 45μm	2	2	2	2	2	3	2	2	3	3	2	3	3	4	5
PLUMBOJAROSITE SO–7B	125–500μm	9	9	9	9	8	3	9	8	7	8	9	8	7	7	4
PREHNITE PS–21A	125–500μm	3	3	1	2	2	3	2	2	3	3	1	1	1	1	1
PYRITE S–02A	< 45μm	7	6	6	5	2	1	7	7	7	8	7	7	6	7	4
PYROLUSITE O–6A	125–500μm	1	1	0	1	1	2	5	6	6	6	6	7	6	7	7
PYROPHYLLITE PS–7A	125–500μm	7	8	7	7	7	7	7	6	6	6	4	3	2	2	2
PYRRHOTITE	< 45μm	7	8	8	8	8	8	8	8	8	8	8	8	8	8	9
QUARTZ, SMOKY TS–1B	125–500μm	5	6	6	7	8	8	7	7	7	7	8	7	8	7	8
REALGAR S–3A	125–500μm	9	9	9	9	9	8	9	9	7	7	9	9	7	7	7
RHODOCHROSITE C–8A	125–500μm	8	8	8	7	7	5	6	6	4	4	5	4	3	3	3
RHODONITE IN–1A	125–500μm	9	9	9	8	7	0	7	6	2	5	7	6	2	5	5

TABLE C.2 *(continued)*

Material Name	Particle Size (μm)	TM Spectral Ratios														
		2,1	3,1	3,2	4,1	4,2	4,3	5,1	5,2	5,3	5,4	7,1	7,2	7,3	7,4	7,5
RIEBECKITE IN-7A	125–500μm	0	1	1	3	6	7	9	9	9	9	9	9	9	9	9
RUTILE O-2A	125–500μm	1	6	7	7	8	9	7	8	8	7	8	8	8	8	9
SANIDINE TS-14A	<45μm	6	7	6	6	6	6	6	5	5	5	6	6	6	6	7
SAPONITE PS-24A	<45μm	6	6	5	6	5	4	4	3	2	2	3	2	2	2	3
SCHEELITE T-1A	125–500μm	7	7	8	7	7	6	7	7	7	7	7	7	7	7	6
SCORODITE A-2A	125–500μm	9	9	9	8	9	8	6	4	1	1	1	1	0	0	0
SEPIOLITE PS-5A	125–500μm	5	5	6	5	5	6	1	1	1	1	1	1	1	1	1
SERPENTINE PS-20A	125–500μm	8	7	3	6	1	1	6	5	6	6	1	1	1	2	1
SIDERITE C-9A	125–500μm	9	9	9	7	1	0	8	7	6	8	9	8	7	8	9
SILLIMANITE NS-08A	125–500μm	6	7	6	7	6	7	5	5	5	3	5	4	3	3	4
SMITHSONITE C-11A	125–500μm	4	3	1	0	0	0	2	2	2	6	1	1	2	3	2
SNOW (COARSE)	Coarse	1	1	1	1	0	0	0	0	0	0	0	0	0	0	0
SNOW (FINE)	Fine	1	2	1	1	1	1	0	0	0	0	0	0	0	0	0
SNOW (FROST)	Frost (finest)	2	2	1	1	1	1	0	0	0	0	2	0	1	1	2
SODALITE TS-10A	125–500μm	0	2	7	8	9	9	7	8	8	6	7	9	6	6	6
SPHALERITE S-1A	125–500μm	8	8	9	9	8	9	8	8	8	7	5	2	2	1	0
SPODUMENE IN-13A	125–500μm	4	5	6	5	5	5	5	5	5	5	5	5	6	6	7
STIBNITE S-6A	125–500μm	0	0	1	9	9	9	9	9	9	9	9	9	9	9	6
STILBITE TS-9A	125–500μm	5	5	5	4	4	3	1	0	1	0	3	3	3	0	0
STRONTIANITE C-1A	125–500μm	5	5	6	5	5	5	5	5	5	5	3	3	3	3	3
SULFUR E-2A	125–500μm	9	8	4	8	3	3	6	3	4	4	7	5	5	6	7
TALC PS-14A	125–500μm	2	1	1	1	0	1	3	4	5	6	2	2	3	4	3
TINCALCONITE B-4A	125–500μm	3	3	4	3	3	2	0	0	0	0	0	0	0	0	1
TITANITE NS-07A	125–500μm	8	8	8	8	8	8	6	3	2	1	6	5	4	3	8
TOPAZ NS-06A	125–500μm	1	2	1	2	3	5	3	4	4	5	1	2	2	2	2
TOURMALINE, DRAVITE-S CS-1A	125–500μm	0	0	2	2	6	8	5	6	7	6	8	9	9	9	9
TREMOLITE IN-5A	125–500μm	5	5	5	4	3	2	4	5	4	5	3	3	3	3	3
TRIPHYLITE P-4A	125–500μm	4	7	9	0	0	0	0	0	0	1	8	8	8	9	9
TRONA C-4A	125–500μm	6	6	6	6	6	7	0	0	0	0	6	6	0	0	0
TSCHERMIGITE SO-6A	125–500μm	2	2	2	1	1	2	0	0	0	0	0	0	0	0	0
ULEXITE B-3A	125–500μm	3	3	3	2	2	2	0	0	0	0	0	0	0	0	0
VERMICULITE PS-18B	<45μm	8	8	8	8	7	7	8	7	7	7	7	7	7	7	5
VESUVIANITE SS-3A	125–500μm	8	8	7	7	6	6	6	6	6	6	3	2	2	2	1
WITHERITE C-2A	125–500μm	5	5	5	5	5	6	5	5	5	4	4	3	4	4	4
WOLLASTONITE IN-2A	125–500μm	3	3	3	3	2	2	2	2	2	3	3	4	4	4	6
ZINCITE O-13A	<45μm	3	3	2	2	4	3	2	3	3	3	4	4	4	5	6
ZIRCON NS-09A	125–500μm	8	9	9	9	9	9	8	8	8	7	9	8	8	7	6

code of actinolite would be $9 - 6 = 3$. Therefore, all reciprocal ratio codes can be simply calculated from the ratio codes given in Table C.2.

If you wanted to make a color ratio composite such that actinolite were red, you might display the $R_{3,1}$ ratio as blue (code of 0), $R_{4,2}$ as green (code of 0), and $R_{5,2}$ as red (code of 9). Only a few other members of the data set would appear red. It may be possible to select a different ratio trio that would make actinolite even more distinctive.

Table C.3 gives the LANDSAT MSS ratio codes for the same data set. There are only six nonreciprocal ratios (with pairs of bands) possible with four spectral bands. The square filter assumption discussed above yields too high a band 5 (0.6–0.7 μm) brightness for vegetation, because of the great increase in reflectance from 0.68–0.70 μm for vigorous vegetation. In reality, the filter for MSS band 5 cuts off most of this increase. Hence, the ratios that use MSS band 5 are too high ($R_{5,4}$) or too low ($R_{6,5}$ and $R_{7,5}$) for vegetation in this table. This is not as serious a problem with the narrower band 3 of TM.

Finally, Table C.4 gives the upper limits of the LANDSAT TM and MSS brightness and ratio codes. For instance, if the brightness of a member of the data set were greater than 5.635% reflectance in TM band 1, but less than 9.920% reflectance, it would have a brightness code of 1 for that spectral band. If reflectance were less than 5.635%, its brightness code for TM band 1 would be zero. If its brightness were greater than 86.69%, its TM band 1 brightness code would be 9. The same reasoning applies to the ratio codes, except that the upper limit for ratio code 9 for any given spectral ratio is the highest value recorded for that ratio in the entire data set, rather than 100%, as it is for the single-band reflectances (brightness codes). The ratio of reflectances is not bounded like the individual reflectances in the numerator or denominator of a spectral ratio.

References

GROVE, C. I., S. J. HOOK, and E. D. PAYLOR II. 1992. *Laboratory Reflectance Spectra of 160 Minerals, 0.4 to 2.5 Micrometers*. JPL Publication 92-2. Pasadena, Calif.: Jet Propulsion Laboratory.

SALISBURY, J. W., D. M. D'ARIA, and A. WALD. 1994. Measurements of Thermal Infrared Spectral Reflectance of Frost, Snow, and Ice. *Journal of Geophysical Research* 99, no. B12: 24235–24240.

TABLE C.3   LANDSAT MSS RATIO CODES (MINERALS, VEGETATION, AND SNOW).

Material	Particle Size	MSS Spectral Ratios					
		R(5,4)	R(6,4)	R(6,5)	R(7,4)	R(7,5)	R(7,6)
ACTINOLITE IN–4A	125–500μm	0	0	4	1	4	4
ALBITE TS–6A	125–500μm	4	4	3	2	1	1
ALBITE TS–6A	125–500μm	4	4	4	2	1	1
ALMANDINE GARNET NS–04A	125–500μm	9	9	9	9	7	0
ALUNITE SO–4A	125–500μm	8	8	7	7	7	5
AMBLYGONITE P–3A	125–500μm	4	4	4	4	4	5
ANALCIME TS–18A	< 45μm	5	5	5	5	5	4
ANATASE, SYNTHETIC O–12A	< 45μm	2	2	2	3	3	3
ANDESINE TS–4A	125–500μm	4	4	4	2	2	1
ANGLESITE SO–10A	125–500μm	7	7	8	8	8	8
ANHYDRITE SO–1A	125–500μm	6	7	8	7	8	7
ANORTHITE TS–5A	125–500μm	4	3	1	1	1	1
ANTHOPHYLLITE IN–8A	125–500μm	2	0	0	0	0	0
ANTLERITE SO–11A	125–500μm	0	0	0	0	0	7
APATITE P–1A	125–500μm	0	0	1	6	8	9
APHTHITALITE SO–9A	125–500μm	4	4	4	4	4	4
ARSENOPYRITE S–5A	125–500μm	4	5	5	5	5	5
ATACAMITE H–4A	125–500μm	0	0	0	0	8	9
AUGITE IN–15A	< 45μm	0	0	0	0	0	6
AZURITE C–12A	125–500μm	0	0	0	0	7	9
BARITE SO–3A	125–500μm	3	3	4	4	4	4
BERYL CS–2A	125–500μm	0	0	0	0	0	8
BIOTITE PS–23A	125–500μm	6	6	7	7	8	9
BORAX B–6A	125–500μm	2	1	1	1	1	1
BORNITE S–9A	125–500μm	3	7	9	9	9	9
BRUCITE OH–1A	125–500μm	6	6	1	3	1	0
BUDDINGTONITE, FELDS TS–11A	125–500μm	8	8	8	8	8	8
BYTOWNITE TS–13A	< 45μm	4	4	4	4	3	2
CALCITE C–3E	125–500μm	4	4	5	3	2	1
CASSITERITE O–3A	125–500μm	2	3	5	5	6	7
CELESTITE SO–5A	125–500μm	3	5	7	5	7	6
CERUSSITE C–10A	125–500μm	4	4	4	5	4	5
CHABAZITE TS–15A	< 45μm	4	4	5	4	4	4
CHALCOCITE S–8A	125–500μm	1	1	1	3	5	7
CHALCOPYRITE S–4A	125–500μm	8	7	5	7	6	7
CHLORITE, RIPIDOLITE PS–12A	125–500μm	0	0	0	0	2	7
CLINOZOISITE SS–4A	125–500μm	9	8	7	8	7	7
COLEMANITE B–1A	125–500μm	3	3	3	3	2	2
COLUMBITE O–7A	125–500μm	1	1	2	5	6	7
CONIFER (AVE.)		9	9	9	9	9	9
COOKEITE PS–9A	125–500μm	7	6	5	6	5	5
CORDIERITE CS–3A	125–500μm	6	6	7	0	0	0
CORRENSITE PS–10A	125–500μm	8	8	8	8	8	8
CORUNDUM, SYNTHETIC O–15A	125–500μm	2	1	2	2	2	2
CRISTOBALITE TS–7A	< 45μm	5	4	4	4	3	3
CRYOLITE H–1A	125–500μm	2	2	2	3	3	4
CUMMINGTONITE IN–6A	125–500μm	2	1	1	1	2	2
DICKITE PS–3A	< 45μm	7	6	6	6	6	6
DIOPSIDE IN–9B	125–500μm	6	7	8	6	6	2

TABLE C.3 *(continued)*

Material	Particle Size	R(5,4)	R(6,4)	R(6,5)	R(7,4)	R(7,5)	R(7,6)
		MSS Spectral Ratios					
DOLOMITE C–5C	125–500μm	6	6	6	6	5	5
ENSTATITE IN–10B	125–500μm	7	0	0	0	0	2
EPIDOTE SS–1C	125–500μm	8	8	9	9	9	8
FAYALITE NS–1A	125–500μm	7	4	0	1	0	0
FERROAXINITE CS–4A	125–500μm	9	9	9	7	1	0
FLUORITE, PURPLE H–2A	125–500μm	6	7	7	7	7	6
FORSTERITE, SYNTHETI NS–2A	< 45μm	2	2	3	3	3	4
GAHNITE O–11A	125–500μm	0	6	9	8	9	9
GALENA S–7A	125–500μm	1	1	1	1	1	1
GIBBSITE, SYNTHETIC OH–3A	< 45μm	2	2	2	2	2	3
GLAUBERITE SO–8A	125–500μm	5	5	6	6	6	6
GLAUCONITE PS–19A	125–500μm	0	0	0	0	0	0
GLAUCOPHANE IN–3A	125–500μm	1	3	7	7	9	9
GOETHITE OH–02A	< 45μm	9	9	9	9	9	8
GRAPHITE E–1A	125–500μm	1	3	7	6	7	8
GRASS (SENESCENT)		9	9	9	9	9	8
GRASS (VIGOROUS)		9	9	9	9	9	9
GROSSULAR GARNET NS–03B	125–500μm	6	6	6	6	6	5
GYPSUM SO–2B	125–500μm	3	2	2	2	2	2
HALITE HO–3A	125–500μm	3	3	3	3	3	4
HEMATITE O–1A	< 45μm	9	9	9	9	9	8
HEMIMORPHITE SS–2A	125–500μm	3	2	3	3	3	3
HOWLITE B–5A	125-500μm	4	3	3	3	2	3
HYDROXYAPOPHYLLITE PS–22A	125–500μm	2	2	2	2	2	2
HYPERSTHENE IN–14A	< 45μm	6	6	6	6	7	7
ICE		1	1	1	1	2	2
ILLITE PS–11A	< 45μm	8	8	8	8	9	8
JAROSITE SO–7A	125–500μm	9	9	8	9	8	6
JOHANNSENITE IN–12A	125–500μm	7	7	7	8	8	8
KAOLINITE, WELL ORDERED PS–01A	< 45μm	5	5	5	5	5	4
KERNITE B–2A	125–500μm	3	2	2	2	2	2
LABRADORITE TS–2A	125–500μm	6	5	5	1	0	0
LEPIDOLITE, YELLOW PS–13A	125–500μm	7	6	3	4	1	0
MAGNESIOCHROMITE O–8A	125–500μm	1	1	4	4	6	6
MAGNESITE C–6A	125–500μm	5	5	4	5	4	4
MAGNETITE O–4A	< 45μm	1	1	0	1	1	2
MALACHITE C–7A	125–500μm	0	0	0	0	0	9
MICROCLINE TS–17A	125–500μm	8	7	6	7	6	6
MIMETITE A–1A	125–500μm	9	8	7	8	7	7
MONTEBRASITE P–2A	125–500μm	3	3	3	3	4	4
MONTMORILLONITE PS–02B	< 45μm	5	5	4	4	4	4
MUSCOVITE PS–16A	125–500μm	7	7	6	7	6	5
NATROJAROSITE SO–7C	125–500μm	9	9	6	8	5	3
NATROLITE TS–8A	125–500μm	4	4	3	4	4	4
NEPHELINE TS–16A	125–500μm	3	3	3	4	5	5
NONTRONITE PS–6A	< 45μm	8	8	7	8	7	6
OLIGOCLASE TS–3A	125–500μm	3	3	3	3	3	3
ORTHOCLASE TS–12A	125–500μm	4	4	3	4	3	3
PALYGORSKITE PS–4A	< 45μm	5	5	5	5	5	5

TABLE C.3 *(continued)*

Material	Particle Size	MSS Spectral Ratios					
		R(5,4)	R(6,4)	R(6,5)	R(7,4)	R(7,5)	R(7,6)
PERICLASE O–14A	< 45μm	2	2	2	2	3	3
PLUMBOJAROSITE SO–7B	125–500μm	9	9	8	9	6	2
PREHNITE PS–21A	125–500μm	2	3	4	3	3	3
PYRITE S–02A	< 45μm	6	5	1	2	1	0
PYROLUSITE O–6A	125–500μm	1	1	1	2	4	6
PYROPHYLLITE PS–7A	125–500μm	7	7	6	7	7	7
PYRRHOTITE	< 45μm	8	8	8	8	8	8
QUARTZ, SMOKY TS–1B	125–500μm	6	7	8	8	8	8
REALGAR S–3A	125–500μm	9	9	9	9	9	7
RHODOCHROSITE C–8A	125–500μm	7	7	6	6	5	3
RHODONITE IN–1A	125–500μm	9	8	8	1	0	0
RIEBECKITE IN–7A	125–500μm	0	1	5	7	8	9
RUTILE O–2A	125–500μm	7	8	9	8	9	9
SANIDINE TS–14A	< 45μm	7	6	6	6	5	5
SAPONITE PS–24A	< 45μm	5	5	5	5	4	3
SCHEELITE T–1A	125–500μm	8	7	6	7	6	6
SCORODITE A–2A	125–500μm	9	9	8	9	9	9
SEPIOLITE PS–5A	125–500μm	5	6	5	5	5	4
SERPENTINE PS–20A	125–500μm	5	3	1	4	4	6
SIDERITE C–9A	125–500μm	9	8	5	0	0	0
SILLIMANITE NS–08A	125–500μm	7	6	6	6	6	6
SMITHSONITE C–11A	125–500μm	2	0	0	0	0	0
SNOW (COARSE)	Coarse	1	0	0	0	0	0
SNOW (FINE)	Fine	1	1	1	1	0	1
SNOW (FROST)	Frost (finest)	1	1	1	1	1	1
SODALITE TS–10A	125–500μm	6	9	9	9	9	9
SPHALERITE S–1A	125–500μm	8	9	9	9	9	9
SPODUMENE IN–13A	125–500μm	5	5	5	5	5	6
STIBNITE S–6A	125–500μm	0	9	9	9	9	9
STILBITE TS–9A	125–500μm	5	4	3	4	3	2
STRONTIANITE C–1A	125–500μm	5	5	5	5	5	5
SULFUR E–2A	125–500μm	6	5	2	5	3	3
TALC PS–14A	125–500μm	1	0	0	0	1	1
TINCALCONITE B–4A	125–500μm	3	3	2	2	2	2
TITANITE NS–07A	125–500μm	8	8	8	7	7	1
TOPAZ NS–06A	125–500μm	1	2	3	3	4	5
TOURMALINE, DRAVITE-S CS–1A	125–500μm	0	1	7	6	8	8
TREMOLITE IN–5A	125–500μm	4	4	4	3	2	1
TRIPHYLITE P–4A	125–500μm	8	8	8	0	0	0
TRONA C–4A	125–500μm	6	6	6	6	6	5
TSCHERMIGITE SO–6A	125–500μm	2	2	1	1	1	1
ULEXITE B–3A	125–500μm	3	2	2	2	2	1
VERMICULITE PS–18B	< 45μm	8	8	7	8	7	7
VESUVIANITE SS–3A	125–500μm	7	7	7	7	7	7
WITHERITE C–2A	125–500μm	5	6	6	6	5	5
WOLLASTONITE IN–2A	125–500μm	3	2	2	1	1	1
ZINCITE O–13A	< 45μm	2	2	2	2	3	3
ZIRCON NS–09A	125–500μm	9	9	9	9	8	0

TABLE C.4 UPPER LIMITS OF LANDSAT TM AND MSS BRIGHTNESS AND RATIO CODES.

Brightness Code	0	1	2	3	4	5	6	7	8	9
TM band 1	5.635	9.92	14.81	30.61	45.48	57.22	68.06	76.49	86.69	100
TM band 2	6.415	12.56	19.07	31.84	52.67	64.71	72.74	80.05	88.35	100
TM band 3	5.37	10.63	26.87	42.77	55.81	69.88	75.31	83.47	89.33	100
TM band 4	5.535	13.09	33.62	48.56	59.82	71.3	76.54	83.14	89.84	100
TM band 5	10.89	17.08	30.99	46.85	57.55	64.05	70.63	77.46	81.93	100
TM band 7	6.445	12.26	22.96	32.53	41.01	55.19	62.7	67.34	72.97	100
MSS band 1	6.04	11.83	18.17	32.19	50.88	64.6	71.88	79.45	87.58	100
MSS band 2	5.44	10.74	26.95	42.49	55.61	69.23	74.95	83.14	89.46	100
MSS band 3	5.31	13.28	32.43	46.92	58.85	71.04	76.73	82.68	90.3	100
MSS band 4	6.205	14.64	31.04	50.34	60.44	71.55	76.6	83.02	89.27	100

Ratio Code	0	1	2	3	4	5	6	7	8	9
TM $R2,1$	0.947	0.998	1.012	1.028	1.049	1.066	1.142	1.216	1.327	3.227
TM $R3,1$	0.894	0.967	1.013	1.041	1.076	1.105	1.225	1.367	1.614	8.304
TM $R3,2$	0.95	0.994	1.002	1.014	1.026	1.038	1.087	1.133	1.236	2.573
TM $R4,1$	0.873	0.993	1.015	1.068	1.089	1.15	1.296	1.541	2.265	9.776
TM $R4,2$	0.892	0.979	1.003	1.02	1.039	1.078	1.151	1.302	1.638	6.648
TM $R4,3$	0.911	0.972	0.992	1.003	1.012	1.028	1.05	1.113	1.204	8.947
TM $R5,1$	0.506	0.865	0.993	1.056	1.151	1.28	1.62	2.007	3.366	12
TM $R5,2$	0.484	0.863	0.956	1.018	1.069	1.163	1.365	1.776	2.502	6.667
TM $R5,3$	0.452	0.808	0.949	0.999	1.039	1.105	1.245	1.598	2.382	8.526
TM $R5,4$	0.47	0.776	0.951	0.992	1.019	1.063	1.172	1.331	1.742	10.09
TM $R7,1$	0.336	0.65	0.848	0.959	1.014	1.174	1.414	1.935	2.611	12.09
TM $R7,2$	0.33	0.643	0.797	0.906	0.977	1.079	1.242	1.644	2.353	12.89
TM $R7,3$	0.256	0.553	0.779	0.872	0.946	1.02	1.187	1.538	2.095	13.16
TM $R7,4$	0.229	0.453	0.756	0.829	0.932	0.998	1.169	1.39	1.916	11.84
TM $R7,5$	0.429	0.61	0.734	0.811	0.864	0.936	0.982	1.01	1.09	5.11
MSS $R5,4$	0.955	0.995	1.005	1.022	1.036	1.053	1.098	1.161	1.32	2.26
MSS $R6,4$	0.934	0.995	1.01	1.029	1.047	1.077	1.154	1.305	1.623	4.972
MSS $R6,5$	0.966	0.994	1	1.007	1.013	1.027	1.051	1.096	1.186	3.742
MSS $R7,4$	0.894	0.965	0.999	1.021	1.057	1.105	1.188	1.421	1.657	8.334
MSS $R7,5$	0.915	0.964	0.992	1.008	1.027	1.056	1.084	1.193	1.346	6.224
MSS $R7,6$	0.938	0.975	0.989	1.002	1.013	1.026	1.053	1.09	1.212	2.354

Index

▼ ▼ ▼ ▼ ▼ ▼ ▼ ▼ ▼ ▼ ▼ ▼ ▼ ▼ ▼ ▼

COLOR PLATES

Color Plate 1: Red Canyon Rim, Wyoming, USA as seen from a color photograph taken from the ground (covering approximately 2 km x 2 km). Referred to in Chapter 1, page 4.

Color Plate 2: Red Canyon Rim, Wyoming, USA as seen from a false color aerial photo of 1:40,000 scale (covering 9.23 km x 9.23 km), with the square showing the approximate location of the ground photo shown in CP 1. Referred to in Chapter 1, page 4.

Color Plate 3: Red Canyon Rim, Wyoming, USA as seen from a LANDSAT MSS false color image (MSS Bands 4, 5, and 7 displayed as blue, green, and red colors, respectively) of one LANDSAT frame (covering 185 km x 185 km), with the rectangle showing the approximate location of the aerial photo in CP 2. Referred to in Chapter 1, page 4.

Color Plate 4: LANDSAT MSS image (covering 185 km x 185 km) of the Wind River Basin and Range, Wyoming, USA. A dark-object-corrected color ratio image with $R'(7,6)$, $R'(6,5)$, and $R'(5,4)$ displayed as blue, green, and red, respectively is shown, where ferric oxides appear orange and vegetation of all types appear blue-green (aqua-colored). Water and shadows are black. Referred to in Chapter 5, page 116.

Color Plate 5: LANDSAT MSS dark-object-corrected color ratio image with $R'(7,6)$, $R'(6,5)$, and $R'(5,4)$ displayed as blue, green, and red, respectively of the Wind River Basin and Range, Wyoming, USA with an overlay showing oil and gas fields in the Wind River Basin. Ferric oxides are displayed as orange in this spectral ratio image. Numbers 1–4 denote fields on classic anticlinal structures, some with Triassic redbeds exposed. Number 5 is the Beaver Creek Field, in Tertiary-aged sediments at the surface. Number 6 is the large gas field near Lost Cabin and numbers 7–8 are oil fields, all three of which were discovered after this ratio image was produced in 1974. North is approximately 7° left of the vertical axis of the image, which covers approximately 185 km x 185 km. Referred to in Chapter 8, page 228.

Color Plate 6: Supervised classification by the parallelepiped method with dark-object-corrected ratio preprocessing for the Wind River Basin and Range, Wyoming, USA from the same LANDSAT MSS frame as shown in CP3. Area shown is approximately 185 km x 185 km. Triassic redbeds are shown in red, predominantly gray shale and claystone (with some fresh granite) in blue, gray-green and gray-yellow arkosic sandstone and conglomerates in black, variegated shale and siltstones in green, buff-colored sandstone in orange, and limestone and granite in brown. Classification with TM data would yield much better results for many more categories owing to the greater number of spectral bands and higher spatial resolution of TM over MSS data. Referred to in Chapter 5, page 140.

Color Plate 7: A natural color image, produced by displaying LANDSAT TM bands 1, 2, and 3 as blue, green, and red, respectively, of a subframe area (approximately 21 km x 19 km) around Santo Domingo, Dominican Republic. Referred to in Chapter 5, page 104.

Color Plate 8: A natural color image, produced by displaying LANDSAT TM bands 1, 2, and 3 as blue, green, and red, respectively, of a whole-frame area (approximately 185 km x 185 km) of the Dominican Republic. The rectangle shows the area covered by the image in CP7. Referred to in Chapter 5, page 106.

Color Plate 9: Four LANDSAT TM quarter-frame images of the Minas Plomosas area (approximately 93 km x 93 km) in the Mexican state of Chihuahua (after Torres et al., 1989). (a) Natural color image of TM bands 1, 2, and 3 displayed in blue, green, and red. (b) TM color ratio image of ratios $R(5,7)$, $R(4,3)$, and $R(3,2)$ displayed in blue, green, and red. (c) Thermal infrared image of TM band 6 shown in black and white, with warmer areas shown as bright. (d) Principal component transformation image of P1, P2, and P3 displayed as blue, green, and red. Referred to in Chapter 5, page 132.

Color Plate 10: Night-time (2:12 AM) thermal infrared image (10-12μm) of the Great Lakes (USA) as recorded by Heat Capacity Mapping Mission (HCMM) on August 22, 1978 (after Vincent et al., 1981). Referred to in Chapter 3, page 33.

Color Plate 11: Color composite image of airborne thermal infrared multispectral scanner (TIMS) bands 1, 3, and 5 (a) and color spectral ratio image (b) of ratios of TIMS bands 2/3, 3/4, and 4/5 shown in blue, green, and red, respectively (both color images) for Death Valley, California, near noon on August 27, 1982. The TIMS spectral bands are centered at 8.3 μm, 8.7 μm, 9.1 μm, 9.8 μm, 10.4 μm, and 11.3 μm. (After Gillespie, 1992. Reprinted by permission of Elsevier Science, Inc., *Remote Sensing of the Environment*.) Referred to in Chapter 5, page 134.

Color Plate 12: Decorrelation contrast stretch image of airborne thermal infrared multispectral scanner (TIMS) data for Death Valley, California near noon on August 27, 1982 (after Gillespie, 1992). The image is approximately 25 km from top to bottom, and North is toward the top. The decorrelation stretch image is of channels 1, 3, and 5 displayed in blue, green, and red, respectively. The TIMS spectral bands are centered at 8.3μm, 8.7μm, 9.1μm, 9.8μm, 10.4μm, and 11.3μm. (Reprinted by permission of Elsevier Science, Inc., *Remote Sensing of the Environment*.) Referred to in Chapter 5, page 134.

Color Plate 13: Spectraseis multifrequency image for a 7.2 km long seismic section of Fraser Township of Bay County, Michigan. Multifrequency channels 3 (30–34 Hz), 4 (34–38 Hz), and 7 (46–50 Hz) are displayed such that their highest amplitudes appear as green, red, and blue, respectively. The Dundee Formation top occurs approximately 1 km below surface and the Trenton Formation top occurs at approximately 3 km below surface. Both the Dundee and Trenton Formations are primarily carbonate rocks. Referred to in Chapter 6, page 159.

Color Plate 14: Color magnetic directional gradient enhancement image of the McDonald Fault, Northwest Territories, Canada. The area covered is 60°–64° N. Lat. and 104°–112° W. Long., with North toward the top. The intensity (brightness) of the image is controlled by directional gradient enhancement of "illumination" from the NE direction and hue is controlled by total magnetic field strength, from red (strongest) to purple (weakest) magnetic field strength. Red areas have more iron content nearer to the surface than other areas. Referred to in Chapter 6, page 159.

Color Plate 15: Color gravity directional gradient enhancement image of the state of Ohio, USA. The intensity (brightness) of the image is controlled by directional gradient enhancement of "illumination" from the East (right of image) and hue is controlled by total gravity field strength, from red (strongest) to purple (weakest) field strength. Red areas are where denser rocks are nearer the surface. Referred to in Chapter 6, page 159.

Color Plate 16: Aerial photo reflectance image of Mt. Brighton, Michigan, USA area, with superimposed color elevation contours derived from automatic processing of the digitized sterio pair by the ATOM software package. Color contour intervals are 10 ft., or 3.1 m. (After James, 1990). Referred to in Chapter 6, page 168.

Color Plate 17: Aerial photo reflectance image of region near Clarkston, Michigan, USA area, with superimposed color elevation contours derived from automatic processing of the digitized stereo pair by the ATOM software package. Color contour intervals are 4 ft., or 1.2 m. Referred to in Chapter 6, page 168.

Color Plate 18: Color contour map of Mt. Ertsberg, Irian Jaya, Indonesia area, produced from 20-m (left image) and 10-m (right image) SPOT satellite stereo pair. Contour interval is 84.685 m. Referred to in Chapter 6, page 170.

Color Plate 19: Simplified geologic map of the Cuprite mining district, Nevada (after Abrams et al., 1977). Referred to in Chapter 7, page 185.

Color Plate 20: Color ratio composite image from MSDS multispectral scanner data, with ratios $1.6\mu m/0.48\mu m$, $1.6\mu m/2.2\mu m$, and $0.6\mu m/1.0\mu m$ spectral ratios displayed as green, red, and blue, respectively (after Abrams et al., 1977). Numbers refer to areas where spectra and/or samples were collected in the field. Referred to in Chapter 7, page 186.

Color Plate 21: Color aerial photo of the Cuprite mining district, Nevada (after Abrams et al., 1977). Referred to in Chapter 7, page 186.

Color Plate 22: Color ratio images of ATM and TIMS data for the Cuprite Mining District, Nevada. The following ratios were displayed as red, green, and blue, respectively, for each of the images as follows: (a) ATM R(3,2), ATM R(5,6), TIMS R(1,2); (b) ATM R(3,2), TIMS R(2,3), TIMS R(1,2); (c) ATM R(3,2), ATM R(5,6), ATM R(5,2),; (d) TIMS R(4,3), TIMS R(2,3), TIMS R(1,2). Pixel size is 30 m, and the area covered is the same as shown in the map of CP19 (after Vincent et al., 1984). Referred to in Chapter 7, page 187.

Color Plate 23: LANDSAT Thematic Mapper image of TM bands 1, 4, and 7 displayed as blue, green, and red, respectively, of the Mountain Pass, California carbonatite region. The width of this figure is 15.36 km, and North is about 7° left of the vertical axis of the image. Referred to in Chapter 7, page 203.

Color Plate 24: Oil slicks in the Aegean Sea from images of specially contrast-stretched, LANDSAT II MSS bands 4, 5, and 7 displayed in blue, green, and red, respectively. These dark slicks are suspected to have originated from tankers that pumped their bilges. Referred to in Chapter 8, page 238.

Color Plate 25: Canonical transform image of the Bendix 24-channel scanner data, June 1976, of the Lisbon Valley, Utah area. Numbers show type areas for image mapping units described in the reference (after Abrams et al., 1984a). Q = Quaternary alluvium; N = Navaho Sandstone; K = Kayenta Formation; Cu = Chinle Formation, Upper Member; Cl = Chinle Formation, Moss Back Member. (M. J. Abrams, J. E. Conel and H. R. Lang, 1984, reprinted by permission of the American Association of Petroleum Geologists.) Referred to in Chapter 8, page 230.

Color Plate 26: Color infrared photomosaic of the area approximately corresponding to CP25 (after Abrams et al., 1984a). (M. J. Abrams, J. E. Conel and H. R. Lang, 1984, reprinted by permission of the American Association of Petroleum Geologists.) Referred to in Chapter 8, page 230.

Color Plate 27: Natural color (a) and color infrared (CIR) photos (b) of two potted house plants (poinsettia on the left and prayer lily on the right), dry leaves (between the potted plants), and two empty Pyrex dishes in a sun porch under natural illumination on a cloudy, winter day (January 9, 1996) in Bowling Green, Ohio. The fallen swamp maple and oak leaves had been collected out-of-doors in the same town. The word MARYLAND is printed on a piece of paper underneath the Pyrex dishes. The same CIR film and similar minus-blue filter were used for (b) as is used for aerial CIR photography. Referred to in Chapter 9, page 273.

Color Plate 28: Color infrared (CIR) photos of two potted house plants (poinsettia on the left and prayer lily on the right), dry leaves (between the potted plants), and two Pyrex dishes containing (a) clear tap water of 2-cm depth in left dish and 4-cm depth in right dish, and (b) maple leaves after 16 days of soaking in 4-cm of water in left dish and oak leaves after 8 days of soaking in water in 4-cm of water in right dish, in a sun porch under natural illumination on a cloudy winter day (January 9, 1996) in Bowling Green, Ohio. The fallen swamp maple and oak leaves had been collected out-of-doors in the same town. The word MARYLAND is printed on a piece of paper underneath the Pyrex dishes. The same CIR film and similar minus-blue filter were used for (b) as is used for aerial CIR photography. Referred to in Chapter 9, page 273.

Color Plate 29: (a) Color infrared (CIR) photo of two potted house plants (poinsettia on the left and prayer lily on the right), dry leaves (between the potted plants), maple leaves (left foreground) and oak leaves (right foreground), one hour after being removed from the water-filled Pyrex dishes in Fig. 9-8 (b). (b) CIR photo of a house in Bowling Green, Ohio, on a cloudy, winter day (January 9, 1996), with snow cover on the ground. The fallen swamp maple and oak leaves of (a) had been collected out-of-doors in the same town. Referred to in Chapter 9, page 273.

Color Plate 30: The main color image shows a 15-km-wide region on Mount Erebus volcano (Ross Island, Antarctica), produced from LANDSAT TM bands 7, 5, and 4 displayed as red, green, and blue, respectively, collected on January 26, 1985. Snow is blue and clouds are white. Hot lava in the summit crater is displayed here as yellow and red. The lower left insert is an enlargement of the summit crater from the main image (same bands and 30 m spatial resolution), whereas on the lower right is a thermal infrared (TM band 6) image of the identically enlarged summit crater area, with 120 m spatial resolution. (After Rothery and Francis, 1990. Reprinted with permission of the International Journal of Remote Sensing. Published by Taylor & Francis. Referred to in Chapter 10, page 303.)

Color Plate 31: Simulated perspective view of Mt. Etna, Sicily, with LANDSAT TM bands 7, 5, and 2 displayed as red, green, and blue, respectively, and draped over DEM produced from digitized 1:50,000 scale topographic maps. The image, collected on June 20, 1984, shows hot lava (displayed as orange) flowing from the summit into the Valle del Bove, a massive sector collapse from an earlier caldera. This is a five-times-vertically-exaggerated view from the southeast toward the summit area. (After Pieri et al, 1990. Reprinted with permission from *Geology,* published by the Geological Society of America.) Referred to in Chapter 10, page 303.

Color Plate 32: SAR interferograms of Eureka Valley, California, USA from ERS-1 radar images collected on (a) September 14 and November 23, 1992, with full color cycle corresponding to an elevation difference of 50 m; (b) November 23, 1992 and November 8, 1993, with full color cycle corresponding to an elevation difference of 78 m, to a 2.8-cm line-of-sight surface displacement, or to a combination of both; (c) Double difference of interferograms shown at left and center, with topography removed and full color cycle corresponding to a displacement of the ground of 2.8-cm in the direction of the satellite motion. An earthquake of magnitude 6.1 occurred on May 17, 1993, causing this displacement. White lines in the right image indicate location of profiles shown in Fig. 10-7 (after Pelter and Rosen, 1995). Referred to in Chapter 10, page 306.

Color Plate 33: Monthly average sea ice concentrations for the south polar region for February, May, August, and November of 1985, from the data of the Nimbus-7 SMMR satellite sensor (after Parkinson and Gloersen, 1993. *Reprinted with permission of Cambridge University Press.*) Referred to in Chapter 10, page 321.

Color Plate 34: Satellite mosaic image (produced by the National Remote Sensing Center, United Kingdom) of Antarctica, compiled from 1-km resolution AVHRR satellite data obtained by NOAA weather satellites (after Merson, 1989). Referred to in Chapter 10, page 322.

Color Plate 35: Sea surface temperatures in the Atlantic and eastern Pacific oceans showing: (a) Conditions during and El Nino event (June, 1983), and (b) During normal conditions (June, 1984) (after Njoku and Brown, 1993). Referred to in Chapter 10, page 323.

Color Plate 36: Differences in 6-year means for 1977–1982, less 1971–1976, for observed sea surface temperatures (SST). Red (blue) denotes increased (decreased) flow of heat into the atmosphere (after Graham, 1995). Referred to in Chapter 10, page 323.

Color Plate 37: Differences in 6-year means for 1977–1982, less 1971–1976, for observed sea surface temperatures (SST), with seismic epicenters that occurred between the years of 1961–1967 overlaid in black. Red (blue) denotes increased (decreased) flow of heat into the atmosphere (underlying figure after Graham, 1995). Overlay shows seismic epicenters for 1961–1967 time period. (After Barazangi and Dorman, 1969. Reprinted with permission of the *Bulletin of the Seismological Society of America.*) Referred to in Chapter 10, page 327.